Numerical Methods for Strong Nonlinearities in Mechanics

SCIENCES

Mechanics, Field Director – Gilles Pijaudier-Cabot

Numerical Methods in Mechanics, Subject Head – Sylvain Drapier

Numerical Methods for Strong Nonlinearities in Mechanics

Contact and Fracture

Coordinated by
Jacques Besson
Frédéric Lebon
Eric Lorentz

iSTE

WILEY

First published 2024 in Great Britain and the United States by ISTE Ltd and John Wiley & Sons, Inc.

Apart from any fair dealing for the purposes of research or private study, or criticism or review, as permitted under the Copyright, Designs and Patents Act 1988, this publication may only be reproduced, stored or transmitted, in any form or by any means, with the prior permission in writing of the publishers, or in the case of reprographic reproduction in accordance with the terms and licenses issued by the CLA. Enquiries concerning reproduction outside these terms should be sent to the publishers at the undermentioned address:

ISTE Ltd
27-37 St George's Road
London SW19 4EU
UK

www.iste.co.uk

John Wiley & Sons, Inc.
111 River Street
Hoboken, NJ 07030
USA

www.wiley.com

© ISTE Ltd 2024
The rights of Jacques Besson, Frédéric Lebon and Eric Lorentz to be identified as the authors of this work have been asserted by them in accordance with the Copyright, Designs and Patents Act 1988.

Any opinions, findings, and conclusions or recommendations expressed in this material are those of the author(s), contributor(s) or editor(s) and do not necessarily reflect the views of ISTE Group.

Library of Congress Control Number: 2024941794

British Library Cataloguing-in-Publication Data
A CIP record for this book is available from the British Library
ISBN 978-1-78945-081-1

ERC code:
PE8 Products and Processes Engineering
PE8_7 Mechanical and manufacturing engineering (shaping, mounting, joining, separation)
PE8_8 Materials engineering (biomaterials, metals, ceramics, polymers, composites, etc.)
PE8_9 Production technology, process engineering

Contents

Preface

Jacques BESSON[1], Frédéric LEBON[2] and Éric LORENTZ[3]

[1] *CNRS, Mines Paris PSL, France*

[2] *LMA, Aix-Marseille Université, France*

[3] *EDF Lab Paris-Saclay, EDF R&D/ERMES, Palaiseau, France*

For 40 years now, numerical simulation has become essential for studying the behavior of structures, from design to operational analyses, and in many industrial fields (building, transport, energy, manufacturing, etc.). Along with the advances in the physics of materials and computer science, it allows for a more realistic description than ever before. Nonetheless, two phenomena raise numerical difficulties to an extent that they still remain, to this day, the main subject of a large body of research: these are the questions concerning contact and friction, on the one hand, and damage mechanics, on the other hand. This is all the more critical considering that these mechanisms are often involved in the response of a structure to external loading. And due to the lack of robust numerical solutions, the design engineer may be compelled to overlook them – or at least to approach them in a crude manner – at the risk of the simulation incurring potential loss of reliability, which in turn requires that more margins of error be taken, usually with an impact on economic and ecological costs.

Although physically distinct, the phenomena of frictional contact and damage both result in a strong nonlinearity of the mechanical problem. They are actually characterized by a threshold behavior (detachment/contact; elasticity/damage) that restricts the regularity of the problem, now non-differentiable (*non-smooth mechanics*). In addition, this loss of regularity has the disadvantage of involving the main variables of the problem (displacement, speed and damage), with two direct consequences: on the one hand, the mathematical characteristics of the problem shift

Numerical Methods for Strong Nonlinearities in Mechanics,
coordinated by Jacques BESSON, Frédéric LEBON and Éric LORENTZ. © ISTE Ltd 2024.

away from well-controlled forms and require in particular innovative discretization schemes, and, on the other hand, the low regularity persists at the scale of the degrees of freedom of the discretized problem. This makes it particularly difficult for solving the corresponding large algebraic systems in a robust and efficient manner. Moreover, neither uniqueness nor even the existence of solutions can be established, which is reflected by the presence of bifurcation points, limit loads and structural instabilities, which prove always challenging to overcome at the numerical level. Finally, the coupling of these phenomena with other nonlinearities such as plasticity or large deformations results in further weakening the entire numerical structure.

As far as couplings are concerned, the phenomena of contact-friction and damage also interact with each other. In fact, during the final stage of damage, macroscopic cracks appear. It is then necessary to solve the problem of frictional contact between the crack lips. This is also the case at the microscopic scale, where damage can be interpreted – depending on the materials – as a network of microcracks or decohesion between particles and matrix: the contact and friction conditions then have a direct impact on the damage model, in particular its unilateral response according to whether the microcracks open or close. Finally, some damage models aim to concentrate degradation phenomena on surfaces rather than distributing them in the bulk. These are then referred to as cohesive models whose shape is closely similar to that of frictional contact laws. If on top of that we add the fact that loading and kinematic relations of mechanical parts generally derive from a frictional contact mechanism, it is thus most of the time all of the nonlinearities of the problem that may simultaneously occur, at the cost of increased numerical complexity.

The collection of numerical tools deployed to address these problems includes some of the topics that will be examined in this book and can already be mentioned:

– alternatives to the traditional Newton method better suited to regular problems;

– methods for solving non-symmetric problems that result from the potentially unassociated nature of damage and friction laws;

– continuation methods for passing critical points and adjusting loading changes in the vicinity of the limit loads;

– geometric methods for surface detection and changes, namely contact surfaces, cracks resulting from the location of the damage field or the boundary of the damaged area;

– spatial discretization schemes based on mixed finite elements for avoiding possible numerical locking problems;

– finite element techniques for representing discontinuity surfaces within or between meshes.

The authors of the book have contributed their particular perspectives on these issues. Therefore, Franz Chouly, Patrick Hild and Yves Renard propose Lagrangian, augmented Lagrangian and Nitsche methods, recently applied, for addressing frictional contact problems. They provide a fine mathematical analysis within the framework of small elastic perturbations. They show the relationships between these methods within the framework of small and large deformations. Pierre Alart is interested in solving qualified systems with diffuse non-regularity, that is, for which non-regularity travels through the system through a multitude of potential contacts. This concerns both granular media and structures (tensegrity systems, for example). Domain decomposition-based techniques are analyzed. Vladislav A. Yastrebov is dedicated to numerical methods for addressing contact problems at small scales, namely for surfaces where roughness is taken into account. He shows the peculiarities of numerical processing of such problems. In the second part, Jacques Besson covers ductile damage. He addresses the issue of coupling between large strain plasticity and damage and that of spatial localization of damage, which often requires the introduction of an additional balance equation deriving from the microstructures of the material. This last issue is also considered by Eric Lorentz for brittle damage, more brutal than the previous one, insisting on the role of this new nonlocal equation and its numerical processing, which can alternatively be interpreted as a regularization of a cracking problem, in the spirit of phase field models. It is also an opportunity to examine numerical methods adapted to the treatment of instabilities as a result of the brutal nature of brittle damage. The nonlocal equation leads to a spatial organization of damage that also could be imposed directly: this is the guiding principle behind the thick level set (TLS) method presented by Nicolas Moës. The problem is rewritten in terms of finding a level surface corresponding to the boundary of the damaged area. This modeling of damage also builds a bridge to cracking models in which the major numerical problem remains the representation of a crack without a priori on the mesh, typically via extended finite element (XFEM) elements that encapsulate the description of a displacement discontinuity (the crack) and the surrounding volume. Finally, Sylvia Feld-Payet focuses on the transition from damage to crack. More particularly, she addresses the issues of geometric crack detection from a damage field, remeshing accordingly to explicitly represent the new crack and transferring fields from the old to the new mesh, which is being computed.

October 2024

PART 1

Contact and Friction

1

Lagrangian and Nitsche Methods for Frictional Contact

Franz CHOULY[1,2], Patrick HILD[3] and Yves RENARD[4,5]

[1] *Université Bourgogne-Franche-Comté, Dijon, France*
[2] *University of Chile, Santiago, Chile*
[3] *CNRS, Université de Toulouse, France*
[4] *Institut Camille Jordan, Université de Lyon, France*
[5] *CNRS, INSA Lyon, France*

1.1. Introduction

Augmented Lagrangians and Lagrangians are constrained optimization tools that very early have naturally been applied to contact problems with deformable solids (see, for example, Rockafellar 1974, 1976). The augmented Lagrangian has since quite widely become established for the approximation and resolution of contact problems in small and large strains, mainly following the research of Curnier and Alart (1988); Alart and Curnier (1991); Simo and Laursen (1992). The method by Nitsche (1971) was originally proposed to allow a Dirichlet-type boundary condition to be weakly taken into account, precisely avoiding the use of Lagrange multipliers. Only recently has it been extended to contact conditions with or without friction in Chouly and Hild (2013a); Annavarapu et al. (2014); Chouly (2014); Chouly et al. (2015). The close connection between Nitsche and Lagrangian methods is however quite clear and it is the objective of this chapter to shed some light on this relationship. This is achieved namely by looking into the mechanisms underlying these methods, and also by way of presenting some recent developments within the framework of small and large elastic strains.

Numerical Methods for Strong Nonlinearities in Mechanics,
coordinated by Jacques BESSON, Frédéric LEBON and Éric LORENTZ. © ISTE Ltd 2024.

Section 1.2 first presents the continuous problem of frictional contact between two elastic solids, within the framework of small strains. Section 1.3 is dedicated to finite element approximation within the framework of small strains, where mathematical analysis of numerical methods is possible. Section 1.4 finally presents the extension of the methods described in previous sections to the regime of large elastic transformations, as well as numerical results related to this context.

1.2. Small-strains frictional contact between two elastic bodies

The problem of frictional contact between two elastic solids is first described in section 1.2.1, and then in section 1.2.2, this problem is reformulated as a quasi variational inequality. Then, section 1.2.3 introduces the weak multiplier form, and section 1.2.4 introduces the proximal augmented Lagrangian formulation. These reformulations are the basis of the numerical approximations presented in sections 1.3 and 1.4.

1.2.1. *Contact between two elastic bodies*

We consider two elastic solids whose respective reference configurations are denoted by Ω^1 and Ω^2 corresponding to two domains of $\mathbb{R}^d$ ($d = 2$ or 3) of regular boundaries (piecewise of class $\mathscr{C}^1$), as shown in Figure 1.1. At the boundaries $\partial\Omega^1$ and $\partial\Omega^2$ of Ω^1 and Ω^2, we can identify the boundaries Γ_D^1 and Γ_D^2 (with non-empty interiors) on which the elastic bodies are clamped, the boundaries Γ_N^1 and Γ_N^2 with an imposed force density and Γ_C^1 and Γ_C^2 which are the potential contact boundaries, slave and master, respectively. We assume that these boundaries form a partition without boundaries overlapping of $\partial\Omega^1$ and $\partial\Omega^2$.

The two elastic bodies are subjected to force densities (volumic forces if $d = 3$) denoted as f^1 and f^2 and on Γ_N^1 and Γ_N^2 to force densities (surface forces if $d = 3$) denoted as ℓ^1 and ℓ^2. The focus is now on expressing the contact condition with Coulomb friction. To this end, we consider the slave surface Γ_C^1. For a point $x \in \Gamma_C^1$, the point $y \in \Gamma_C^2$ that potentially comes into contact therewith must be determined. This is called contact pairing. In the contact condition of small–strain approximation, this correspondence is determined on the reference configuration and is not questioned during deformation. In general, a projection is used, but it is not the only possible choice. Let us consider this correspondence:

$$\begin{aligned} \Pi : \Gamma_C^1 &\to \Gamma_C^2 \\ x &\mapsto y = \Pi(x) \end{aligned}$$

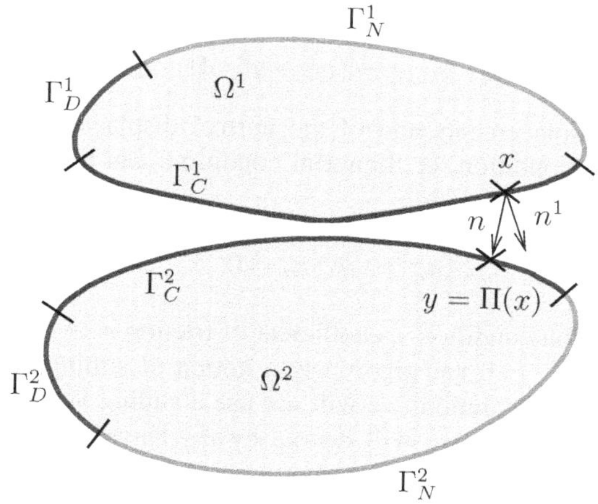

Figure 1.1. *Two bodies with their respective potential contact boundaries. For a color version of this figure, see www.iste.co.uk/lebon/mechanics.zip*

There are then two outward vectors of interest at point x (see Figure 1.1): the outward unit normal vector of Ω^1, which we will denote by n^1, and the outward unit vector in the direction of $y = \Pi(x)$, which we will denote by n and which can be defined by:

$$n = \begin{cases} \dfrac{y-x}{|y-x|} & \text{if } (y-x)\cdot n^1 > 0 \\ n^1 & \text{if } (y-x)\cdot n^1 = 0 \\ -\dfrac{y-x}{|y-x|} & \text{if } (y-x)\cdot n^1 < 0 \end{cases}$$

since the last two cases (i.e. for $(y-x)\cdot n^1 \leq 0$) are expected, either when contact is established in the reference configuration, or if both domains are overlapping, which is a priori not prohibited. There is no reason that these two vectors n^1 and n should be equal, in general, except by using the "ray tracing" strategy exposed in section 1.4.1. The vector n is usually called the contact normal. To express the contact condition, we need to determine what is the normal component of the stress. Let $u^1 : \Omega^1 \to \mathbb{R}^d$ be the displacement of the first body and $\sigma(u^1)$ its Cauchy stress tensor. So, we will denote:

$$\sigma_n = (\sigma(u^1)n^1)\cdot n, \; \sigma_t = (I - n \otimes n)(\sigma(u^1)n^1)$$

the normal and tangential component decomposition of the stress on the slave contact boundary. We will also denote:

$$g_0 = (y-x)\cdot n$$

the initial *gap* between the two potential contact surfaces as well as:

$$[\![u]\!] = (u^1 - u^2 \circ \Pi), \ [\![u_n]\!] = (u^1 - u^2 \circ \Pi) \cdot n$$

the jumps of the displacements and of the normal displacements. Therefrom, the non-interpenetration condition, or Signorini condition, can be written on Γ_C^1 as the following complementarity relation:

$$[\![u_n]\!] \leq g_0, \ \sigma_n \leq 0, \ ([\![u_n]\!] - g_0)\sigma_n = 0 \quad [1.1]$$

To write the friction condition, a coefficient of friction is of course needed, which will be denoted by $\mathscr{F} \geq 0$ and rigorously a notion of sliding velocity. Here, in a supposedly quasi-static evolution, we will not use a sliding velocity but a tangential displacement increment that we will denote by d_t. Duvaut and Lions (1972) and Kikuchi and Oden (1988) use the expression $d_t(u_1, u_2) = (I - n \otimes n)[\![u]\!]$, leading to a problem which, although artificial, exhibits the same characteristics as that obtained for an expression of d_t that would derive from a time discretization that can be written as:

$$d_t(u_1, u_2) = (I - n \otimes n)([\![u]\!] - [\![u^0]\!])$$

where $[\![u^0]\!]$ is the displacement jump at the previous time step. The friction condition is then written as:

$$|\sigma_t| \leq -\mathscr{F}\sigma_n, \text{ if } d_t \neq 0, \text{ then } \sigma_t = \mathscr{F}\sigma_n \frac{d_t}{|d_t|} \quad [1.2]$$

The second Newton law, or action-reaction principle, imposes that:

$$\sigma(u^1)n^1 = \sigma(u^2 \circ \Pi)n^2 \circ \Pi J_\Pi \quad [1.3]$$

where n^2 is the outward unit normal to Γ_C^2 at point $y = \Pi(x)$ and J_Π is the Jacobian of the transformation Π between the two surfaces Γ_C^1 and Γ_C^2.

The description of the linearized elasticity law is carried out by the intermediate of the small strain tensor $\varepsilon(u) = (\nabla u + \nabla u^T)/2$. The Cauchy stress tensor is then connected to the strain tensor by the fourth-order elasticity tensor $\mathcal{A}$ with the usual symmetry and coerciveness properties. This relationship is written as $\sigma(u) = A\,\varepsilon(u)$. The displacements u^1, u^2 of the two elastic bodies are then subjected to the following equations on Ω^i, $i = 1, 2$ in addition to the contact and friction equations [1.1], [1.2] and [1.3]:

$$\begin{cases} -\text{div}\,\sigma(u^i) = f^i \text{ in } \Omega^i \\ \sigma(u^i) = A\,\varepsilon(u^i) \text{ in } \Omega^i \\ u^i = 0 \text{ on } \Gamma_D^i \\ \sigma(u^i)n^i = \ell^i \text{ on } \Gamma_N^i \end{cases} \quad [1.4]$$

1.2.2. *The classical weak inequality form*

The weak formulation in the form of inequality that can be found in Duvaut and Lions (1972) and Kikuchi and Oden (1988) can be constructed by introducing the following spaces:

$$V = H^1(\Omega^1; \mathbb{R}^d) \times H^1(\Omega^2; \mathbb{R}^d)$$

$$V_0 = \{v = (v^1, v^2) \in V : v^1 = 0 \text{ on } \Gamma_D^1 \text{ and } v^2 = 0 \text{ on } \Gamma_D^2\}$$

and the set of admissible displacements:

$$K = \{v = (v^1, v^2) \in V_0 : [\![v_n]\!] - g_0 \in K_0\}$$

$$K_0 = \{v \in L^2(\Gamma_C^1) : v \leq 0\}$$

The spaces of normal and tangential traces on Γ_C^1 are also introduced as:

$$X_N = \left\{w \in L^2(\Gamma_C^1; \mathbb{R}) : \exists v \in H^1(\Omega^1; \mathbb{R}^d), v{=}0 \text{ on } \Gamma_D^1, w{=}v_{|\Gamma_C^1} \cdot n\right\}$$

$$X_T = \left\{w \in L^2(\Gamma_C^1; \mathbb{R}^{d-1}) : \exists v \in H^1(\Omega^1; \mathbb{R}^d), v{=}0 \text{ on } \Gamma_D^1, w{=}(I{-}n{\otimes}n)v_{|\Gamma_C^1}\right\}$$

as well as their respective topological duals X_N' and X_T'. Let (f_1, f_2) be in $L^2(\Omega^1; \mathbb{R}^d) \times L^2(\Omega^2; \mathbb{R}^d)$, (ℓ^1, ℓ^2) in $L^2(\Gamma_N^1; \mathbb{R}^d) \times L^2(\Gamma_N^2; \mathbb{R}^d)$, the following bilinear and linear forms on V are defined:

$$a(u, v) = \sum_{i=1,2} \int_{\Omega^i} \sigma(u^i) : \varepsilon(v^i)\, d\Omega$$

$$L(v) = \sum_{i=1,2} \int_{\Omega^i} f^i \cdot v^i\, d\Omega + \sum_{i=1,2} \int_{\Gamma_N^i} \ell^i \cdot v^i\, d\Gamma$$

as well as the functional corresponding to the virtual work of the frictional force:

$$j(s, v) = \langle s, |d_t(v)| \rangle_{X_N', X_N}$$

where $s = -\mathscr{F}\sigma_n(u)$ is the friction threshold. The notation $\langle \cdot, \cdot \rangle_{X_N', X_N}$ denotes the product of duality between the spaces X_N' and X_N. When s is regular, this product is reduced to the integral $\int_{\Gamma_C^1} s|d_t(v)|d\Gamma$. Therefore, the classical weak form associated with [1.1]–[1.4] is written as:

$$\begin{cases} \text{Find } u \in K \text{such that} \\ a(u, v - u) - j(\mathscr{F}\sigma_n(u), v) + j(\mathscr{F}\sigma_n(u), u) \geq L(v - u), \quad \forall v \in K \end{cases} \qquad [1.5]$$

When there is no friction ($\mathscr{F} = 0$), the weak problem [1.5] is a variational inequality of the first kind. Then, the Stampacchia theorem allows us to conclude that it admits a unique solution, which moreover is the only minimizer of the functional $\mathcal{J}(u) = \frac{1}{2}a(u,u) - L(u)$ on the convex set K. In the presence of friction, results of existence could be shown, for example, in Eck et al. (2005), given that the coefficient of friction $\mathscr{F}$ is small. A recent result in Ballard and Iurlano (2023) claims that existence holds for any friction coefficient. Regarding the uniqueness of the solution, counterexamples were presented for large friction coefficients in Hild (2003, 2004) and a criterion for characterizing the uniqueness of the solution was presented in Renard (2006). The uniqueness of the solution for a sufficiently small coefficient is still an open problem.

It should be noted that when the friction threshold $s \in X'_N$ is known (this is then referred to as Tresca friction), the weak problem, which is then a variational inequality of the second kind, is equivalent to the minimization of the non-regular functional $\mathcal{J}_s(u) = \mathcal{J}(u) + j(s,u)$ on the convex K. The existence and uniqueness of the solution are also guaranteed, since the functional $j(\cdot,\cdot)$ has the property of being convex and lower semi-continuous compared to its second argument (see, for example, Glowinski 1984; Kikuchi and Oden 1988).

1.2.3. *The principle of duality and the weak form with multipliers*

The use of Lagrange multipliers leads to transforming a constrained minimization problem into a problem with simple constraints on multipliers. In this case, it also allows the non-regular nature of the norm involved in the expression of the functional $j(\cdot,\cdot)$ to be taken into consideration. Given that Coulomb friction does not naturally derive from a potential, we look into this matter with the problem involving Tresca friction. By using $I_{K_0}(\cdot)$ the characteristic function for the set K_0, a convex function equal to 0 in K_0 and $+\infty$ elsewhere, the solution to the Tresca problem minimizes on V_0 the functional:

$$\tilde{\mathcal{J}}_s(u) = \mathcal{J}(u) + j(s,u) + I_{K_0}([\![u_n]\!] - g_0)$$

The principle of dualization consists in introducing an auxiliary variable, here $\mu = (\mu_n, \mu_t) \in X_N \times X_T$, and set:

$$\mathcal{W}(u,\mu) = \mathcal{J}(u) + \langle s, |d_t(u) - \mu_t| \rangle_{X'_N, X_N} + I_{K_0}([\![u_n]\!] - g_0 - \mu_n) \qquad [1.6]$$

so that $\mathcal{W}(u,0) = \tilde{\mathcal{J}}_s(u)$ and that $\mathcal{W}(u,\cdot)$ be convex. Therefore, the solution to the Tresca problem is the minimizer of $\mathcal{W}(u,0)$ on V_0 and applying the Fenchel-Legendre conjugate to the functional $\mathcal{W}(u,\cdot)$ (see, for instance, Rockafellar and Wets 1998), this will be the saddle point of the Lagrangian:

$$\mathscr{L}(u,\lambda) = -\sup_{\mu} \left(\int_{\Gamma_C^1} \lambda \cdot \mu \, d\Gamma - W(u,\mu) \right).$$

In practice, the computation of this conjugate yields the sets:

$$\Lambda_N = \{\lambda_n \in X'_N : \langle \lambda_n, v_n \rangle_{X'_N, X_N} \geq 0, \forall v_n \in X_N, v_n \leq 0\}$$
$$\Lambda_T(s) = \{\lambda_t \in X'_T : -\langle \lambda_t, v_t \rangle_{X'_T, X_T} - \langle s, |v_t| \rangle_{X'_N, X_N} \leq 0, \forall v_t \in X_T\}$$

which are, respectively, the sets of admissible normal and tangential stresses. This leads to obtaining the Lagrangian expression:

$$\mathscr{L}(u, \lambda) = \mathcal{J}(u) - \langle \lambda_n, [\![u_n]\!] - g_0 \rangle_{X'_N, X_N}$$
$$-\langle \lambda_t, d_t(u) \rangle_{X'_T, X_T} - I_{\Lambda_N}(\lambda_n) - I_{\Lambda_T(s)}(\lambda_t)$$

which is a Lagrangian with constraints on the multiplier that are represented by the presence of characteristic functions $I_{\Lambda_N}(\lambda_n)$ and $I_{\Lambda_T(s)}(\lambda_t)$. This Lagrangian has then simply to be differentiated to obtain the optimality system. Bearing in mind the fact that the subdifferential of the characteristic function of a set is its normal cone (we still refer to Rockafellar and Wets 1998), the formulation of the Tresca friction problem can be derived as follows (see also Laborde and Renard 2008):

$$\begin{cases} \text{Find } u \in V_0, \lambda_n \in \Lambda_N \text{ and } \lambda_t \in \Lambda_T(s) \text{ such that:} \\ a(u, v) = L(v) + \langle \lambda_n, [\![v_n]\!] \rangle_{X'_N, X_N} + \langle \lambda_t, [\![v_t]\!] \rangle_{X'_T, X_T}, \quad \forall v \in V_0 \\ \langle \lambda_n - \mu_n, [\![u_n]\!] - g_0 \rangle \leq 0, \quad \forall \mu_n \in \Lambda_N \\ \langle \lambda_t - \mu_t, d_t(u) \rangle \leq 0, \quad \forall \mu_t \in \Lambda_T(s). \end{cases} \qquad [1.7]$$

A weak form of the problem with Coulomb friction can be then easily obtained by replacing $\Lambda_T(s)$ by $\Lambda_T(-\mathscr{F}\sigma_n(u))$.

1.2.4. *Proximal augmented Lagrangian: principle and use*

A drawback of the simple Lagrangian presented in section 1.2.3 is that removing constraints and non-regular terms on u induces some constraints on the Lagrange multipliers. An augmented Lagrangian makes it possible to obtain an optimality system without constraints. To this end, the formula in [1.6] has to be modified keeping in mind that any convex quantity can be added to the variable μ provided that the latter cancels out for $\mu = 0$. The proximal augmented Lagrangian (see Rockafellar and Wets 1998) consists of introducing a norm of the additional variable μ, that is, in this case:

$$\mathcal{W}_\gamma(u, \mu) = \mathcal{W}(u, \mu) + \frac{\gamma}{2} \int_{\Gamma_C^1} |\mu|^2 d\Gamma$$

where $\gamma > 0$ is the augmentation parameter. It should be noted that by using a norm in $L^2(\Gamma_C^1)$, only the cases where the multiplier is in this space are taken into consideration. This will be assumed hereafter, and will make it possible to replace duality products by integrals. Using again the Fenchel–Legendre conjugate with respect to the variable μ, the augmented Lagrangian is obtained:

$$\mathscr{L}_\gamma(u,\lambda) = -\sup_{\mu}\left(\left(\int_{\Gamma_C^1} \lambda \cdot \mu \, d\Gamma - W_\gamma(u,\mu)\right)\right)$$

which gives after computation:

$$\begin{aligned}
\mathscr{L}_\gamma(u,\lambda) &= \mathcal{J}(u) - \int_{\Gamma_C^1} \lambda_n(\llbracket u_n \rrbracket - g_0) d\Gamma - \int_{\Gamma_C^1} \lambda_t \cdot d_t(u) d\Gamma \\
&\quad - \int_{\Gamma_C^1} \frac{1}{2\gamma}\left[\lambda_n - \gamma(\llbracket u_n \rrbracket - g_0) + (\lambda_n - \gamma(\llbracket u_n \rrbracket - g_0))_-\right]^2 d\Gamma \\
&\quad + \int_{\Gamma_C^1} \frac{\gamma}{2}(\llbracket u_n \rrbracket - g_0)^2 d\Gamma \\
&\quad - \int_{\Gamma_C^1} \frac{1}{2\gamma}\left|\lambda_t - \gamma d_t(u) - P_{B(s)}(\lambda_t - \gamma d_t(u))\right|^2 d\Gamma \\
&\quad + \int_{\Gamma_C^1} \frac{\gamma}{2}|d_t(u)|^2 d\Gamma
\end{aligned} \qquad [1.8]$$

where $(x)_- = (|x| - x)/2$ designates the negative part of x and $P_{B(s)}$ is the projection onto the ball of center 0 and radius s. The optimality system of this augmented Lagrangian is the weak form of the following Tresca friction contact problem:

$$\left\{\begin{aligned}
&\text{Find } u \in V_0,\ \lambda \in L^2(\Gamma_C^1;\mathbb{R}^d) \text{ such that for any } v \in V_0 \\
&\text{and any } \mu \in L^2(\Gamma_C^1;\mathbb{R}^d) \text{ one has:} \\
&a(u,v) = L(v) - \int_{\Gamma_C^1} (\lambda_n - \gamma(\llbracket u_n \rrbracket - g_0))_- \llbracket v_n \rrbracket d\Gamma \\
&\qquad + \int_{\Gamma_C^1} P_{B(s)}(\lambda_t - \gamma d_t(u)) \llbracket v_t \rrbracket d\Gamma \\
&\qquad - \frac{1}{\gamma}\int_{\Gamma_C^1} \left[\lambda_n + (\lambda_n - \gamma(\llbracket u_n \rrbracket - g_0))_-\right] \mu_n d\Gamma \\
&\qquad - \frac{1}{\gamma}\int_{\Gamma_C^1} \left[\lambda_t - P_{B(s)}(\lambda_t - \gamma d_t(u))\right] \cdot \mu_t \, d\Gamma.
\end{aligned}\right. \qquad [1.9]$$

A weak form of the Coulomb friction problem can then be obtained by replacing the threshold s by the friction threshold from Coulomb's law, that is, either with $-\mathscr{F}\lambda_n$, or with $\mathscr{F}(\lambda_n - \gamma(\llbracket u_n \rrbracket - g_0))_-$.

The mechanism by which the augmented Lagrangian provides an optimality condition that is not a constrained condition is related to the Moreau–Yosida regularization. So to quickly introduce the principle, we examine the optimality condition on the multiplier λ which is obtained by canceling out:

$$\partial_\lambda \mathscr{L}_\gamma(u, \lambda) = \partial_\lambda(\mathcal{W}_\gamma^{*\mu}(u, \lambda))$$

where $\mathcal{W}_\gamma^{*\mu}(u, \lambda)$ is the Fenchel–Legendre conjugate of $\mathcal{W}_\gamma(u, \mu)$ with respect to the variable μ in $L^2(\Gamma_C^1)$. Now, for $\varphi : X \to \mathbb{R}$ a convex functional on a Hilbert space X, from the definitions of the conjugate and the subderivative, it follows that: $x \in \partial\varphi(u) \Leftrightarrow -u \in \partial\varphi^*(x)$. Here, for $X = L^2(\Gamma_C^1)$, this gives:

$$0 \in \partial_\lambda \mathscr{L}_\gamma(u, \lambda) \Leftrightarrow -\lambda \in \partial_\mu \mathcal{W}_\gamma(u, 0) \Leftrightarrow 0 \in (\partial_\mu \mathcal{W}(u, \cdot) + \gamma I)^{-1}(-\lambda)$$

where I is the identity. The term $(\partial_\mu \mathcal{W}(u, \cdot) + \gamma I)^{-1}$ is called the Moreau–Yosida resolvent of $\partial_\mu \mathcal{W}(u, \cdot)$. If $\partial_\mu \mathcal{W}(u, \cdot)$ is indeed maximal monotone, that is, if $\mathcal{W}(u, \mu)$ is indeed semi-continuous convex with respect to μ, then the resolvent will be single valued and a contraction for all $\gamma > 0$ (see Brézis 1973). This implies at least a Lipschitz-continuous regularity to the optimality condition on the multiplier.

As an example, consider the normal part $\mathcal{W}^n(u, \mu_n) = I_{K_0}(\llbracket u_n \rrbracket - g_0 + \mu_n)$. The subderivative of a characteristic function I_C of a convex subset C of a Hilbert space X is $\partial I_C = N_C$ where N_C, the normal cone to C is defined by:

$$N_C(x) = \begin{cases} \{w \in X : \langle w, v - x \rangle \leq 0 \ \ \forall v \in C\} & \text{if } x \in \overline{C} \\ \emptyset & \text{otherwise} \end{cases}$$

and in particular satisfies $N_C(x) = \{0\}$ for x in the interior of C. This set is nontrivial only for $x \in \partial C$. It thus follows here that $\partial_{\mu_n} \mathcal{W}^n(u, \mu_n) = N_{K_0}(\llbracket u_n \rrbracket - g_0 + \mu_n)$, and for the standard Lagrangian, we then have:

$$0 \in (\partial_\mu \mathcal{W}^n(u, \cdot))^{-1}(-\lambda_n) \Leftrightarrow g_0 - \llbracket u_n \rrbracket \in N_{\Lambda_N}(\lambda_n)$$

because it can be easily verified that $N_{K_0}^{-1} = N_{\Lambda_N}$ (the two sets K_0 and Λ_N being two mutually polar cones). The condition $g_0 - \llbracket u_n \rrbracket \in N_{\Lambda_N}(\lambda_n)$ is indeed the one represented in the second line of the optimality system of the simple Lagrangian [1.7] and it can be seen that this condition remains constrained because subjected to a normal cone. Conversely, for the augmented Lagrangian, we take $\mathcal{W}^n_\gamma(u,\mu_n) = \mathcal{W}^n(u,\mu_n) + \frac{\gamma}{2}\int_{\Gamma_C^1} |\mu_n|^2 d\Gamma$, and it follows that:

$$0 \in (N_{K_0}([\![u_n]\!] - g_0 + \cdot) + \gamma I)^{-1}(-\lambda_n).$$

As the resolvent of a normal cone can be written as a projection (see Rockafellar and Wets 1998), it yields here:

$$(N_{K_0}([\![u_n]\!] - g_0 + \cdot) + \gamma I)^{-1}(-\lambda_n) = \frac{1}{\gamma}(\lambda_n + (\lambda_n - \gamma([\![u_n]\!] - g_0))_-)$$

and the normal part of the second equation in [1.9], whose regularity is Lipschitz, finally can be found. In section 1.3, it will also be shown how to more directly make use of the relation $\lambda_n + (\lambda_n - \gamma([\![u_n]\!] - g_0))_- = 0$, which is a reformulation of the contact conditions [1.1], weakly imposed in [1.9], in order to obtain a Nitsche-type formulation.

1.3. Finite element approximation in small deformations

To achieve a finite element approximation of the contact problem of two elastic bodies, consider two meshes $\mathcal{T}_h^1$ and $\mathcal{T}_h^2$, regular in the sense of Ciarlet (1991), for the two domains Ω^1 and Ω^2. These meshes are made of geometric elements of desired shapes and sizes, such that:

$$\Omega^i = \bigcup_{T \in \mathcal{T}_h^i} T.$$

For the sake of simplicity, it is therefore considered that the domains Ω^1 and Ω^2 are exactly covered, which may not be the case when these domains present curved boundaries that are impossible to cover with the geometric elements under consideration. Moreover, it is also assumed that the meshes comply with the partition of the boundaries into Γ_D^i, Γ_N^i and Γ_C^i.

By V_0^h, we will denote the subspace of V_0 of piecewise regular functions (usually polynomial) on the two meshes $\mathcal{T}_h^1$ and $\mathcal{T}_h^2$. For instance, if the choice is a P_1 Lagrange element, the meshes will consist of triangles in dimension $d = 2$ and tetrahedra in dimension $d = 3$ and the unknowns will be continuous piecewise affine functions.

There are many different ways for approaching the frictional contact problem. An important feature from the practical point of view is whether we are considering matching (or compatible) meshes or not. The meshes will be said to be coinciding (or compatible) if for any finite element node on the slave boundary $a_i \in \Gamma_C^1$, then

$\Pi(a_i)$ is also a finite element node of the mesh $\mathcal{T}_h^2$. In the latter case, more direct approximations can be considered (see, for example, Wriggers 2002).

It should also be noted that the application of the techniques presented here for the account of the contact condition is not restricted to finite element approximations only, despite that this choice remains one of the most widespread in structural mechanics. They can actually be combined with any other type of discretization based on the weak formulation of the problem (Galerkin methods) and which then allows for an integral formulation of the contact condition. It is particularly possible to use recent NURBS-based (*Non-Uniform Rational B-Splines*) isogeometric-type discretizations, which make it easier to link with computer-aided design (see, for example, Temizer et al. (2011, 2012); De Lorenzis et al. (2012); Kim and Youn (2012); Seitz et al. (2016); Hu et al. (2018); Antolin et al. (2019)) or even polytopal methods, which allow meshes consisting of polygons or polyhedra (see, for example, Wriggers et al. (2016); Cascavita et al. (2020); Chouly et al. (2020)).

The following section will present a state of the art with regard to the results of existence-uniqueness of discrete solutions and their convergence, and the multiplier methods seen in section 1.2.3 are revisited. Then, section 1.3.2 presents a stabilized method that allows more flexibility in choosing the finite element spaces for the displacement fields and the multiplier. From this stabilized method, a primal (without multiplier) and consistent method can be derived, the so called Nitsche's method, which is covered in section 1.3.3. The existing relationship between this method and the augmented Lagrangian presented in section 1.2.4 is detailed in section 1.3.4, then the penalty method is finally found in section 1.3.5 as an approximation of Nitsche's method.

1.3.1. *State of the art, methods with multipliers*

First, it should be understood that each continuous weak formulation, such as those described in section 1.2 or as penalized or regularized formulations, has its finite element discretized analogue. The state of the art regarding these discrete versions is described below:

1) For frictionless contact, the existence and uniqueness of solutions are guaranteed for any type of "reasonable" discretization. The convergence of discrete solutions with an optimal rate remained an open problem since the 1970s and was only established recently, first for the Nitsche method in 2013 (Chouly and Hild 2013a), and then for the weak inequality in 2015 (Drouet and Hild 2015) (the latter result could then be applied to other discretizations). The analyses published earlier were in fact suboptimal or otherwise involved additional artificial hypotheses concerning the behavior of the solution on the contact boundary (see, in particular, Wohlmuth (2011); Drouet and Hild (2015); Chouly et al. (2017, 2023) for a detailed review of the results related to this matter).

2) For Tresca friction, the existence and uniqueness of solutions is also ensured for any type of "reasonable" discretization. As for the convergence of discrete solutions, it is generally established with suboptimal convergence rates, without knowing whether this suboptimality is due to analysis defect or not (see Wohlmuth (2011); Chouly and Hild (2013a); Chouly (2014); Chouly et al. (2017, 2023) for a detailed state of the art). In dimension two, there is an optimal result established in Wohlmuth (2011) but which involves a technical assumption on the behavior of the solution in the contact and friction zone. In dimension three, there is no optimal result for most methods to our knowledge. However, the recent introduction of the Nitsche method (see section 1.3.3) has made it possible to produce discrete solutions that optimally converge toward the solution of the continuous problem, both in dimensions two and three (see Chouly 2014). Later on, an optimal result has been obtained for the penalty formulation in Chouly et al. (2023), for low-order Lagrange finite elements, and in dimension two and three, that combines both the techniques of Drouet and Hild (2015) and Chouly and Hild (2013b).

3) With regard to Coulomb friction, independently of the discretization chosen (weak inequality, multiplier-based method, Nitsche method, etc.), it can generally be shown with fixed-point arguments that the discrete problem admits (at least) one solution, regardless of the value of the coefficient of friction $\mathscr{F}$. In addition, it can be shown that the solution is unique if the coefficient of friction is less than $Ch^{1/2}$, where $C > 0$ is a constant depending on the geometry but not on the mesh used. It is still unclear whether the term in $Ch^{1/2}$ corresponds either to real cases (with multiple solutions that appear when the mesh is refined) or if it is simply a defect of mathematical analysis. Studies concerning the non-uniqueness of discrete solutions were conducted and simple explicit examples of non-uniqueness obtained (see, for example, Hild (2002) for examples of multiple solutions with a coefficient of friction > 1). In addition, examples of non-uniqueness were obtained numerically (since finite element solutions cannot be explained by hand) in Hassani et al. (2003) for given and arbitrarily small friction coefficients. The more sensitive issue of finding a discrete (explicit or not) example with which multiple solutions are obtained for any arbitrarily small coefficient of friction remains open.

To date, the additional results (compared to the above) have mainly been obtained for multiplier methods that have been the subject of most publications in recent decades. These methods consist of a discretization of [1.7] which, in the case of Coulomb friction, is written as:

$$\begin{cases} \text{Find } u^h \in V_0^h,\ \lambda_n^h \in \Lambda_N^h \text{and } \lambda_t^h \in \Lambda_T^h(-\mathscr{F}\lambda_N^h) \text{ such that:} \\ a(u^h, v^h) = L(v^h) + \displaystyle\int_{\Gamma_C^1} \lambda_n^h [\![v_n^h]\!] + \lambda_t^h \cdot [\![v_t^h]\!] d\Gamma, \quad \forall v^h \in V_0^h \\ \displaystyle\int_{\Gamma_C^1} (\lambda_n^h - \mu_n^h)([\![u_n^h]\!] - g_0^h) d\Gamma \leq 0, \quad \forall \mu_n^h \in \Lambda_N^h \\ \displaystyle\int_{\Gamma_C^1} (\lambda_t^h - \mu_t^h) \cdot d_t(u^h) d\Gamma \leq 0, \quad \forall \mu_t^h \in \Lambda_T^h(-\mathscr{F}\lambda_N^h) \end{cases} \qquad [1.10]$$

where $V_0^h, \Lambda_N^h, \Lambda_T^h$ are discrete versions of $V_0, \Lambda_N, \Lambda_T$ (several choices are possible). Historically, the first convergence results were obtained for the formulation type [1.10] (with displacements approximated with a P_1-Lagrange and piecewise constant multipliers) in Haslinger (1983). In this reference, convergence is proven under the assumption of the existence of a solution to the continuous problem (i.e. if the coefficient of friction is sufficiently small) and the author establishes the existence of a subsequence of discrete solutions converging toward a solution to the continuous problem. Using piecewise linear continuous displacements and multipliers, Hild and Renard (2007) obtain the convergence of solutions with a velocity in $Ch^{1/2}$ under the assumptions ensuring the existence of a unique solution established in Renard (2006). The authors further assume that the solution of the continuous problem is in $H^{3/2+\varepsilon}(\Omega)$ ($\varepsilon > 0$) (see also Chouly et al. 2023 for a similar result, but with a method that involves different discrete convex sets). All previous results were obtained in space dimension two and some extend to dimension three. In addition, almost all of these results were obtained for compatible meshes, and it is reasonable to consider generalizing them to noncompatible meshes.

Within this context, there are two effective approaches that have already proven successful for frictionless problems with noncompatible meshes: the first approach is the standard "mortar" approach in which the displacements (in each solid) are continuous and piecewise P_k ($k = 1.2$ in practice) and the multipliers are chosen in the trace space of one of the two meshes and thereby are continuous and piecewise P_k just as displacements (see Ben Belgacem et al. 1999; Hild 2000). The second, more recent approach, called LAC (*local average contact*) differs from the first only in the multipliers that are chosen piecewise P_0 (independently of the degree k chosen for displacements) on macro-meshes (see Drouet and Hild 2017; Abbas et al. 2018). Displacement-multiplier pairs satisfy an independent inf-sup condition of h for both mortar and LAC approaches.

1.3.2. *Absence of inf-sup condition and stabilized methods*

An important aspect in the formulation [1.10] is the need to have an inf-sup condition between the displacement space and the multiplier space in order for the problem to be well posed (at least in the frictionless case). If this condition is not available (not satisfied or too difficult to demonstrate), it is possible to add a term to the previous formulation and this is then referred to as a stabilized multiplier method. For the frictional problem, such a method is written as:

$$
\left\{
\begin{aligned}
&\text{Find } u^h \in V_0^h,\ \lambda_n^h \in \Lambda_N^h \text{ and } \lambda_t^h \in \Lambda_T^h(-\mathscr{F}\lambda_N^h) \text{ such that:}\\
&a(u^h, v^h) - \int_{\Gamma_C^1} \frac{1}{\gamma}(\sigma(u^h)n)\cdot(\sigma(v^h)n)d\Gamma\\
&= L(v^h) + \int_{\Gamma_C^1} \lambda_n^h \left([\![v_n^h]\!] - \frac{1}{\gamma}\sigma_n(v^h)\right) d\Gamma\\
&\quad + \int_{\Gamma_C^1} \lambda_t^h \cdot \left([\![v_t^h]\!] - \frac{1}{\gamma}\sigma_t(v^h)\right) d\Gamma, \quad \forall v^h \in V_0^h\\
&\int_{\Gamma_C^1} (\lambda_n^h - \mu_n^h)([\![u_n^h]\!] - g_0^h)d\Gamma\\
&\quad + \int_{\Gamma_C^1} \frac{1}{\gamma}(\lambda_n^h - \mu_n^h)(\lambda_n^h - \sigma_n(u^h))d\Gamma \le 0, \quad \forall \mu_n^h \in \Lambda_N^h\\
&\int_{\Gamma_C^1} (\lambda_t^h - \mu_t^h)\cdot d_t(u^h)d\Gamma\\
&\quad + \int_{\Gamma_C^1} \frac{1}{\gamma}(\lambda_t^h - \mu_t^h)\cdot(\lambda_t^h - \sigma_t(u^h))d\Gamma \le 0, \forall \mu_t^h \in \Lambda_T^h(-\mathscr{F}\lambda_N^h)
\end{aligned}
\right.
\qquad [1.11]
$$

where $\gamma = \gamma_0/h_T$ is the stabilization parameter in which $\gamma_0 > 0$ is chosen large enough and h_T denotes the diameter of the element T. This is the stabilization originally proposed by Barbosa and Hughes (1992b) for variational inequalities, and adapted to the contact framework in Hild and Renard (2010). Although numerical analysis was only carried out for frictionless contact and with compatible meshes (see Hild and Renard 2010), it is reasonable to assume that all of the results obtained in unstabilized cases with an inf-sup condition can be redemonstrated for stabilized problems. Recently, in Beaude et al. (2023), (suboptimal) error estimates have been derived for Tresca friction, but that are robust with respect to the value of the stabilization parameter for inf-sup compatible pairs. Still in the case of inf-sup compatible pairs, Beaude et al. (2023) establish the convergence to the mixed formulation when the stabilization parameter is large. Extension for Coulomb friction have also been studied in Beaude et al. (2023); Lleras (2009).

1.3.3. *Nitsche's method seen as a limit stabilized method model*

In Stenberg (1995) (see also Juntunen 2015; Chouly 2023), for Dirichlet boundary conditions, we can find an explanation of the relationship between the stabilized method of Barbosa and Hughes (1992a) and that previously proposed by Nitsche (1971), which can be found from that of Barbosa and Hughes through the local elimination of the multiplier. A similar approach can be followed within the framework of contact and the Nitsche method can be obtained from the previous stabilized formulation. The last two equalities in [1.11] can indeed be expressed as follows:

$$\begin{cases} \displaystyle\int_{\Gamma_C^1} (\lambda_n^h - \mu_n^h)(\lambda_n^h - \sigma_n(u^h) + \gamma([\![u_n^h]\!] - g_0^h))d\Gamma \le 0, \quad \forall \mu_n^h \in \Lambda_N^h \\ \displaystyle\int_{\Gamma_C^1} (\lambda_t^h - \mu_t^h) \cdot (\lambda_t^h - \sigma_t(u^h) + \gamma d_t(u^h))d\Gamma \le 0, \quad \forall \mu_t^h \in \Lambda_T^h(-\mathscr{F}\lambda_N^h) \end{cases}$$

or equivalently using projection operators in terms of $L^2(\Gamma_C^1)$:

$$\begin{cases} \lambda_n^h = \mathrm{Proj}_{\Lambda_N^h} \left(\sigma_n(u^h) - \gamma([\![u_n^h]\!] - g_0^h)\right) \\ \lambda_t^h = \mathrm{Proj}_{\Lambda_T^h(-\mathscr{F}\lambda_N^h)} \left(\sigma_t(u^h) - \gamma d_t(u^h)\right) \end{cases}$$

Since the formulation in [1.11] does not require an inf-sup condition and is well posed irrespective of the choice of the multiplier space (at least for frictionless contact), it can be considered that the discrete space of the multipliers should "tend" to the continuous, that is, $\Lambda_N^h \to \Lambda_N$ and $\Lambda_T^h(-\mathscr{F}\lambda_N^h) \to \Lambda_T(-\mathscr{F}\lambda_N)$, which is equivalent to looking for a first multiplier $\lambda_n^h \le 0$ and a second multiplier $\lambda_t^h \in B(-\mathscr{F}\lambda_n^h)$. It is therefore formally obtained that:

$$\begin{cases} \lambda_n^h = -\left(\sigma_n(u^h) - \gamma([\![u_n^h]\!] - g_0^h)\right)_- \\ \lambda_t^h = \mathrm{P}_{B(-\mathscr{F}\lambda_n^h)} \left(\sigma_t(u^h) - \gamma d_t(u^h)\right) \end{cases} \qquad [1.12]$$

We can note that the formal expressions of [1.12] are the discrete counterparts of the relations:

$$\begin{aligned} \sigma_n(u) &= -\left(\sigma_n(u) - \gamma([\![u_n]\!] - g_0)\right)_- \\ \sigma_t(u) &= \mathrm{P}_{B(-\mathscr{F}\sigma_n(u))} \left(\sigma_t(u) - \gamma d_t(u)\right) \end{aligned} \qquad [1.13]$$

which are strictly equivalent to the contact [1.1] and friction [1.2] conditions reformulated as equations (see a formal proof in Chouly (2014); Chouly et al. (2023), for example). These reformulations of contact and friction conditions can be directly obtained by deriving the proximal augmented Lagrangian associated with the contact problem, as discussed in section 1.2.4. Moreover, it should be noted that the discrete character of multipliers in [1.12] no longer results from belonging to a finite-dimensional space but from the fact that they can be expressed according to discrete unknowns u^h. The two expressions of [1.12] are then replaced in the first equation of [1.11] to obtain Nitsche's formulation. By setting the approximate friction threshold as:

$$s(u^h) = \mathscr{F}(\sigma_n(u^h) - \gamma([\![u_n^h]\!] - g_0^h))_-$$

we get:

$$\begin{cases} \text{Find } u^h \in V_0^h \text{ such that for all } v^h \in V_0^h \text{ one has:} \\ a(u^h, v^h) - \displaystyle\int_{\Gamma_C^1} \frac{1}{\gamma}(\sigma(u^h)n) \cdot (\sigma(v^h)n) d\Gamma \\ = L(v^h) \\ \quad + \displaystyle\int_{\Gamma_C^1} \frac{1}{\gamma} \left(\sigma_n(u^h) - \gamma(\llbracket u_n^h \rrbracket - g_0^h)\right)_- \left(\sigma_n(v^h) - \gamma\llbracket v_n^h \rrbracket\right) d\Gamma \\ \quad - \displaystyle\int_{\Gamma_C^1} \frac{1}{\gamma} P_{B(s(u^h))} \left(\sigma_t(u^h) - \gamma d_t(u^h)\right) \cdot \left(\sigma_t(v^h) - \gamma\llbracket v_t^h \rrbracket\right) d\Gamma \end{cases} \quad [1.14]$$

The formulation in [1.14] admits a solution if γ_0 is large enough. In addition, this solution is unique if $\mathscr{F}^2\gamma_0 h^{-1}$ is sufficiently small (Chouly et al. 2019a, 2019b). An alternative formulation, called mean-Nitsche, can be obtained using static condensation and elimination of the multiplier, when spaces of piecewise constant multipliers are considered (Beaude et al. 2023). This formulation is primal, but involves local projection operators of the contact and friction condition on each contact edge or face.

Further analysis in the frictionless case (see Chouly et al. 2015), or with Tresca friction (see Chouly 2014), made it possible to generalize the Nitsche method by introducing an additional parameter denoted by $\theta \in \mathbb{R}$ as formulation [1.14] corresponds to $\theta = 1$. This generalization has resulted in distinguishing three cases of interest: when $\theta = 1$ which corresponds on the one hand to the natural case resulting from the stabilized method and which, by its symmetry properties, derives from the energy function:

$$\begin{aligned} \mathcal{J}_{\mathrm{N}}(u^h) = \mathcal{J}(u^h) &- \int_{\Gamma_C^1} \frac{1}{2\gamma}\sigma_n(u^h)^2 \, d\Gamma \\ &+ \int_{\Gamma_C^1} \frac{1}{2\gamma}(\sigma_n(u^h) - \gamma(\llbracket u_n^h \rrbracket - g_0^h))_-^2 \, d\Gamma \end{aligned} \quad [1.15]$$

(see Chouly et al. 2017 for proof and extension to Tresca friction). The second case of interest corresponds to $\theta = 0$ which leads to a particularly simple formulation, close to an augmented Lagrangian formulation (see section 1.3.4) or also to a penalized formulation (see section 1.3.5). As for the third case obtained for $\theta = -1$, the condition on γ_0 (quite large) vanishes (see Chouly et al. 2015), and the method obtained is more robust (the reader can also refer to Burman et al. (2017) for a "penalty free" variant inspired by the method proposed in Burman (2012) for the Dirichlet condition). When Coulomb friction is added, the analyses become more complicated: these generalized formulations are being studied in Chouly et al. (2022) (see Chouly et al. 2019 for a summary of the results) and are written as:

$$\begin{cases} \text{Find } u^h \in V_0^h \text{ such that } v^h \in V_0^h \text{ one has:} \\ a(u^h, v^h) - \displaystyle\int_{\Gamma_C^1} \frac{\theta}{\gamma}(\sigma(u^h)n) \cdot (\sigma(v^h)n) d\Gamma \\ = L(v^h) \\ \quad + \displaystyle\int_{\Gamma_C^1} \frac{1}{\gamma} \left(\sigma_n(u^h) - \gamma(\llbracket u_n^h \rrbracket - g_0^h)\right)_- \left(\theta\sigma_n(v^h) - \gamma\llbracket v_n^h \rrbracket\right) d\Gamma \\ \quad - \displaystyle\int_{\Gamma_C^1} \frac{1}{\gamma} P_{B(s(u^h))} \left(\sigma_t(u^h) - \gamma d_t(u^h)\right) \cdot \left(\theta\sigma_t(v^h) - \gamma\llbracket v_t^h \rrbracket\right) d\Gamma \end{cases} \quad [1.16]$$

Finally, it should be noted that Nitsche's method, similarly to Barbosa and Hughes' stabilization, only makes sense for the discretized problem and does not admit a counterpart at the continuous level, other than purely formal, unlike most other methods (multipliers, augmented Lagrangian, penalty) that can be written for the continuous problem (see section 1.2). On the other hand, it is a consistent method, which does not induce an additional approximation of contact and friction conditions, and it is also a primal method, where the only unknown is the discrete displacement field.

1.3.4. *Relationship between Nitsche and proximal augmented Lagrangian*

For problems, where $\theta = 0$, Nitsche's method [1.16] can be rewritten more simply as:

$$\begin{cases} \text{Find } u^h \in V_0^h \text{ such that:} \\ a(u^h, v^h) = L(v^h) - \displaystyle\int_{\Gamma_C^1} \left(\sigma_n(u^h) - \gamma(\llbracket u_n^h \rrbracket - g_0^h)\right)_- \llbracket v_n^h \rrbracket d\Gamma \\ \qquad + \displaystyle\int_{\Gamma_C^1} P_{B(s(u^h))} \left(\sigma_t(u^h) - \gamma d_t(u^h)\right) \cdot \llbracket v_t^h \rrbracket d\Gamma \end{cases}$$

Through the introduction of multipliers, as new unknowns, instead of normal and tangential stresses, and by weakly placing contact and friction conditions [1.13] on these multipliers, we obtain from the previous formulation:

$$
\left\{
\begin{aligned}
&\text{Find } (u^h, \lambda_n^h, \lambda_t^h) \in V_0^h \times X_N^h \times X_T^h \\
&\text{such that for } (v^h, \mu_n^h, \mu_t^h) \in V_0^h \times X_N^h \times X_T^h \text{ we have:} \\
&a(u^h, v^h) = L(v^h) - \int_{\Gamma_C^1} (\lambda_n^h - \gamma([\![u_n^h]\!] - g_0^h))_- [\![v_n^h]\!] d\Gamma \\
&\qquad + \int_{\Gamma_C^1} P_{B(s(u^h))}(\lambda_t^h - \gamma d_t(u^h)) \cdot [\![v_t^h]\!] d\Gamma \\
&\qquad - \frac{1}{\gamma} \int_{\Gamma_C^1} (\lambda_n^h + (\lambda_n^h - \gamma([\![u_n^h]\!] - g_0^h))_-) \, \mu_n^h \\
&\qquad + (\lambda_t^h - P_{B(s(u^h))}(\lambda_t^h - \gamma d_t(u^h))) \cdot \mu_t^h \, d\Gamma = 0
\end{aligned}
\right. \tag{1.17}
$$

Here, unlike the previous formulations, the discrete spaces for multipliers, X_N^h, respectively X_T^h, are vector subspaces (without inequality constrains) of X_N, respectively X_T. We then find the discrete version of the proximal augmented Lagrangian formulation [1.9] presented in section 1.2.4. The use of augmented Lagrangian as an approximation technique has been revisited recently (Burman et al. 2019; Burman and Hansbo 2017; Burman et al. 2023), by establishing the close relationship with the Nitsche method (see also Chouly et al. 2017, 2023). For frictionless contact, in particular, in Burman et al. (2019), the identification of the close connection with the Nitsche method resulted in establishing an optimal convergence result for formulations such as [1.17].

1.3.5. *The connection between Nitsche and penalty*

Still for $\theta = 0$, Nitsche's method [1.16] can also be written as:

$$
\left\{
\begin{aligned}
&\text{Find } u^h \in V_0^h \text{ such that for any } v^h \in V_0^h \text{ one has:} \\
&a(u^h, v^h) = L(v^h) - \int_{\Gamma_C^1} \gamma \left(\frac{1}{\gamma} \sigma_n(u^h) - ([\![u_n^h]\!] - g_0^h) \right)_- [\![v_n^h]\!] d\Gamma \\
&\qquad + \int_{\Gamma_C^1} \gamma \, P_{B(\frac{1}{\gamma} s(u^h))} \left(\frac{1}{\gamma} \sigma_t(u^h) - d_t(u^h) \right) \cdot [\![v_t^h]\!] d\Gamma
\end{aligned}
\right. \tag{1.18}
$$

Assuming that γ is large enough, the terms in $1/\gamma$ are neglected in the previous formulation [1.18], which gives the following penalized model:

$$
\left\{
\begin{aligned}
&\text{Find } u^h \in V_0^h \text{ such that } \forall \, v^h \in V_0^h \\
&a(u^h, v^h) = L(v^h) - \int_{\Gamma_C^1} \gamma \left([\![u_n^h]\!] - g_0^h \right)_+ [\![v_n^h]\!] d\Gamma \\
&\qquad - \int_{\Gamma_C^1} \gamma \, P_{B(\mathscr{F}([\![u_n^h]\!] - g_0^h)_+)} \left(d_t(u^h) \right) \cdot [\![v_t^h]\!] d\Gamma
\end{aligned}
\right.
$$

where $(x)_+ = (|x| + x)/2$ is the positive part of x. It can be verified that this formulation is not fully consistent. This penalized formulation admits a continuous counterpart (see, for example, Kikuchi and Oden (1988)), which can be interpreted as a regularization of the contact and friction conditions, whose approximation is more accurate for large γ. This formulation allows in particular an interpenetration proportional to the contact pressure and inversely proportional to the value of γ (see among others the numerical examples presented in Wohlmuth (2011)). The numerical analysis of this formulation has apparently not been carried out in cases with Coulomb friction. For frictionless problems or involving Tresca friction, the initial analyses of Kikuchi and Oden (1988) were improved in Chouly and Hild (2013b) and more recently in Chouly et al. (2023); Dione (2020).

1.4. Large strain finite element approximation

We now consider how the augmented Lagrangian and Nitsche methods can be extended within the framework of large elastic strains, and also examine the challenges that this implies. Consider the Lagrangian description, with materials supposed to follow a hyperelastic law (the extension to other constitutive laws can be taken into account as, for example, in Seitz (2019); Seitz et al. (2019) for plasticity and thermoelasticity). We will denote by Ω^1 and Ω^2 the reference configurations of the two elastic solids and $u^i : \Omega^i \to \mathbb{R}^d$, $i = 1, 2$ their displacements. To differentiate between deformed and reference configurations, we adopt the notation conventions shown in Figure 1.2. Let:

$$\begin{aligned} \varphi^i : \Omega^i &\longrightarrow \mathbb{R}^d \\ X &\longmapsto x = X + u^i(X) \end{aligned}$$

the transformation associated with the elastic solid of index i. We denote also Γ_D^1 and Γ_D^2, the respective boundaries where a clamped condition is prescribed, as well as Γ_N^1 and Γ_N^2 the free boundaries.

The mapping:

$$\begin{aligned} \Pi : \Gamma_C^1 &\longrightarrow \Gamma_C^2 \\ X &\longmapsto Y = \Pi(X) \end{aligned}$$

which connects the points on the slave surface Γ_C^1 to their contact candidate on the master surface Γ_C^2 is obviously no longer a given problem as in the case of small strains, but a mapping that depends on the two displacements u^1 and u^2. We give in section 1.4.1 two conventional strategies for defining Π.

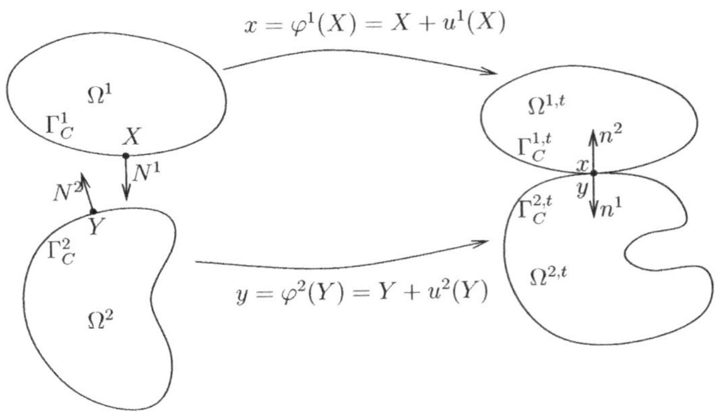

Figure 1.2. *Large-strain Lagrangian description*

For each solid i, the strains are described by the Cauchy–Green tensor $C^i = (F^i)^T F^i$, where $F^i = \nabla\varphi^i = I + \nabla_X u^i$ is the strain gradient, and $J^i = \det(F^i)$ the associated Jacobian. The Green–Lagrange strain tensor $E^i = (C^i - I)/2$ will also be used. The Cauchy stress tensor is always denoted by σ^i, and $\hat{\sigma}^i = J^i\sigma(F^i)^{-T}$, $S^i = J^i(F^i)^{-1}\sigma(F^i)^{-T}$ will be the first and second Piola–Kirchhoff stress tensor, respectively. For a hyperelastic law, there exists a potential W^i which depends on the strain through E^i or C^i (see, for example, Gurtin 1981; Ogden 1984; Ciarlet 1988), such as:

$$S^i = \frac{\partial W^i}{\partial E}(E^i) = 2\frac{\partial W^i}{\partial C}(C^i)$$

To focus on the description of the contact and friction terms, in the following the potential energy of the system $\mathcal{J}(u)$ will be adjusted, and is written for example, in the presence of gravitational forces:

$$\mathcal{J}(u) = \sum_{i=1}^{2} \left(\int_{\Omega^i} W^i(E^i)\, dX - \int_{\Omega^i} \rho_0^i\, g \cdot u^i\, dX \right)$$

where ρ_0^i is the density in the reference configuration of the body i and g is the gravitational acceleration vector. Depending on the models, this potential energy may contain other terms, such as terms representing the potential energies of forces at the boundaries Γ_N^1 and Γ_N^2.

The directional derivative of a quantity A with respect to a displacement $u = (u^1, u^2)$ and in the direction $\delta u = (\delta u^2, \delta u^2)$ will be denoted by:

$$\mathcal{D}A(u)[\delta u]$$

where even $\mathcal{D}A[\delta u]$ if there is no ambiguity about the argument of the quantity A. This directional derivative is defined by:

$$\mathcal{D}A(u)[\delta u] = \lim_{\varepsilon \to 0} \frac{A(u^1 + \varepsilon\, \delta u^1, u^2 + \varepsilon\, \delta u^2) - A(u^1, u^2)}{\varepsilon}$$

when this limit exists.

We will first see, in section 1.4.1, how the *gap* function can be defined, and then explain the formulation of the contact conditions in section 1.4.2. We will then show in sections 1.4.3 and 1.4.4 how the augmented Lagrangian and Nitsche methods seen previously can be adapted to the framework of large strains. We will conclude with practical considerations in section 1.4.5 by considering the choice of the numerical parameter γ, which can be seen in the different methods, and then by presenting some numerical examples in section 1.4.6.

1.4.1. *About contact pairing and* gap *function*

The function Π associates a point X on the slave surface with a point Y on the master surface, facing each other. This association can be achieved in different ways. The most conventional strategy consists in using the orthogonal projection of $x = \varphi(X)$ onto the deformed master surface as illustrated in Figure 1.3a (see, for example, Laursen 2002).

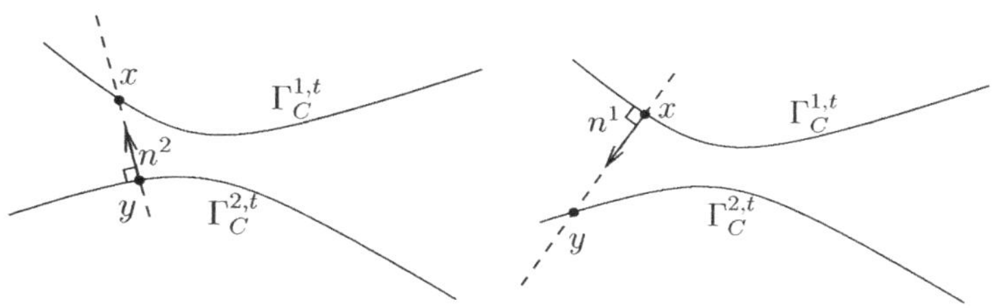

a) Projection strategy b) *Raytracing* strategy

Figure 1.3. *Illustration of projection and raytracing strategies*

An alternative strategy, called *raytracing*, corresponding to the description of Figure 1.3(b), consists of defining y as the closest intersection between the master surface and the straight line passing through the point x and carried by the normal n^1 (see, for example, Poulios and Renard 2015). The *gap* functions that correspond to these two policies are then defined by:

$g = n^1 \cdot (y - x)$ for raytracing

and $g = n^2 \cdot (x - y)$ for projection

This also gives:

$$y = x + g\, n^1 \text{ for raytracing} \qquad [1.19]$$

$$\text{and } y = x - g\, n^2 \text{ for projection} \qquad [1.20]$$

It is useful for what follows to get the directional derivative of the gap. First of all, we get (see Poulios and Renard 2015):

$$\mathcal{D}y[\delta u] = \delta u^2(Y) + F^2\, \mathcal{D}Y[\delta u]. \qquad [1.21]$$

Since $\mathcal{D}Y[\delta u]$ is tangent to Γ^2_C, the vector $F^2\, \mathcal{D}Y[\delta u]$ is tangent to $\Gamma^{2,t}_C$ and thus:

$$n^2 \cdot F^2\, \mathcal{D}Y[\delta u] = 0. \qquad [1.22]$$

Using [1.19] and [1.20], it can also be rewritten that:

$$\mathcal{D}y[\delta u] = \delta u^1(X) + \mathcal{D}g[\delta u]\, n^1 + g\, \mathcal{D}n^1[\delta u] \text{ for raytracing} \qquad [1.23]$$

$$\mathcal{D}y[\delta u] = \delta u^1(X) - \mathcal{D}g[\delta u]\, n^2 - g\, \mathcal{D}n^2[\delta u] \text{ for projection} \qquad [1.24]$$

and by combining the expressions [1.21]–[1.24], the following derivatives of the gap can be obtained:

$$\mathcal{D}g[\delta u] = -\frac{n^2}{n^1 \cdot n^2} \cdot \left(\delta u^1(X) - \delta u^2(Y) + g\, \mathcal{D}n^1[\delta u]\right) \text{ for raytracing} \qquad [1.25]$$

$$\mathcal{D}g[\delta u] = n^2 \cdot \left(\delta u^1(X) - \delta u^2(Y)\right) \text{ for projection.} \qquad [1.26]$$

Computing the derivative of the normal n^1 does not raise any particular problem and gives:

$$\mathcal{D}n^1[\delta u] = -(I - n^1 \otimes n^1)(F^1)^{-T}\, (\nabla \delta u^1(X))^T\, n^1.$$

On the other hand, the derivative of the normal n^2 is much more complex because it depends on the variation of the point y. The computation of the derivative $\mathcal{D}Y[\delta u]$ for the projection gives according to $\mathcal{D}n^2[\delta u]$:

$$\mathcal{D}Y[\delta u] = (F^2)^{-1}\, \left(I - n^2 \otimes n^2\right) \left(\delta u^1(X) - \delta u^2(Y) - g\mathcal{D}n^2[\delta u]\right)$$

whereas for raytracing, with the use of the expression of $\mathcal{D}n^1[\delta u]$, it follows that:

$$\mathcal{D}Y[\delta u] =$$
$$(F^2)^{-1}\left(I - \frac{n^1 \otimes n^2}{n^1 \cdot n^2}\right)\left(\delta u^1(X) - \delta u^2(Y) - g\,(F^1)^{-T}\,(\nabla \delta u^1(X))^T\, n^1\right).$$

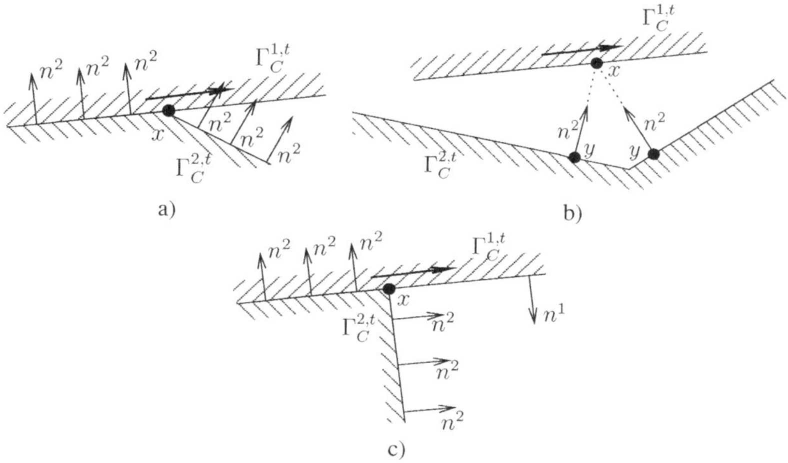

Figure 1.4. *Example of discontinuity of the normal n^2 with respect to x: (a) when crossing an element boundary whether for projection or raytracing; (b) in the presence of concave boundaries for projection (in this case, even y is discontinuous); and (c) when raytracing fails (in this case, y becomes undefined)*

It can thus be seen that for the projection, it will be necessary to have an expression of the derivative of the normal n^2 with respect to displacement, which, on the one hand, is very complex because these derivatives depend in particular on the curvature of the master surface (see the expressions in Laursen (2002) for example). On the other hand, it is challenging since the normal n^2 is not continuous with respect to displacement in a number of situations and particularly in the projection case. Non-continuity examples can be seen described in Figure 1.4. These discontinuities create issues when attempting to make a Newton method converge. We can avoid problem (a) in Figure 1.4 by using finite elements $\mathscr{C}^1$, which ensure the continuity of the normal when crossing the boundary between the elements as in Padmanabhan and Laursen (2001) with Hermite elements, or as in Krstulović-Opara et al. (2002); Stadler et al. (2003); Lengiewicz et al. (2011) with splines, Bézier curves or NURBS, respectively. However, the difficulty remains when modeling angular surfaces. Problem (b) is more difficult to avoid and can cause Newton's method to oscillate between two possible positions y. Strategies are then sometimes employed to avoid these oscillations (see, for example, Alart 1997). When using the raytracing strategy, priority is given to the use of the normal n^1, which generally does not raise any discontinuity problem. A example of discontinuity of point y with raytracing is however depicted in illustration (c) in Figure 1.4.

1.4.2. *Formulation of contact and friction conditions*

The non-interpenetration condition is simply expressed as $g(u) \geq 0$. The frictional contact condition can either be described with the Cauchy stress or in Piola stress. We choose here a constrained Piola description. The Piola contact stress is therefore decomposed at point X (slave boundary) into normal and tangential part as follows:

$$\hat{\sigma}^1(u^1)N^1 = \underbrace{(\hat{\sigma}^1(u^1)N^1 \cdot n)\, n}_{= \hat{\sigma}_n(u)} + \underbrace{(I - n \otimes n)\, \hat{\sigma}^1(u^1)N^1}_{= \hat{\sigma}_t(u)}$$

where n is either n^1 for raytracing or $-n^2$ for the projection such that:

$$y = x + gn. \qquad [1.27]$$

It should be noted that when the contact is effective ($\hat{\sigma}_n < 0$), the outward unit normal vectors n^1 and n^2 are opposite. The quantity $\hat{\sigma}_n$ represents the contact pressure (represented in the reference configuration) in X, and must be nonpositive. Contact conditions can be written as:

$$g(u) \geq 0 \qquad [1.28a]$$

$$\hat{\sigma}_n(u) \leq 0 \ \text{ on } \Gamma_C^1 \qquad [1.28b]$$

$$\hat{\sigma}_n(u) g(u) = 0. \qquad [1.28c]$$

For any $\gamma > 0$, contact conditions [1.28a]–[1.28c] can be rewritten using the following reformulation that is derived from the proximal augmented Lagrangian (see section 1.2.4, and see also equation [1.13] in section 1.3.3):

$$\hat{\sigma}_n(u) = -(\hat{\sigma}_n(u) + \gamma g(u))_-$$

With friction, normal and tangential stresses are coupled to the relative sliding velocity. Relative velocity can be simply defined as:

$$v_r(X) = \dot{\varphi}^1(X) - \dot{\varphi}^2(Y)$$

but this relative velocity does not satisfy the principle of objectivity when the gap is nonzero (see Curnier et al. 1995). As in Curnier et al. (1995), we can then use the definition:

$$v_r(X) = \dot{\varphi}^1(X) - \dot{\varphi}^2(Y) + g\,\dot{n}$$

which is an objective quantity and which coincides with the sliding velocity when contact is established. This, of course, indicates that we are dealing with a evolution problem. It is possible to revert to a problem without time derivative by achieving a temporal discretization, we will then set:

$$v_r(X) = \frac{1}{\Delta t}(\varphi^1(X) - \varphi^2(Y)) - \frac{1}{\Delta t}(\varphi_0^1(X) - \varphi_0^2(Y))$$

with simple relative velocity and:

$$\begin{aligned} v_r(X) &= \frac{1}{\Delta t}(\varphi^1(X) - \varphi^2(Y) + g\, n\,) - \frac{1}{\Delta t}(\varphi_0^1(X) - \varphi_0^2(Y) + g\, n_0\,) \\ &= -\frac{1}{\Delta t}(\varphi_0^1(X) - \varphi_0^2(Y) + g\, n_0\,) \end{aligned}$$

for objective relative velocity, where the formula in [1.27] was used to obtain the second line. The notations φ_0^1, φ_0^2 and n_0, respectively, denote the deformed and the normal to the previous time step (always depending on the chosen strategy), and Δt is the time step associated with discretization.

Despite that this relative velocity is a priori tangential when the contact is persistent in the continuous problem, this will not exactly be the case in the approximate problem, and it may seem more intuitive to define the tangential displacement increment by $d_t = \Delta t(I - n \otimes n)v_r$. For more simplicity, we will hereafter choose:

$$d_t = \Delta t\, v_r$$

The Coulomb friction conditions are then written as ($\mathscr{F}$ still denotes the coefficient of friction):

$$\begin{cases} \|\hat{\sigma}_t(u)\| \leq -\mathscr{F}\hat{\sigma}_n(u) \text{ if } d_t = 0 \\ \hat{\sigma}_t(u) = \mathscr{F}\hat{\sigma}_n(u)\dfrac{d_t}{\|d_t\|} \text{ otherwise.} \end{cases} \qquad [1.29]$$

Using expressions derived from the augmented Lagrangian (see also equation [1.13]), the Coulomb friction condition can be reformulated based on the projection $P_{B(\tau)}$. Still for $\gamma > 0$, the friction condition is then equivalent to the non-regular equation:

$$\hat{\sigma}_t(u) = P_{B(-\mathscr{F}\hat{\sigma}_n(u))}(\hat{\sigma}_t(u) - \gamma d_t)$$

1.4.3. *Augmented Lagrangian and penalization*

We present here the augmented Lagrangian technique for the contact problem in large transformations previously described. An initial formulation thereof will be first given. It will then be seen that a simpler formulation can be reached if we accept to lose symmetry. The associated finite element formulation will then be described, as well as the Uzawa algorithm for the effective resolution of the problem obtained. Finally, it will be seen how it is possible to find the penalized formulation from the Uzawa algorithm.

1.4.3.1. *Augmented Lagrangian and weak form of the friction problem*

It is less direct to obtain a large-strain augmented Lagrangian type formulation for the frictional problem than in the framework of small strains. It is possible to give an analogue of the proximal augmented Lagrangian [1.8] for large strains that would be for a given frictional threshold problem s:

$$
\begin{aligned}
&\mathscr{L}_\gamma(u,\lambda_n,\lambda_t) = \mathcal{J}(u)\\
&+\int_{\Gamma_C^1} \frac{1}{2\gamma}\left((\lambda_n+\gamma g(u))_-^2 - \lambda_n^2\right) d\Gamma - \int_{\Gamma_C^1} \lambda_t\cdot d_t(u) d\Gamma\\
&-\int_{\Gamma_C^1} \frac{1}{2\gamma}\left|\lambda_t - \gamma d_t(u) - P_{B(s)}(\lambda_t-\gamma d_t(u))\right|^2 d\Gamma + \int_{\Gamma_C^1} \frac{\gamma}{2}|d_t(u)|^2 d\Gamma
\end{aligned}
$$

Here, compared to [1.8], the property $x(x)_- = (x)_-^2$ was used to simplify the expression associated with the normal stress. Large-strain Tresca friction may seem even more artificial than in small strains since not only is friction allowed when there is no contact but the relative velocity between two potentially very distant points cannot be interpreted as a sliding velocity and highly depends on the chosen contact pairing strategy. If we write the optimality system of this Lagrangian, it follows that:

$$
\left\{
\begin{aligned}
&\mathcal{D}\mathcal{J}(u)[\delta u] - \int_{\Gamma_C^1} (\lambda_n + \gamma g(u))_- \mathcal{D}g(u)[\delta u]\, d\Gamma\\
&-\int_{\Gamma_C^1} P_{B(s)}(\lambda_t - \gamma d_t(u))\cdot \mathcal{D}d_t(u)[\delta u]\, d\Gamma\\
&= 0, \quad \forall\, \delta u,\ (\delta u = 0 \text{ on } \Gamma_D^1 \cup \Gamma_D^2)\\
&-\int_{\Gamma_C^1} \frac{1}{\gamma}\left(\lambda + (\lambda_n+\gamma g)_- n - P_{B(s)}(\lambda_t - \gamma d_t(u))\right)\cdot \delta\lambda\, d\Gamma\\
&= 0, \quad \forall\, \delta\lambda
\end{aligned}
\right.
\qquad [1.30]
$$

with $\lambda = \lambda_n n + \lambda_t$. If the second equation of [1.30] is indeed a weak equivalent of the contact [1.28] and friction [1.29] conditions, the interpretation of λ_n and λ_t as a constraint is ensured only if the terms $\mathcal{D}g(u)[\delta u]$ and $\mathcal{D}d_t(u)[\delta u]$ are expressed in terms of relative displacement. This is the case for $\mathcal{D}g(u)[\delta u]$ when using the projection (see formula [1.26]) but additional terms are already present when using raytracing (see formula [1.25]). Regarding the derivative $\mathcal{D}d_t(u)[\delta u]$, it is written as:

$$\mathcal{D}d_t(u)[\delta u] = \delta u^1(X) - \delta u^2(Y) + (F_0^2 - F^2)\mathcal{D}Y[\delta u]$$

with simple relative velocity and:

$$\mathcal{D}d_t(u)[\delta u] = F_0^2 \mathcal{D}Y[\delta u] - \mathcal{D}g[\delta u]n_0$$

with objective relative velocity. In both cases, the computation involves $\mathcal{D}Y[\delta u]$, which supposes the computation of the second derivative of Y to obtain the tangent problem, as part of Newton's method solution.

1.4.3.2. *Simplified weak form and non-preservation of symmetry*

To simplify the expression of [1.30], we replace $\mathcal{D}g(u)[\delta u]$ with $-n \cdot (\delta u^1(X) - \delta u^1(Y))$ and $\mathcal{D}d_t(u)[\delta u]$ with $(\delta u^1(X) - \delta u^2(Y))$. This changes the interpretation of the multiplier λ, but not the fact that a unilateral friction contact condition is correctly imposed. To take into account Coulomb friction, we replace the threshold s either by $-\mathscr{F}\lambda_n$ or by $\mathscr{F}(\lambda_n + \gamma g)_-$. The second expression is preferred, for example, in Curnier and Alart (1988) and its main advantage is that its sign is always positive irrespective of the value of λ_n during solver iterations. If this second expression is adopted, the weak form of the Coulomb friction problem is written as:

$$\begin{cases} \mathcal{D}\mathcal{J}(u)[\delta u] + \displaystyle\int_{\Gamma_C^1} ((\lambda_n + \gamma g(u))_- n \\ -P_{B(\mathscr{F}(\lambda_n+\gamma g(u))_-)}(\lambda_t - \gamma d_t(u)) \cdot (\delta u^1(X) - \delta u^2(Y)) d\Gamma \\ = 0, \ \forall \delta u, \ (\delta u = 0 \text{ on } \Gamma_D^1 \cup \Gamma_D^2) \\ -\displaystyle\int_{\Gamma_C^1} \frac{1}{\gamma}(\lambda + (\lambda_n + \gamma g(u))_- n \\ -P_{B(\mathscr{F}(\lambda_n+\gamma g(u))_-)}(\lambda_t - \gamma d_t(u))) \cdot \delta\lambda \, d\Gamma = 0, \ \forall \delta\lambda \end{cases} \quad [1.31]$$

However, the modification carried out makes it so the tangent problem loses symmetry, even when the coefficient of friction $\mathscr{F}$ is zero.

The conservation of symmetry can be a challenge to use potentially more efficient linear solvers. In our case, the dilemma is between using formulation [1.30] (previously adapted to Coulomb friction) and using the simplified formulation in [1.31]. The formulation in [1.31], in addition to its simplicity, has the advantage of providing a simple interpretation of multipliers in terms of contact force density and its redifferentiation is also simpler for obtaining the tangent problem. Nonetheless, its disadvantage resides in the loss of symmetry of the tangent problem. This is not a disadvantage if the loss of symmetry has already been identified, for example, due to intrinsically non-symmetric constitutive laws (nonassociated plasticity for example), but it can be a disadvantage if it is the only cause of non-symmetry and that symmetric linear solvers are to be used.

In the latter case, the use of formulation [1.30] may seem more appealing. However, as mentioned in section 1.4.1, in addition to its additional complexity, the derivative of the normal n^2 and the second derivative of Y are also extremely difficult to obtain, because it depends on the curvature of the master surface, and often impossible to compute because of the discontinuities in n^2 and Y. This is the reason why many implementations that use the projection strategy adopt simplified versions of the tangent problem that do not involve the curvature of the master surface (see, for example, Konyukhov and Schweizerhof 2012) at the cost of possibly losing the second-order convergence of the Newton method.

Moreover, because of the unassociated nature of Coulomb's law of friction, the global tangent system will be non-symmetric. Relatively to the Uzawa algorithm (see section 1.4.3.4), strategies have been developed, for example, in Laursen and Simo (1993); Konyukhov and Schweizerhof (2012), to recover the symmetry of the tangent problem in displacement. These strategies mainly consist of taking the non-symmetrical parts to the previous iteration of Uzawa's algorithm.

The method chosen in this chapter is to use the simplified formulation of [1.31], even if it implies losing the symmetry of the tangent problem. This choice, which can also be found in Puso and Laursen (2004); Popp et al. (2013); Poulios and Renard (2015), is driven by the concern to provide a weak formulation as regular as possible in relation to the displacement. Actually, the non-continuity cases of the normal n^2 and the position Y are potentially problematic when using the Newton method.

1.4.3.3. *Finite element approximation*

In order to write the finite element approximation of problem [1.31], we consider V_0^h and X^h as two finite element spaces. The first space V_0^h is intended to approximate the displacement (as in the previous sections, the index 0 means that the clamped conditions are integrated therein) and the second X^h the multiplier. The latter is therefore restricted to the slave boundary Γ_C^1. As in small strains, the relative choice of these two spaces is subject to whether an inf-sup condition is satisfied (see section 1.3.1) in order to guarantee the uniqueness of the multiplier and not to over-constrain the problem. After finite element discretization, the problem [1.31] is written as:

$$\left\{\begin{array}{l}\text{Find } u^h = (u^{1,h}, u^{2,h}) \in V_0^h \text{ and } \lambda^h \in X^h \\ \text{such that } \forall\, \delta u^h \in V_0^h \text{ and } \forall\, \delta\lambda^h \in X^h \\ \mathcal{DJ}(u^h)[\delta u^h] \\ +\displaystyle\int_{\Gamma_C^1} (\lambda_n^h + \gamma g(u^h))_- n \cdot (\delta u^{1,h}(X) - \delta u^{2,h}(Y))\, d\Gamma \\ -\displaystyle\int_{\Gamma_C^1} P_{B(\mathscr{F}(\lambda_n^h+\gamma g(u^h))_-)}(\lambda_t^h - \gamma d_t(u^h)) \\ \cdot(\delta u^{1,h}(X) - \delta u^{2,h}(Y))\, d\Gamma = 0 \\ -\displaystyle\int_{\Gamma_C^1} \frac{1}{\gamma}(\lambda^h + (\lambda_n^h + \gamma g(u^h))_- n \\ P_{B(\mathscr{F}(\lambda_n^h+\gamma g(u^h))_-)}(\lambda_t^h - \gamma d_t(u^h))) \cdot \delta\lambda^h\, d\Gamma = 0 \end{array}\right. \qquad [1.32]$$

The numerical resolution of this highly nonlinear discrete problem is usually done using a generalized Newton method (Curnier and Alart 1988; Renard 2013) (also called non-regular Newton's method (Christensen 2002)). This requires us to calculate the tangent problem associated with the previous problem.

The tangent problem for solving by a Newton method is obtained by redifferentiating [1.32] with respect to displacement and multiplier. This results in a rather complex tangent system due to the large number of terms, including when using the simplifications introduced above. A computer implementation can quickly prove difficult to realize as the slightest omission or error can degrade the convergence speed of Newton's method.

An alternative to direct implementation in a compiled language is to use automation strategies such as those found in the AceGen (Lengiewicz et al. 2011) or GetFEM (Renard and Poulios 2020) software program for example, which makes it possible to automatically derive the tangent problem and obtain an efficient computation from a description of the weak form [1.32].

1.4.3.4. *Numerical resolution using Uzawa's algorithm*

For solving problem [1.32] by a Newton method, a simultaneous displacement and Lagrange multiplier solution has to be performed. An alternative widely employed in structural mechanics software is based on Uzawa's algorithm (see, for example, Simo and Laursen 1992, in this context). We refer to Bertsekas (1982); Kunisch and Stadler (2005); Laborde and Renard (2008) for the principle of the Uzawa algorithm for a Lagrangian or an augmented Lagrangian. In this case, the algorithm can be written as:

0) choose an initial value λ_0^h for the multiplier and a forward step $r > 0$;

1) solve the first equation of [1.32] in displacement, for the multiplier set to the value $\lambda^{h,k}$. Let $u^{h,k}$ be the solution obtained;

2) calculate $\lambda^{h,k+1}$ based on the following updates:

$$\lambda_n^{h,k+1} = \lambda_n^{h,k} + \frac{r}{\gamma}\left(-(\lambda_n^{h,k} + \gamma g(u^{h,k}))_- - \lambda_n^{h,k}\right) \quad [1.33]$$

$$\lambda_t^{h,k+1} = \lambda_t^{h,k} + \frac{r}{\gamma}\left(P_{B(\mathscr{F}(\lambda_n^{h,k} + \gamma g(u^{h,k}))_-)}(\lambda_t^{h,k} - \gamma d_t(u^{h,k})) - \lambda_t^{h,k}\right) \quad [1.34]$$

3) loop in step 1 by incrementing k until a convergence criterion is satisfied.

Algorithm 1.1. *The Uzawa algorithm*

Since step 1 of the algorithm consists of solving a nonlinear problem, it is usually performed using a Newton method. Step 2 corresponds to a fixed-step gradient method $r > 0$ on the multiplier λ^h. For the continuous problem, not approximated by finite elements, the update given by equations [1.33] and [1.34] gives a fixed point in $L^2(\Gamma_C^1)$, which in Stadler (2004); Kunisch and Stadler (2005) was shown to be always convergent for the Tresca friction problem (in small strains), when $r = \gamma$. In practice, this choice is the one that is always made. Moreover, the fixed point admits a contraction constant that is all the smaller as $r = \gamma$ is chosen large (and therefore converges all the faster). On the other hand, it is a first-order convergence to be compared with the 2nd-order convergence of Newton's method.

With regard to the finite element problem, equations [1.33] and [1.34] are obviously not satisfactory as such because the result of the right-hand side calculation are not necessarily in the finite element space X^h. A first solution is to project them again, which yields (for $r = \gamma$): find $\lambda^{h,k+1} \in X^h$ such that for all $\delta\lambda^h \in X^h$ we have:

$$\int_{\Gamma_C^1} (\lambda^{h,k+1} + (\lambda_n^{h,k} + \gamma g(u^{h,k})_- n$$
$$- P_{B(\mathscr{F}(\lambda_n^{h,k} + \gamma g(u^{h,k}))_-)}(\lambda_t^{h,k} - \gamma d_t(u^{h,k}))) \cdot \delta\lambda^h \, d\Gamma = 0$$

This is equivalent to considering the second equation of problem [1.32] by taking all of the terms at iteration k except the first λ^h, taken at iteration $k+1$. The price to pay is an inversion of an additional linear system reduced to the slave contact boundary. The cost of this resolution, however, is mostly negligible compared to the Newton method when solving the first equation.

The additional cost of this resolution is avoided in many references by updating the degrees of freedom of the multiplier (see, for example, Simo and Laursen 1992).

Denoting by $\lambda_i^{h,k+1}$ the contact constraint at the finite element node $a_i \in \Gamma_C^1$ and $\lambda_{t,i}^{h,k+1}$, $\lambda_{n,i}^{h,k+1}$ the normal and tangential parts, this update is written as:

$$\lambda_{n,i}^{h,k+1} = -(\lambda_{n,i}^{h,k} + \gamma g(u^{h,k}(a_i)))_-$$

$$\lambda_{t,i}^{h,k+1} = P_{B(\mathscr{F}(\lambda_{n,i}^{h,k}+\gamma g(u^{h,k}(a_i)))_-)}(\lambda_{t,i}^{h,k} - \gamma d_t(u^{h,k}(a_i)))$$

This usually represent an additional approximation and means that the contact and friction condition is locally imposed on the finite element node of the multiplier. The use of bi-orthogonal bases in Popp et al. (2012, 2013) makes it possible to perform this operation without additional approximation.

1.4.3.5. *Connection to penalty approximation*

In addition to its good convergence properties, one of the reasons for using the Uzawa algorithm for augmented Lagrangian resides in its proximity to the problem with penalization. Indeed, if we take $\lambda^{h,0} = 0$, the first iteration of the Uzawa algorithm consists of finding $u^h = (u^{1,h}, u^{2,h})$ as a solution of:

$$\begin{cases} D\mathcal{J}(u^h)[\delta u^h] + \displaystyle\int_{\Gamma_C^1} (\gamma g(u^h))_- n \cdot (\delta u^{1,h}(X) - \delta u^{2,h}(Y))\, d\Gamma \\ + \displaystyle\int_{\Gamma_C^1} P_{B(\mathscr{F}(\gamma g(u^h))_-)}(\gamma d_t(u^h)) \cdot (\delta u^{1,h}(X) - \delta u^{2,h}(Y))\, d\Gamma \\ = 0, \quad \forall\, \delta u^h \in V_0^h \end{cases} \qquad [1.35]$$

which is precisely a problem with penalized contact and friction conditions; γ is now the penalty parameter (see section 1.3.5). Therefore, it is possible to solve the problem with penalization [1.35], then to perform, on a poorly satisfied contact condition or friction condition criterion one or more iterations of the Uzawa algorithm to obtain a better solution.

1.4.4. *Nitsche's method*

We saw in section 1.3.3 that the Nitsche method could be interpreted as a stabilized Lagrangian method in which a multiplier condensation operation was performed. Although the stabilized Lagrangian method makes it possible to perform this condensation quite simply, in special cases of multiplier spaces (see Chouly 2023; Gustafsson et al. 2017a, 2019), it should be noted that such strategies also exist on the augmented Lagrangian for constant multipliers or bi-orthogonal bases in Popp et al. (2012, 2013); Seitz (2019), the condensation being however limited to the tangent problem.

The Nitsche method will be first presented, deriving it from an energy functional to find its symmetric variant $\theta = 1$. We will then see that in the context of contact between several elastic solids (or self-contact), it is possible to write an unbiased variant of this method.

1.4.4.1. *Differentiation from a potential*

For the frictionless problem, it is possible to obtain a Nitsche method starting from the potential [1.15] obtained in small strains (see Chouly et al. 2017; Mlika et al. 2017), which is transposed here to the framework of large strains:

$$\mathcal{J}_{\mathrm{N}}(u^h) = \mathcal{J}(u^h) - \int_{\Gamma_C^1} \frac{1}{2\gamma} \hat{\sigma}_n(u)^2 \, d\Gamma + \int_{\Gamma_C^1} \frac{1}{2\gamma} (\hat{\sigma}_n(u) + \gamma g(u))_-^2 \, d\Gamma$$

where $\gamma > 0$ is again the Nitsche parameter. The optimality system of this potential is written as:

$$\begin{cases} \mathcal{D}\mathcal{J}(u^h)[\delta u^h] \\ -\displaystyle\int_{\Gamma_C^1} \frac{1}{\gamma} \hat{\sigma}_n(u^h) \mathcal{D}\hat{\sigma}_n(u^h)[\delta u^h] \, d\Gamma \\ -\displaystyle\int_{\Gamma_C^1} \frac{1}{\gamma} (\hat{\sigma}_n(u^h) + \gamma g(u^h))_- \mathcal{D}(\hat{\sigma}_n(u^h) + \gamma g(u^h))[\delta u^h] \, d\Gamma = 0, \\ \forall \, \delta u^h \in V_0^h \end{cases}$$

The term $\mathcal{D}\hat{\sigma}_n[\delta u]$ appears here in which can be seen the derivative of the normal:

$$\mathcal{D}\hat{\sigma}_n[\delta u] = \mathcal{D}((\hat{\sigma} N) \cdot n)[\delta u] = (\mathcal{D}\hat{\sigma}[\delta u] N) \cdot n + (\hat{\sigma} N) \cdot \mathcal{D}n[\delta u]$$

This term may obviously seem problematic since it involves the tangent rigidity $\mathcal{D}\hat{\sigma}[\delta u]$ in the weak form, which implies, rather unusually, the computation of the derivative of this tangent rigidity to get the expression of the tangent problem.

Still for the frictionless problem, the different variants of the Nitsche method are then obtained quite naturally for $\theta \in \mathbb{R}$:

$$\begin{cases} \mathcal{D}\mathcal{J}(u^h)[\delta u^h] \\ -\displaystyle\int_{\Gamma_C^1} \frac{\theta}{\gamma} \hat{\sigma}_n(u^h) \mathcal{D}\hat{\sigma}_n(u^h)[\delta u^h] \, d\Gamma \\ -\displaystyle\int_{\Gamma_C^1} \frac{1}{\gamma} (\hat{\sigma}_n(u^h) + \gamma g(u^h))_- \mathcal{D}(\theta\hat{\sigma}_n(u^h) + \gamma g(u^h))[\delta u^h] \, d\Gamma = 0, \\ \forall \, \delta u^h \in V_0^h \end{cases}$$

In this context, the variant $\theta = 0$ seems especially interesting because it makes the term $\mathcal{D}\hat{\sigma}[\delta u]$ vanish, of course always at the cost of a non-symmetric tangent problem.

To obtain the Nitsche method in problems with Coulomb friction, it seems preferable to start from the simplified augmented Lagrangian formulation [1.32]. Actually, replacing the multiplier λ by $\hat{\sigma}N$ directly yields the variant $\theta = 0$, then by generalizing for any $\theta \in \mathbb{R}$, leads then to obtaining (see also Mlika et al. 2017):

$$\left\{\begin{array}{l} \mathcal{D}\mathcal{J}(u^h)[\delta u^h] - \displaystyle\int_{\Gamma_c^1} \frac{\theta}{\gamma}(\hat{\sigma}^h N) \cdot (\mathcal{D}\hat{\sigma}^h[\delta u^h]N)\, d\Gamma \\ - \displaystyle\int_{\Gamma_c^1} \frac{1}{\gamma} \left((\hat{\sigma}_n^h + \gamma g^h)_- n - P_{B(\mathscr{F}(\hat{\sigma}_n^h + \gamma g^h)_-)}(\hat{\sigma}_t^h - \gamma d_t^h)\right) \\ \cdot \left(\theta \mathcal{D}\hat{\sigma}^h[\delta u^h]N - \gamma(\delta u^{1,h}(X) - \delta u^{2,h}(Y))\right) d\Gamma = 0, \quad \forall\, \delta u^h \in V_0^h \end{array}\right.$$

As in the case of the augmented Lagrangian, this formulation avoids normal differentiation terms but at the cost of losing the symmetry of the tangent problem, even when there is no friction ($\mathscr{F} = 0$).

It should also be noted that other variants of the Nitsche method can be obtained through slightly different principles. It can be seen for example in Burman et al. (2017) in small strains, and also within the framework of large strains in Seitz (2019); Seitz et al. (2019), where a family is introduced with an additional parameter covering the method presented here.

1.4.4.2. *Unbiased variant*

An approximation method is said to be unbiased if no difference is made between master surface and slave surface (see, for example, Sauer and De Lorenzis 2015). In this case, we denote by $\Gamma_C = \Gamma_C^1 \cup \Gamma_C^2$ the total potential contact surface. We will always denote by X the current point of Γ_C and $Y \in \Gamma_C$ the point found by way of the projection strategy or raytracing, which is obviously no longer necessarily on the master surface Γ_C^2. It is then easy to see that an unbiased formulation is simply obtained by adding factors $1/2$ and extending the integrals over any Γ_C (see Mlika et al. 2017):

$$\left\{\begin{array}{l} \mathcal{D}\mathcal{J}(u^h)[\delta u^h] - \displaystyle\frac{1}{2}\int_{\Gamma_C} \frac{\theta}{\gamma}(\hat{\sigma}^h N) \cdot (\mathcal{D}\hat{\sigma}^h[\delta u^h]N)\, d\Gamma \\ - \displaystyle\frac{1}{2}\int_{\Gamma_C} \frac{1}{\gamma} \left((\hat{\sigma}_n^h + \gamma g^h)_- n - P_{B(\mathscr{F}(\hat{\sigma}_n^h + \gamma g^h)_-)}(\hat{\sigma}_t^h - \gamma d_t^h)\right) \\ \cdot \left(\theta \mathcal{D}\hat{\sigma}^h[\delta u^h]N - \gamma(\delta u^{1,h}(X) - \delta u^{2,h}(Y))\right) d\Gamma = 0, \quad \forall\, \delta u^h \in V_0^h \end{array}\right.$$

The advantage of this formulation, in addition to providing additional symmetry, lies mainly when searching for self-contact. The lack of separation in the contact surface into master part and slave part implies that no a priori at the actual point where the contact will occur is necessary. There is of course an additional cost related to the need to integrate over the entire contact surface.

1.4.5. *About the value of the parameter* γ

The parameter γ that appears as a penalization parameter in the penalized formulation in [1.35], as an augmentation parameter in the augmented Lagrangian formulation of [1.32] and as a Nitsche parameter in the formulation in [1.35], plays nonetheless a rather similar role in all three approaches. Although the numerical solution is relatively unaffected by the value of the parameter γ in the augmented Lagrangian and Nitsche formulations, a minimum value must be respected with regard to the Nitsche method to preserve the problem coercivity. For the penalty method, this will mean that a good compromise will have to be found between a large value of γ that will ensure a good approximation of the contact and friction conditions and a moderate value that does not affect the convergence of the Newton method. Despite that the concerns are different in the three approaches, the optimal values of the parameter γ are yet similar. Actually, a priori error analyses within the framework of small strains, both for the Nitsche method (Chouly et al. 2015) and for the first-order convergence of the penalty method (Chouly and Hild 2013b), bring forward a dependency on $\gamma = \gamma_0/h$ where h is the mesh size. In addition, the study by Renard (2013) and the dimension analysis carried out in Poulios and Renard (2015) in the context of the augmented Lagrangian lead to the conclusion that γ_0 has the dimension of an elastic modulus. It is therefore quite natural to choose for γ:

$$\gamma = K/h \text{ where } K = \frac{1}{3}\frac{E}{1-2\nu} = \lambda + \frac{2}{3}\mu$$

denotes the *bulk modulus*, which is equal to Young's modulus E when the Poisson coefficient ν is equal to $1/3$ (λ and μ denote here the Lamé coefficients).

Naturally, and mainly for the Nitsche method for large strains, this value may not be sufficient to ensure coercivity. Indeed, when the deformation is very important, the value of γ_0 should rather be connected to the maximum value of the tangent moduli of elasticity. One way to proceed consists then, as proposed in Seitz et al. (2019), in adapting the value of γ_0 to the maximum eigenvalues of the elementary tangent stiffness matrices.

1.4.6. *Numerical tests*

We reproduce here numerical tests that were performed using GetFEM, a public domain software library (Renard and Poulios 2020) both in Poulios and Renard (2015) following an augmented Lagrangian strategy and in Mlika et al. (2017) with the unbiased Nitsche method previously presented. These results obtained with both methods are very similar. They are further corroborated by Seitz (2019) where a similar study is presented.

The concern for the numerical integration on the contact boundary can be included in the practical aspects for the implementation of these methods, whether for the augmented Lagrangian or the Nitsche method. This is common to all mortar-type methods, where an integration of quantities between two incompatible meshes (those of the slave surface and the master surface) is necessary. Given that integrals involve piecewise polynomials, optimal integration a priori requires sub-slicing the interface that complies with both meshes. When performed, this is usually achieved by sub-slicing the faces of the elements of the slave surface in accordance with a projection of the elements of the master surface in a deformed configuration as in Puso and Laursen (2004). This sub-slicing is not only quite complex but must a priori be performed at each iteration of the Newton method. In addition, when curved elements (isoparametric or isogeometric) are employed, or simply for elements with a degree greater than one, the intersections of the element faces become excessively complex and then need to be simplified.

In a linear case with transmission conditions, a comparison is achieved in Lacour et al. (1997). It also proposed non-symmetric integration method, which was then adapted in Mlika (2018) for the Nitsche method. Although it is clear that the sub-slicing of slave faces for integration provides additional accuracy, it is not clear whether the gain obtained is large enough to justify a step that remains computationally expansive. Therefore, in Farah et al. (2015), a comparison is made between an integration without slicing, with full slicing and with partial slicing. The partial slicing strategy consists of slicing only those elements that satisfy a discontinuity of the master surface. In the tests performed, partial slicing offers approximately the same accuracy as total slicing. In Mlika (2018), tests were carried out with total slicing and no slicing at all. In most of these tests performed with linear elements, total slicing does not lead to substantial gain in accuracy. However, a greater gain is reported for the use of quadratic elements. The tests presented in the following were carried out without slicing.

1.4.6.1. *Elastic half-ring*

This test was introduced in Fischer and Wriggers (2005). An elastic ring composed of two layers of neo-Hookean hyperelastic materials has the following strain energy:

$$W(C) = \frac{\mu}{2}(i_1(C) - 3) + \frac{\lambda}{4}(i_3(C) - 1) - (\frac{\mu}{2} - \frac{\lambda}{4}) \ln(i_3(C))$$

where λ and μ are the Lamé coefficients of the material, $C = F^T F$ is still the Cauchy–Green tensor and $i_1(C) = \text{trace}\ (C)$ and $i_3(C) = \det\ (C)$ are the first and third invariants of C. The outer ring has a Young's modulus of 10^3 MPa and the inner ring of 10^5 MPa. The Poisson coefficient is 0.3 for both materials. This significant difference in stiffness creates difficulties for the penalty method (see Fischer and Wriggers 2005). This half ring is pushed against an elastic block with a Young modulus of 300 MPa and a Poisson coefficient of 0.3. The dimensions are shown on the drawing at the top of Figure 1.5, which also specifies the geometry of the bodies in contact. The ends of the half ring are fixed and moved vertically 70 mm downwards in 140 steps of 0.5 mm each.

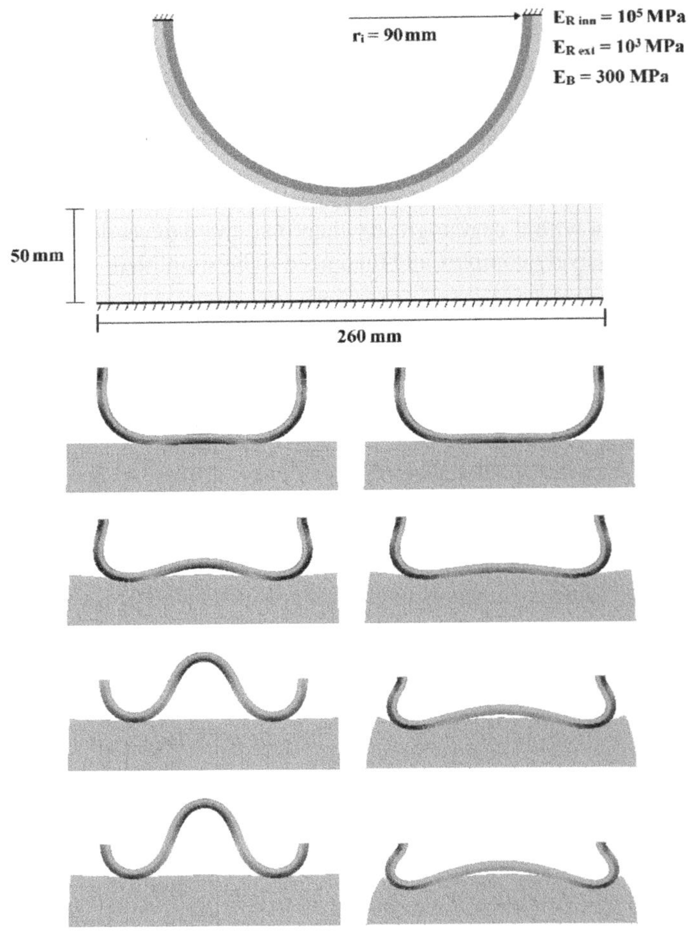

Figure 1.5. *Deformation of the elastic half-ring coming into contact without (a) and with friction for $\mathcal{F}$=0.5 (b) for quadratic quadrilateral finite elements, at loading steps corresponding to 25, 45, 60 and 70 mm of penetration. For a color version of this figure, see www.iste.co.uk/lebon/mechanics.zip*

Figure 1.5 shows four strain states at different loading steps. Graphs (a) correspond to frictionless contact, while graphs (b) correspond to Coulomb friction contact and

$\mathscr{F} = 0.5$. The graphs presented correspond to the use of the variant $\theta = 0$ of the unbiased Nitsche method for a Nitsche parameter $\gamma = \gamma_0/h$ with h the mesh size, γ_0 equal to the Young modulus of the material (10^3 MPa for the half ring and 300 MPa for the block) and the use of the raytracing pairing strategy. In order to be able to compare the results obtained to those of the existing literature, the vertical displacement of the midpoint of the ring is plotted in Figure 1.6 in both cases, with or without friction. The results obtained with Nitsche are in very good agreement with those produced in Poulios and Renard (2015) with an augmented Lagrangian formulation.

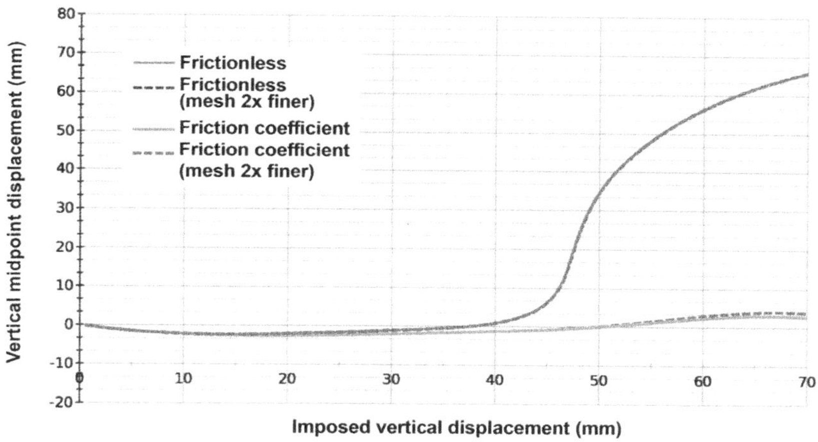

Figure 1.6. *Vertical displacement of the half-ring midpoint for different mesh sizes. For a color version of this figure, see www.iste.co.uk/lebon/mechanics.zip*

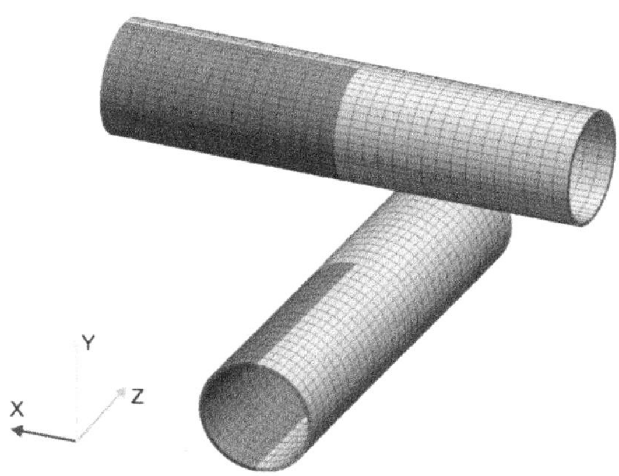

Figure 1.7. *Geometry and mesh of hollow tubes in their undeformed configurations. For a color version of this figure, see www.iste.co.uk/lebon/mechanics.zip*

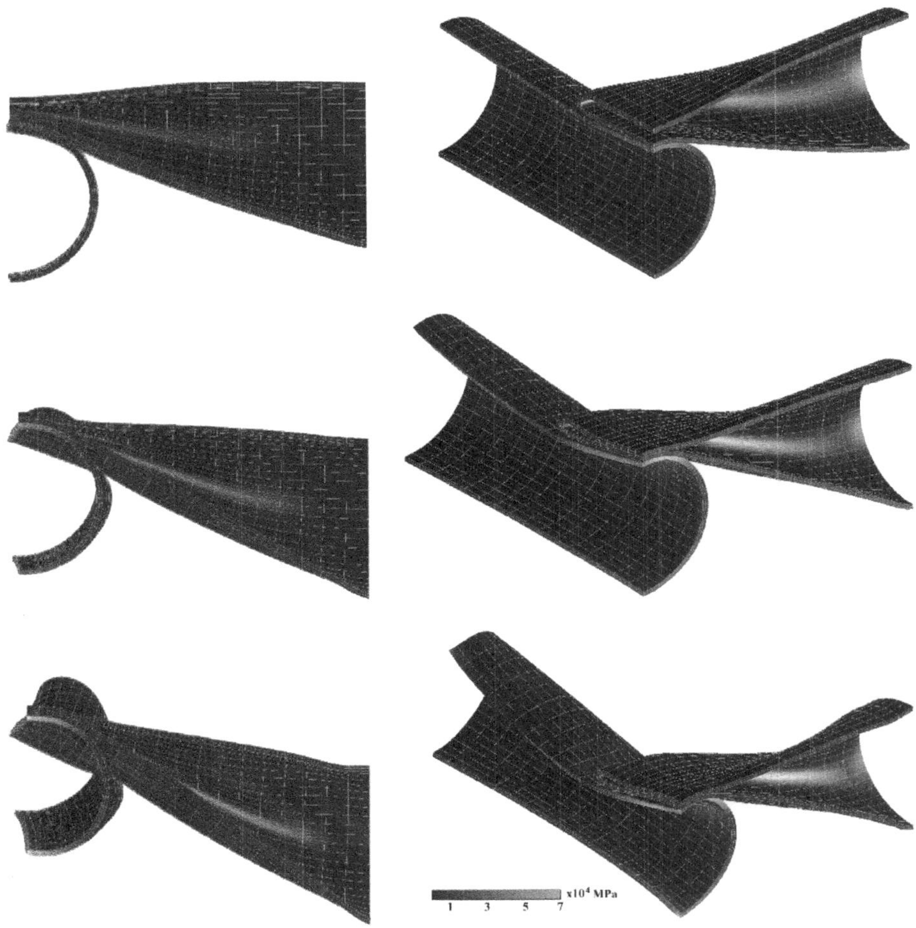

Figure 1.8. *Von-Mises strain and stress of the two crossed tubes. Frictionless example for a penetration of 20 mm, 30 mm and 40 mm, respectively. For a color version of this figure, see www.iste.co.uk/lebon/mechanics.zip*

1.4.6.2. *Crossing hollow tubes with self-contact*

In this example, we study the contact between two crossing hollow tubes leaning on each other. The tubes are 100 mm long, with an outer diameter of 24 mm and wall thickness of 0.8 mm. The material is considered to be neo-Hookean with a Young modulus $E_1 = 10^5$ MPa for the bottom tube and $E_2 = 10^4$ MPa for the top tube. The Poisson coefficient is 0.3 for both tubes. Quadratic hexahedral elements are used with two elements in the thickness.

The meshes that were used are visible in Figure 1.7 and are represented in green the parts where the computation actually takes place; the symmetry of the problem is used to reduce the computation area. The ends of the tubes are subject to imposed displacement conditions, zero for the bottom tube and downward for the top tube with a total imposed displacement of 40 mm divided into 80 equal loading steps. This brings the two tubes into contact and induces strains large enough to observe self-contact inside the top tube, as shown in Figure 1.8. The variant $\theta = 0$ of the unbiased Nitsche method is used with all the inner and outer surfaces of the tubes as contact surface. The Nitsche parameter employed is still $\gamma = \gamma_0/h$ with γ_0 the Young modulus of the solid on which we integrate and h the mesh size.

In Figure 1.8, it can be seen that the self-contact condition is correctly taken into account despite the very large deformations that are involved. Here too, the results are in very strong agreement with those obtained for the augmented Lagrangian on the same test case in Poulios and Renard (2015).

1.5. Acknowledgments

We would like to thank our collaborators Guillaume Drouet, Mathieu Fabre, Vanessa Lleras, Rabii Mlika and Konstantinos Poulios who participated, often decisively, in the research described in this chapter.

1.6. References

Abbas, M., Drouet, G., Hild, P. (2018). The local average contact (LAC) method. *Computer Methods in Applied Mechanics and Engineering*, 339, 488–513.

Alart, P. (1997). Méthode de Newton généralisée en mécanique du contact. *J. Math. Pures Appl.*, 76(1), 83–108.

Alart, P. and Curnier, A. (1991). A mixed formulation for frictional contact problems prone to Newton like solution methods. *Computer Methods in Applied Mechanics and Engineering*, 92(3), 353–375.

Annavarapu, C., Hautefeuille, M., Dolbow, J.E. (2014). A Nitsche stabilized finite element method for frictional sliding on embedded interfaces. Part I: Single interface. *Computer Methods in Applied Mechanics and Engineering*, 268, 417–436.

Antolin, P., Buffa, A., Fabre, M. (2019). A priori error for unilateral contact problems with Lagrange multipliers and isogeometric analysis. *IMA J. Numer. Anal.*, 39(4), 1627–1651.

Ballard, P. and Iurlano, F. (2023). Optimal existence results for the 2d elastic contact problem with Coulomb friction. *arXiv:2303.16298*.

Barbosa, H.J.C. and Hughes, T.J.R. (1992a). Boundary Lagrange multipliers in finite element methods: Error analysis in natural norms. *Numer. Math.*, 62(1), 1–15.

Barbosa, H.J.C. and Hughes, T.J.R. (1992b). Circumventing the Babuška-Brezzi condition in mixed finite element approximations of elliptic variational inequalities. *Computer Methods in Applied Mechanics and Engineering*, 97(2), 193–210.

Beaude, L., Chouly, F., Laaziri, M., Masson, R. (2023). Mixed and Nitsche's discretizations of Coulomb frictional contact-mechanics for mixed dimensional poromechanical models. *Comput. Methods Appl. Mech. Engrg.*, 413, 116124.

Ben Belgacem, F., Hild, P., Laborde, P. (1999). Extension of the mortar finite element method to a variational inequality modeling unilateral contact. *Math. Models Methods Appl. Sci.*, 9(2), 287–303.

Bertsekas, D.P. (1982). *Constrained Optimization and Lagrange Multiplier Methods*. Academic Press, New York/London.

Brézis, H. (1973). *Opérateurs maximaux monotones et semi-groupes de contractions dans les espaces de Hilbert*. Elsevier, New York.

Burman, E. (2012). A penalty-free nonsymmetric Nitsche-type method for the weak imposition of boundary conditions. *SIAM J. Numer. Anal.*, 50(4), 1959–1981.

Burman, E. and Hansbo, P. (2017). Deriving robust unfitted finite element methods from augmented Lagrangian formulations. In *Geometrically Unfitted Finite Element Methods and Applications*, Bordas, S.P.A., Burman, E., Larson, M.G., Olshanskii, M.A. (eds). Springer, Cham, 1–24.

Burman, E., Hansbo, P., Larson, M.G. (2017). The penalty-free Nitsche method and nonconforming finite elements for the Signorini problem. *SIAM J. Numer. Anal.*, 55(6), 2523–2539.

Burman, E., Hansbo, P., Larson, M.G. (2019). Augmented Lagrangian finite element methods for contact problems. *ESAIM, Math. Model. Numer. Anal.*, 53(1), 173–195.

Burman, E., Hansbo, P., Larson, M.G. (2023). The augmented Lagrangian method as a framework for stabilised methods in computational mechanics. *Arch. Comput. Methods Eng.*, 30(4), 2579–2604.

Cascavita, K.L., Chouly, F., Ern, A. (2020). Hybrid high-order discretizations combined with Nitsche's method for Dirichlet and Signorini boundary conditions. *IMA J. Numer. Anal.*, 40(4), 2189–2226.

Chouly, F. (2014). An adaptation of Nitsche's method to the Tresca friction problem. *J. Math. Anal. Appl.*, 411(1), 329–339.

Chouly, F. (2023). A review on some discrete variational techniques for the approximation of essential boundary conditions. hal-03741809.

Chouly, F. and Hild, P. (2013a). A Nitsche-based method for unilateral contact problems: Numerical analysis. *SIAM J. Numer. Anal.*, 51(2), 1295–1307.

Chouly, F. and Hild, P. (2013b). On convergence of the penalty method for unilateral contact problems. *Appl. Numer. Math.*, 65, 27–40.

Chouly, F., Hild, P., Renard, Y. (2015). Symmetric and non-symmetric variants of Nitsche's method for contact problems in elasticity: Theory and numerical experiments. *Math. Comp.*, 84(293), 1089–1112.

Chouly, F., Fabre, M., Hild, P., Mlika, R., Pousin, J., Renard, Y. (2017). An overview of recent results on Nitsche's method for contact problems. In *Geometrically Unfitted Finite Element Methods and Applications,* Bordas, S.P.A., Burman, E., Larson, M.G., Olshanskii, M.A. (eds). Springer, Cham, 93–141.

Chouly, F., Hild, P., Lleras, V., Renard, Y. (2019). Nitsche-based finite element method for contact with Coulomb friction. In *Numerical Mathematics and Advanced Applications ENUMATH 2017*, Radu, F.A., Kumar, K., Berre, I. et al. (eds). Springer, Cham, 839–847.

Chouly, F., Ern, A., Pignet, N. (2020). A hybrid high-order discretization combined with Nitsche's method for contact and Tresca friction in small strain elasticity. *SIAM J. Sci. Comput.*, 42(4), A2300–A2324.

Chouly, F., Hild, P., Lleras, V., Renard, Y. (2022). Nitsche method for contact with Coulomb friction: Existence results for the static and dynamic finite element formulations. *J. Comput. Appl. Math.*, 416, 18.

Chouly, F., Hild, P., Renard, Y. (2023). *Finite Element Approximation of Contact and Friction in Elasticity*. Birkhäuser, Cham. doi: 10.1007/978-3-031-31423-0.

Christensen, P.W. (2002). A semi-smooth Newton method for elasto-plastic contact problems. *Int. J. Solids Struct.*, 39(8), 2323–2341.

Ciarlet, P.G. (1988). *Mathematical Elasticity. Volume 1: Three-Dimensional Elasticity*. North-Holland Publishing Co., Amsterdam.

Ciarlet, P.G. (1991). The finite element method for elliptic problems. In *Handbook of Numerical Analysis, II*, Ciarlet, P.G. and Lions, J.L. (eds). North Holland, Amsterdam, 17–352.

Curnier, A. and Alart, P. (1988). A generalized Newton method for contact problems with friction. *J. Méc. Théor. Appl.*, 7(suppl. 1), 67–82.

Curnier, A., He, Q.C., Klarbring, A. (1995). Continuum mechanics modelling of large deformation contact with friction. In *Contact Mechanics*, Raous, M., Jean, M., Moreau, J.J. (eds). Springer, Boston, MA, 145–158.

De Lorenzis, L., Wriggers, P., Zavarise, G. (2012). A mortar formulation for 3D large deformation contact using NURBS-based isogeometric analysis and the augmented Lagrangian method. *Comput. Mech.*, 49(1), 1–20.

Dione, I. (2020). Optimal error estimates of the unilateral contact problem in a curved and smooth boundary domain by the penalty method. *IMA J. Numer. Anal.*, 40(1), 729–763.

Drouet, G. and Hild, P. (2015). Optimal convergence for discrete variational inequalities modelling Signorini contact in 2D and 3D without additional assumptions on the unknown contact set. *SIAM J. Numer. Anal.*, 53(3), 1488–1507.

Drouet, G. and Hild, P. (2017). An accurate local average contact method for nonmatching meshes. *Numer. Math.*, 136(2), 467–502.

Duvaut, G. and Lions, J.L. (1972). *Les inéquations en mécanique et en physique*. Dunod, Paris.

Eck, C., Jarušek, J., Krbec, M. (2005). *Unilateral Contact Problems*. CRC Press, Boca Raton, FL.

Farah, P., Popp, A., Wall, W.A. (2015). Segment-based vs. element-based integration for mortar methods in computational contact mechanics. *Comput. Mech.*, 55(1), 209–228.

Fischer, K.A. and Wriggers, P. (2005). Frictionless 2D contact formulations for finite deformations based on the mortar method. *Comput. Mech.*, 36(3), 226–244.

Glowinski, R. (1984). *Numerical Methods for Nonlinear Variational Problems*. Springer-Verlag, New York.

Gurtin, M.E. (1981). *An Introduction to Continuum Mechanics*. Academic Press, New York/London.

Gustafsson, T., Stenberg, R., Videman, J. (2017a). Mixed and stabilized finite element methods for the obstacle problem. *SIAM Journal on Numerical Analysis*, 55(6), 2718–2744.

Gustafsson, T., Stenberg, R., Videman, J. (2017b). A stabilised finite element method for the plate obstacle problem. arXiv:1711.04166.

Gustafsson, T., Stenberg, R., Videman, J. (2019). A stabilised finite element method for the plate obstacle problem. *BIT*, 59(1), 97–124. doi: 10.1007/s10543-018-0728-7.

Haslinger, J. (1983). Approximation of the Signorini problem with friction, obeying the Coulomb law. *Math. Methods Appl. Sci.*, 5(3), 422–437.

Hassani, R., Hild, P., Ionescu, I.R., Sakki, N. (2003). A mixed finite element method and solution multiplicity for Coulomb frictional contact. *Comput. Methods Appl. Mech. Engrg.*, 192(41–42), 4517–4531.

Hild, P. (2000). Numerical implementation of two nonconforming finite element methods for unilateral contact. *Comput. Methods Appl. Mech. Engrg.*, 184(1), 99–123.

Hild, P. (2002). On finite element uniqueness studies for Coulomb's frictional contact model. *Appl. Math. Comp. Sci.*, 12(1), 41–50.

Hild, P. (2003). An example of nonuniqueness for the continuous static unilateral contact model with Coulomb friction. *C. R. Math. Acad. Sci. Paris*, 337(10), 685–688.

Hild, P. (2004). Non-unique slipping in the Coulomb friction model in two-dimensional linear elasticity. *Quart. J. Mech. Appl. Math.*, 57(2), 225–235.

Hild, P. and Renard, Y. (2007). An error estimate for the Signorini problem with Coulomb friction approximated by finite elements. *SIAM J. Numer. Anal.*, 45(5), 2012–2031.

Hild, P. and Renard, Y. (2010). A stabilized Lagrange multiplier method for the finite element approximation of contact problems in elastostatics. *Numer. Math.*, 115(1), 101–129.

Hu, Q., Chouly, F., Hu, P., Cheng, G., Bordas, S.P.A. (2018). Skew-symmetric Nitsche's formulation in isogeometric analysis: Dirichlet and symmetry conditions, patch coupling and frictionless contact. *Computer Methods in Applied Mechanics and Engineering*, 341, 188–220.

Juntunen, M. (2015). On the connection between the stabilized Lagrange multiplier and Nitsche's methods. *Numer. Math.*, 131(3), 453–471.

Kikuchi, N. and Oden, J.T. (1988). *Contact Problems in Elasticity: A Study of Variational Inequalities and Finite Element Methods*. SIAM, Philadelphia.

Kim, J.Y. and Youn, S.K. (2012). Isogeometric contact analysis using mortar method. *Numerical Methods in Engineering*, 89(12), 1559–1581.

Konyukhov, A. and Schweizerhof, K. (2012). *Computational Contact Mechanics: Geometrically Exact Theory for Arbitrary Shaped Bodies*. Springer Science & Business Media, New York.

Krstulović-Opara, L., Wriggers, P., Korelc, J. (2002). A C^1-continuous formulation for 3D finite deformation frictional contact. *Comput. Mech.*, 29(1), 27–42.

Kunisch, K. and Stadler, G. (2005). Generalized Newton methods for the 2D-Signorini contact problem with friction in function space. *M2AN Math. Model. Numer. Anal.*, 39(4), 827–854.

Laborde, P. and Renard, Y. (2008). Fixed point strategies for elastostatic frictional contact problems. *Math. Methods Appl. Sci.*, 31(4), 415–441.

Lacour, C., Maday, Y., Cazabeau, L. (1997). Numerical quadratures and mortar methods. In *Computational Science for the 21st Century*, Bristeau, M.-O., Etgen, G.J., Fitzgibbon, W. et al. (eds). John Wiley & Sons, Chichester, 119–128.

Laursen, T.A. (2002). *Computational Contact and Impact Mechanics*. Springer-Verlag, Berlin.

Laursen, T.A. and Simo, J.C. (1993). Algorithmic symmetrization of Coulomb frictional problems using augmented lagrangians. *Computer Methods in Applied Mechanics and Engineering*, 108(1–2), 133–146.

Lengiewicz, J., Korelc, J., Stupkiewicz, S. (2011). Automation of finite element formulations for large deformation contact problems. *Numerical Methods in Engineering*, 85(10), 1252–1279.

Lleras, V. (2009). A stabilized Lagrange multiplier method for the finite element approximation of frictional contact problems in elastostatics. *Math. Model. Nat. Phenom.*, 4(1), 163–182.

Mlika, R. (2018). Nitsche method for frictional contact and self-contact: Mathematical and numerical study. PhD Thesis, Institut National des Sciences Appliquées de Lyon.

Mlika, R., Renard, Y., Chouly, F. (2017). An unbiased Nitsche's formulation of large deformation frictional contact and self-contact. *Computer Methods in Applied Mechanics and Engineering*, 325, 265–288.

Nitsche, J. (1971). Über ein Variationsprinzip zur Lösung von Dirichlet-Problemen bei Verwendung von Teilräumen, die keinen Randbedingungen unterworfen sind. *Abhandlungen aus dem Mathematischen Seminar der Universität Hamburg*, 36, 9–15.

Ogden, R.W. (1984). *Nonlinear Elastic Deformations*. Dover Publications Inc., New York.

Padmanabhan, V. and Laursen, T.A. (2001). A framework for development of surface smoothing procedures in large deformation frictional contact analysis. *Finite Elem. Anal. Des.*, 37(3), 173–198.

Popp, A., Wohlmuth, B.I., Gee, M.W., Wall, W.A. (2012). Dual quadratic mortar finite element methods for 3D finite deformation contact. *SIAM J. Sci. Comput.*, 34(4), B421–B446.

Popp, A., Seitz, A., Gee, M.W., Wall, W.A. (2013). Improved robustness and consistency of 3D contact algorithms based on a dual mortar approach. *Computer Methods in Applied Mechanics and Engineering*, 264, 67–80.

Poulios, K. and Renard, Y. (2015). An unconstrained integral approximation of large sliding frictional contact between deformable solids. *Comput. Struct.*, 153, 75–90.

Puso, M.A. and Laursen, T.A. (2004). A mortar segment-to-segment frictional contact method for large deformations. *Computer Methods in Applied Mechanics and Engineering*, 193(45–47), 4891–4913.

Renard, Y. (2006). A uniqueness criterion for the Signorini problem with Coulomb friction. *SIAM J. Math. Anal.*, 38(2), 452–467.

Renard, Y. (2013). Generalized Newton's methods for the approximation and resolution of frictional contact problems in elasticity. *Computer Methods in Applied Mechanics and Engineering*, 256, 38–55.

Renard, Y. and Poulios, K. (2020). GetFEM: Automated FE modeling of multiphysics problems based on a generic weak form language. *ACM Trans. Math. Softw.*, 47(1), 1–31.

Rockafellar, R.T. (1974). Augmented Lagrange multiplier functions and duality in nonconvex programming. *SIAM J. Control*, 12, 268–285.

Rockafellar, R.T. (1976). Augmented Lagrangians and applications of the proximal point algorithm in convex programming. *Math. Oper. Res.*, 1(2), 97–116.

Rockafellar, R.T. and Wets, R.J.B. (1998). *Variational Analysis*. Springer-Verlag, Berlin.

Sauer, R.A. and De Lorenzis, L. (2015). An unbiased computational contact formulation for 3D friction. *Numerical Methods in Engineering*, 101(4), 251–280.

Seitz, A. (2019). Computational methods for thermo-elasto-plastic contact. PhD Thesis, Technische Universität München.

Seitz, A., Farah, P., Kremheller, J., Wohlmuth, B.I., Wall, W.A., Popp, A. (2016). Isogeometric dual mortar methods for computational contact mechanics. *Computer Methods in Applied Mechanics and Engineering*, 301(1), 259–280.

Seitz, A., Wall, W.A., Popp, A. (2019). Nitsche's method for finite deformation thermomechanical contact problems. *Comput. Mech.*, 63, 1091–1110.

Simo, J.C. and Laursen, T.A. (1992). An augmented Lagrangian treatment of contact problems involving friction. *Comput. Struct.*, 42(1), 97–116.

Stadler, G. (2004). Infinite-dimensional semi-smooth Newton and augmented Lagrangian methods for friction and contact problems in elasticity. PhD Thesis, University of Graz.

Stadler, M., Holzapfel, G.A., Korelc, J. (2003). C^n continuous modelling of smooth contact surfaces using NURBS and application to 2D problems. *Numerical Methods in Engineering*, 57(15), 2177–2203.

Stenberg, R. (1995). On some techniques for approximating boundary conditions in the finite element method. *J. Comput. Appl. Math.*, 63(1–3), 139–148.

Temizer, I., Wriggers, P., Hughes, T.J.R. (2011). Contact treatment in isogeometric analysis with NURBS. *Computer Methods in Applied Mechanics and Engineering*, 200(9–12), 1100–1112.

Temizer, I., Wriggers, P., Hughes, T.J.R. (2012). Three-dimensional mortar-based frictional contact treatment in isogeometric analysis with NURBS. *Computer Methods in Applied Mechanics and Engineering*, 209/212, 115–128.

Wohlmuth, B.I. (2011). Variationally consistent discretization schemes and numerical algorithms for contact problems. *Acta Numer.*, 20, 569–734.

Wriggers, P. (2002). *Computational Contact Mechanics*. Wiley, Berlin.

Wriggers, P., Rust, W.T., Reddy, B.D. (2016). A virtual element method for contact. *Comput. Mech.*, 58(6), 1039–1050.

2

High-performance Computing in Multicontact Mechanics: From Elastostatics to Granular Dynamics

Pierre ALART
LMGC, Université de Montpellier, France

2.1. Introduction

Contact problems are known for being highly complex in the field of numerical mechanics for several reasons. Contact occurs at the interface of bodies, either as boundary conditions in the framework of continuous modeling of solids, or as an interaction law between bodies modeled by rigid solids, or even particles. In the first case, the numerical modeling of body boundaries is already challenging if the goal is to accurately account for curvatures, corners, surface imperfections, CAD models, etc. Modeling the interaction of two boundaries increases this complexity. A fine modeling of the boundaries therefore leads to no longer considering contact between two boundaries, but a multitude of potential contacts between contact control entities, as, for example, Gauss points on the edges of finite elements facing each other. In that respect, such a problem may be characterized as a multicontact problem. In the second case, unless only two particles in contact are being considered, a system comprising a large number of rigid grains, even convex ones, leads to a problem with many simultaneous potential contacts, which can thereby be called a *multi-contact* problem.

Numerical Methods for Strong Nonlinearities in Mechanics,
coordinated by Jacques BESSON, Frédéric LEBON and Eric LORENTZ. © ISTE Ltd 2024.

Contact problems are highly nonlinear. As such, their solution requires an iterative algorithm whose convergence becomes erratic with stronger nonlinearity and involves a large number of variables and equations. The studies developed here particularly concern this algorithmic aspect. However, contact relationships are not only nonlinear, but also *non-smooth*. The term "non-smooth" can be given several interpretations. Mathematically, the relationships involved are not only equations, but also inequalities.

More finely, Moreau (1999, 2003) makes a distinction between three types of non-smoothness. *Non-smoothness* in space arises from the geometric conditions of non-interpenetration of bodies, which are conditions expressed by inequalities involving the quantities describing the boundaries. Non-interpenetrability bonds are mechanically achieved involving contact forces (or reactions), governed by very irregular laws that also resort to inequalities, whether it is Signorini's law for unilateral contact or Coulomb friction. We are therefore in the presence of *non-smoothness in laws*. Finally, if a collision occurs between bodies treated as being rigid, speed jumps are expected, representative of *non-smoothness in time*. This last non-smoothness will be addressed in section 2.4 concerning the dynamics of granular media.

Regarding multi-contact situations, we focus on systems that we refer to as *diffuse non-smoothness* systems, that is, for which non-smoothness enters the system through a large number of potential contacts. The purpose of iterative algorithms is to determine, on the one hand, effective contacts, but also the intensity of the efforts passing through these contacts. While granular media clearly fall into this category, with contacts or interactions playing a major role in the overall behavior of the system, there are comparable situations existing in structural calculus, as discussed in section 2.2.

It is then interesting, from a numerical strategy point of view, that these problems be addressed, which are not resulting from contact, but from another diffuse non-smoothness. This is the case of tensegrity systems addressed in section 2.3, which are structural problems (as in section 2.2), yet discrete and with very diffuse non-smoothness, such as the granular media of section 2.4.

Given the diversity of the problems considered, only the structure and mathematical properties of the different problems will be specified, but it will not be possible to explain in detail all equations on volume or interface behavior laws. Strategies motivated by mechanics rather than computer science are also treated. Therefore, all sections of this chapter present studies on domain decomposition methods for which problems homogenized by subdomain can function as accelerators of convergence.

2.2. Multicontact in elastostatics

2.2.1. *Development framework*

The problems considered in this section concern contacts between deformable bodies and are developed within the framework of finite element. The simplest problem consists of balancing an elastic body modeled under the hypothesis of small perturbations in grazing contact, without friction, on part of its boundary with a rigid obstacle. Once discretized into finite elements, the mathematical problem consists of finding the assembled nodal displacements in a vector u of size $n_u := n_n \times d$, where n_n is the number of nodes and $d = 2.3$ is the dimension of the modeling space, as well as the normal contact reactions assembled in the vector λ of size n_c, namely the number of contact control points. Control points can be mesh nodes such as Gauss points on the regions of the finite elements of the potential contact surface. The relative displacements normal to the contact surface of the control points are for a grazing contact expressed using a linear operator denoted by H^t, such that the Signorini contact law on the potential contact surface is expressed using the following *complementarity relations*, at each control point indexed by α:

$$(H^t u)_\alpha \geq 0, \lambda_\alpha \geq 0, (H^t u)_\alpha \lambda_\alpha = 0, \alpha = 1, n_c \qquad [2.1]$$

or globally on assembled vectors:

$$H^t u \geq 0, \lambda \geq 0, \lambda . H^t u = 0 \Leftrightarrow 0 \leq H^t u \perp \lambda \geq 0 \qquad [2.2]$$

The problem to be solved is therefore written synthetically as:

$$\text{Find } (u, \lambda) \in \mathbb{R}^{n_u \times n_c} \text{ such that } \begin{cases} Ku - H\lambda = f \\ 0 \leq H^t u \perp \lambda \geq 0 \end{cases} \qquad [2.3]$$

where the first equation is the equilibrium equation: K is the stiffness matrix of the elastic body and f is the vector of the external nodal forces. As it appears in [2.3], the problem is a *linear complementarity problem* (LCP) for which there are various efficient solvers available as long as the system is not too large.

This formulation can be generalized to more complex situations: elasticity is modeled in finite transformation (internal forces depend nonlinearly on u); the contact is not grazing but the contact kinematic variable is a contact distance or gap ($d(u)$ such that $\nabla d(u) = H^t$); the contact can occur on a non-flat surface (H^t depends on u); the contact occurs between several elastic bodies (u is the concatenation of the displacements of each component); the contact relations at each control point not only include the Signorini law, but also Coulomb's friction, or even more complex interaction laws (adhesion, wear), expressed as relations between $d_\alpha(u)$ and λ_α, formally synthesized as $\mathcal{R}(d(u), \lambda) = 0$. The notation λ covers the vector of contact or interaction reactions with not only normal, but also tangential components. The system to be solved then assumes the following, more general form:

$$\begin{cases} F_{int}(u) - H(u)\lambda = f \\ \mathcal{R}(d(u), \lambda) = 0 \end{cases} \qquad [2.4]$$

Such a system can more generally constitute the solution of the equilibrium of a structure at each time or a loading step in an incremental process. For different situations, refer to Wriggers (2012). The system thus formulated can be linear for the first equation (equilibrium), but is nonlinear, and even non-smooth for interaction relations; it is even undefined for some values of λ or $d(u)$ if the inequality constraints are not satisfied. If we now only consider relations of contact and Coulomb friction at the interfaces, the *mixed formulation* of Alart and Curnier (1991) leads to obtaining a system definite and continuous everywhere, but not differentiable everywhere:

$$\begin{cases} Ku - Hp(d(u), \lambda) = f \\ \lambda - p(d(u), \lambda) = 0 \end{cases} \qquad [2.5]$$

The operator $p(.,.)$ includes projections on convex sets (see Alart and Curnier 1991) ensuring continuity. Differentiability can only be weakly defined, as the generalized *Jacobian* of Clarke (1990). Therefore, by setting $\tilde{u} := (u, \lambda)^t$, the previous equations constitute a non-differentiable system $G(\tilde{u}) = 0$, which can be solved using a *generalized Newton method* introduced by Curnier and Alart (1988) and studied by Alart (1997). At each Newton iteration, a linear system with the following form has to be solved:

$$\tilde{K}\tilde{u} = \tilde{f}, \text{ with } \tilde{u} := (u^t, \lambda^t)^t \qquad [2.6]$$

where the matrix $\tilde{K}$ is an "augmented" stiffness, potentially having *zeros on the diagonal* and *non-symmetric* if addressing Coulomb friction as developed in Alart and Curnier (1991). Generally in structural problems, the bodies are fixed on a portion of their boundary in such a way that rigid solid movements are excluded; the stiffness matrix K is therefore invertible, and the "augmented" matrix $\tilde{K}$ is also invertible. When this is not the case, only the effectiveness of the contact (active contact) eliminates rigid solid motion; the matrix K is no longer invertible, but the matrix $\tilde{K}$, corresponding to enough active contacts, is invertible. Somehow, we then "manage" to start the Newton algorithm in such a configuration.

In structural computation, a *multicontact* problem refers to situations where the number of contacts is high compared to the number of degrees of freedom; however, without the former exceeding the latter. The situation where the number of contacts exceeds, sometimes significantly, the number of degrees of freedom mainly concerns granular media modeled as a collection of rigid grains, which are covered in section 2.4. However, as recalled in the introduction, we are mainly focusing on problems *with diffuse non-smoothness*.

Two examples illustrate this concept in structural calculus in a somewhat "dual" way. The first is a *collection of deformable grains* as shown in Figure 2.1. In this case, each grain is a body that is in equilibrium only by contacts with its neighbors; the contacts are initially grazing, the number of contact areas is significant, but the number of contacts is much less than the number of degrees of freedom.

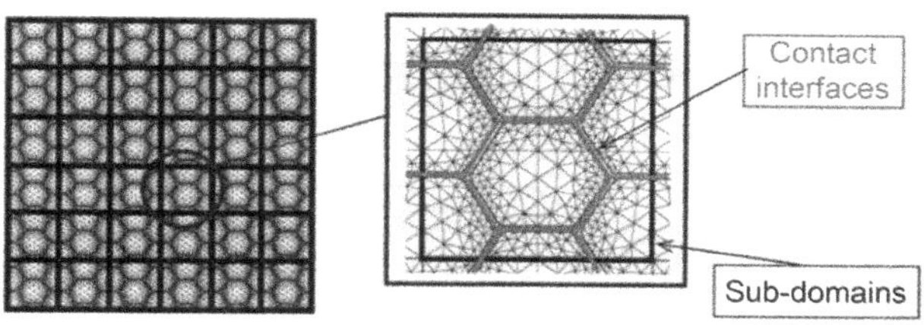

Figure 2.1. *Subdomain decomposition of a collection of deformable hexagonal grains (source: Alart et al. 2001). For a color version of this figure, see www.iste.co.uk/lebon/mechanics.zip*

The second example is a honeycomb *cell structure* shown in Figure 2.2. In this case, the deformable structure is connected and fixing the structure to its edges is sufficient to eliminate any rigid solid motion. There are many areas of potential contact in each cell, but the number of active contacts will remain very small. Finally, the study of buckling in such structures carried out by Alart et al. (2004) requires modeling in finite transformations, which adds a strong nonlinearity. These two examples are discussed in section 2.2.3.

2.2.2. *Parallel solver preconditioning*

It is quite natural to try to generalize the efficient computational approaches dedicated to large-sized structural problems to contact and friction ones. Within the finite element context, one of the approaches developed during the 1980s was the *element-by-element* (EBE) method by Hughes et al. (1983) consisting of, for the context of a conjugate gradient method dedicated to solving a linear system generated by a finite element method, constructing a preconditioner based on the said

Winget factorization of elementary matrices. It is therefore no longer necessary to assemble either the initial stiffness matrix or its approximation by the EBE method. The solution is thus highly parallelizable. The management of contact using the approach of section 2.2.1 is achieved through the notion of *contact element* by analogy with finite elements. This thus makes it possible to generalize the EBE method to contact elements (see Alart et al. 1997). It was first necessary to use a double conjugate gradient method, take the characteristics of matrices resulting from a mixed formulation into account, improve the EBE method for elastostatic problems (Barboteu et al. 1998) and define new approximate factorizations for contact type elementary matrices (see Alart et al. 1997). The tests carried out perform similarly in terms of iterations and computation time or even better than ILU (*Incomplete LU factorization*). Table 2.1 provides one of the comparative results developed in Alart et al. (1997) with a benchmark set by Raous et al. (1988).

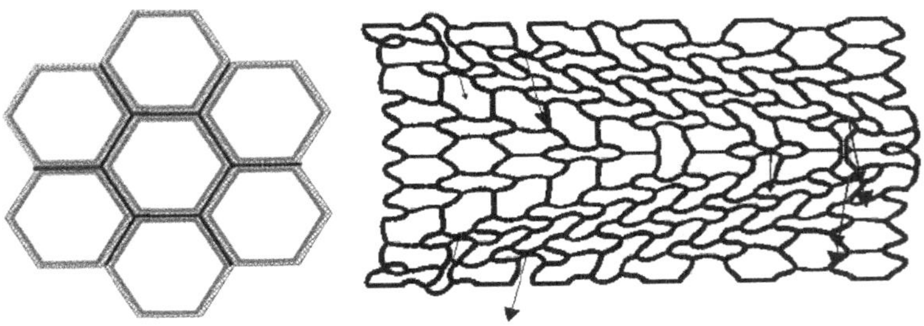

Figure 2.2. *Honeycomb cell medium decomposition into one subdomain per cell (interface at the center of the cell wall) and emergence of two shear bands for high compression with self-contact for a sample of 105 subdomains (cells) and a coefficient of friction of 0.5 (source: Alart et al. 2001). For a color version of this figure, see www.iste.co.uk/lebon/mechanics.zip*

This verifies that the generalized EBE method is faster both in terms of the average number of iterations of the double conjugate gradient (the average is achieved over all Newton iterations necessary for convergence) and computational time. This illustrates the advantage of adapting numerical approaches to the mechanical problem under consideration. Therefore, while the ILU preconditioner is applicable to any type of problem, the EBE preconditioner was specially designed for elastostatic contact problems. In Barboteu's thesis, the same study is developed with a problem of roller shutters with clearance and friction for which can be obtained a ratio of contacts to degrees of freedom all the more significant when considering an increasing number of articulated blades. Such a problem addresses the characteristic of diffuse non-smoothness, such as the problems mentioned in section 2.2.1. While the performances are comparable to those obtained with the example of Table 2.1, the double conjugate gradient proved to be very sensitive to the coefficient of friction, at least with the problem considered. Finally, EBE methods are well suited to vector computers developed in the 1980s and 1990s; they are less so on the various architectures that were subsequently developed, namely with shared or distributed memory or a combination of both. This avenue has therefore not been further investigated.

Indicator	EBE	ILU
Average number of iterations	740	884
CPU(s) time	763	1045

Table 2.1. *Performance of EBE and ILU preconditioners (source: Alart et al. 2001)*

2.2.3. *Domain decomposition: Newton–Schur solver*

The strategy combining mixed formulation and the generalized Newton method is tantamount to solving *in fine* a system such as [2.6]. The only difference with a linear elastostatic problem lies in the unknown x, which involves nodal displacements but also contact reactions. It is only natural to want to extend the domain decomposition methods developed for elastostatics. The work developed from Barboteu's thesis favored the primal approach developed by Le Tallec (1994) and Mandel (1993) with the dual approach of Farhat and Roux (1991), in which a decomposition of the geometric domain that excludes contact interfaces from decomposition interfaces is preferred. In other words, (numerical) decomposition interfaces can be differentiated from potential contact (physical) interfaces. This takes into account the diffuse nature of contact, whose processing is not solely confined to decomposition interfaces, but concerns the interior of the subdomains themselves, unlike the strategies adopted in Dureisseix and Farhat (2001) and Ladevèze et al. (2002). This strategy is illustrated in Figures 2.1 and 2.2 with the two examples shown in section 2.2.1. In the example using deformable grains, the decomposition interfaces slice the grains, the contact interfaces being internal to the subdomains with the possibility of intersecting the numerical interfaces. Concerning the honeycomb structure, numerical interfaces separate cell walls into a connected network without intersection with contact interfaces. The following presentation of the primal approach therefore remains formally very similar to that developed for elastostatics, the variable $\tilde{u}$ sometimes replacing the variable u.

The global domain Ω is therefore decomposed into n_s subdomains without overlapping $\Omega^{(s)}$, $\Omega = \bigcup_{s=1}^{n_s} \Omega^{(s)}$. The global interface Γ is the assembly of local interfaces:

$$\Gamma = \bigcup_{s=1}^{n_s} \Gamma^{(s)} \;\; with \;\; \Gamma^{(s)} = \partial\Omega^{(s)} \cap \left(\bigcup_{p=1,p\neq s}^{n_s} \partial\Omega^{(p)} \right) - \partial\Omega \qquad [2.7]$$

The vectors $\tilde{u}$ and $\tilde{f}$ are decomposed into internal components $\tilde{u}_I$ and $\tilde{f}_I$ and interface components u_Γ and f_Γ. The same procedure shall be applied to subdomains:

$$\tilde{u} = \left\{ \begin{array}{c} \tilde{u}_I \\ u_\Gamma \end{array} \right\}, \; \tilde{f} = \left\{ \begin{array}{c} \tilde{f}_I \\ f_\Gamma \end{array} \right\}, \; \tilde{u}^{(s)} = \left\{ \begin{array}{c} \tilde{u}_i^{(s)} \\ u_b^{(s)} \end{array} \right\}, \; \tilde{f}^{(s)} = \left\{ \begin{array}{c} \tilde{f}_i^{(s)} \\ f_b^{(s)} \end{array} \right\} \qquad [2.8]$$

Numerical interface variables therefore do not have an accent "~" because they are not concerned with contact management. As for a contactless elastostatic problem, the primal approach consists of imposing the continuity of displacement on the numerical interfaces – they are called *perfect* because they are not concerned by non-smoothness – by introducing the restriction operators of the global interface on the local interfaces:

$$R^{(s)} : \Gamma \to \Gamma^{(s)}, \; R^{(s)} : u_\Gamma \to u_b^{(s)} = R^{(s)} u_\Gamma \qquad [2.9]$$

The reference problem similar to [2.6] is then reformulated per subdomain:

$$\left\{ \begin{array}{l} \tilde{K}_{ii}^{(s)} \tilde{u}_i^{(s)} + K_{ib}^{(s)} R^{(s)} u_\Gamma = \tilde{f}_i^{(s)} \;, \; s = 1, n_s \\ \sum_s R^{(s)t} K_{bi}^{(s)} \tilde{u}_i^{(s)} + \sum_s R^{(s)t} K_{bb}^{(s)} R^{(s)} u_\Gamma = f_\Gamma = \sum_s R^{(s)t} f_b^{(s)} \end{array} \right. \qquad [2.10]$$

The interface problem, via a blockwise Gaussian elimination, thus restricted to (numeric) interface displacements, makes use of a global Schur matrix $\tilde{S}$ by assembly of local Schur matrices:

$$\begin{array}{l} \tilde{S} u_\Gamma \; = \; \tilde{b}, \; with \; \tilde{S} = \sum_s R^{(s)t} \tilde{S}^{(s)} R^{(s)}. \\ \tilde{S}^{(s)} = K_{bb}^{(s)} - K_{bi}^{(s)} \tilde{K}_{ii}^{(s)-1} K_{ib}^{(s)}, \; \tilde{b} = \sum_s R^{(s)t} (f_b^{(s)} - K_{bi}^{(s)} \tilde{K}_{ii}^{(s)-1} \tilde{f}_i^{(s)}) \end{array} \qquad [2.11]$$

The Schur matrix is not explicitly constructed and the system $\tilde{S} u_\Gamma = \tilde{b}$ is solved by a preconditioned conjugate gradient algorithm. Each iteration requires, as the most computationally time-consuming procedure, a matrix-vector product $\tilde{S}v$ that amounts to solving a linear system on each subdomain. This system formulates an elastostatic problem with contact conditions on the contact interface, set by the contact status of the current Newton iteration and with Dirichlet condition on the (numeric) domain boundary. These solutions can be performed in parallel, so here the advantage of domain decomposition methods for distributed memory computers can be clearly seen. Nevertheless, such algorithms are only effective with a suitable preconditioner consisting of a preconditioning matrix $\bar{S}^{-1}$, whose computation is easy and that approximates at best the inverse of $\tilde{S}$. Therefore, a parallel solver is said to be *scalable* if the number of iterations to converge does not increase when the number of subdomains is increased. The Neumann–Neumann preconditioner, associated with the primal approach, is tantamount to approaching the inverse of the

Schur matrix by a weighted sum of the inverses of the local Schur matrices, each local inversion amounting to a Neumann problem that is not well posed if the subdomain is "floating" (without sufficient Dirichlet conditions). This preconditioner is not scalable. In elastostatics, Mandel (1993) defined a scalable preconditioner called *balancing method* consisting of solving a coarse problem on the entire structure, based on the rigid solid modes of the subdomains. The choice of differentiating physical contact interfaces from numerical decomposition interfaces does not introduce any additional difficulty compared to an elastostatic problem in the implementation of a Schur solver. However, the *balancing method* preconditioner must be adapted to contact problems in order to regain scalability. Without going into the technical details available elsewhere in previous studies (Alart et al. 2000, 2001), the preconditioner is summarized in the following expression:

$$\bar{S}^{-1} = \left(I - \tilde{G}_*(\tilde{G}_*^t \tilde{S} \tilde{G}_*)^{-1} \tilde{G}_*^t \tilde{S}\right) \sum_s R^{(s)t} D^{(s)} \tilde{S}^{(s)+} D^{(s)} R^{(s)} \qquad [2.12]$$

in which we emphasize, in the following, the adaptations with respect to the elastostatic case contained in the expressions of the generalized inverse $\tilde{S}^{(s)+}$ and in the definition of the operator G_*.

Since the Schur matrix $\tilde{S}^{(s)}$, the generalized inverse $\tilde{S}^{(s)+}$ is not explicitly calculated, but amounts to solving a Neumann problem associated with the matrix $\tilde{K}^{(s)}$ on each subdomain. For a "floating" subdomain, the generalized inverse is defined as the inverse of a matrix regularized by the rigid solid modes – referred to here as *standard rigid modes* – contained in the kernel of $\tilde{K}^{(s)}$, denoted by $\tilde{N}^{(s)}$:

$$\tilde{K}^{(s)+} = (\tilde{K}^{(s)} + \tilde{N}^{(s)} \tilde{N}^{(s)t})^{-1} \qquad [2.13]$$

The first adaptation consists of taking into account possible additional rigid modes arising from the choice of substructuring. Such a situation is illustrated in Figures 2.3 and 2.4 for a collection of sheared deformable grains. Figure 2.3 shows a subdomain during a Newton iteration characterized by a global contact status of the different contact interfaces. Figure 2.4 then represents the additional rigid modes induced by this status; they consist of rigid solid movements of grains or pieces of grains authorized by this status. The detection of these modes is done automatically by the local solver, but the number of kernels collected must be correctly sized to the maximum number of possible modes.

The second adaptation is imposed by friction which induces non-symmetric $\tilde{K}^{(s)}$ matrices. Adopting the concept of additive Schwarz method, it is then necessary to define the operator $\tilde{G}_*$ from *dual rigid modes*, kernels $\tilde{N}_*^{(s)}$ of the transposed matrices $\tilde{K}^{(s)t}$ of $\tilde{K}^{(s)}$:

$$\tilde{G}_* = \left[R^{(1)t} D^{(1)} \tilde{N}_*^{(1)} \;\; ... \;\; R^{(n_s)t} D^{(n_s)} \tilde{N}_*^{(n_s)}\right] \qquad [2.14]$$

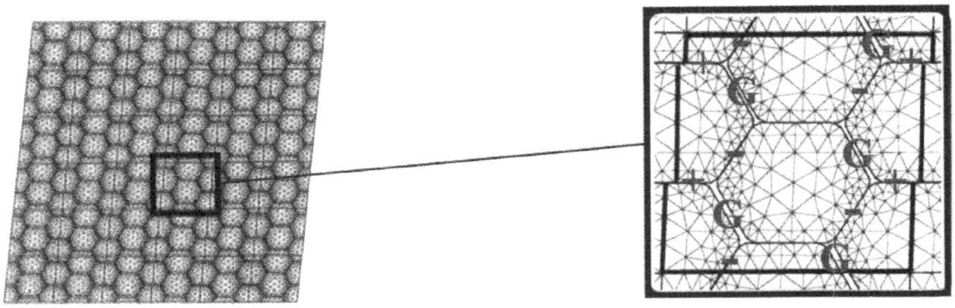

Figure 2.3. *Overall contact status at the internal contact interfaces to a subdomain during a given Newton iteration: G (gap) for non-contact, + and – for forward and backward sliding (source: Alart 2007). For a color version of this figure, see www.iste.co.uk/lebon/mechanics.zip*

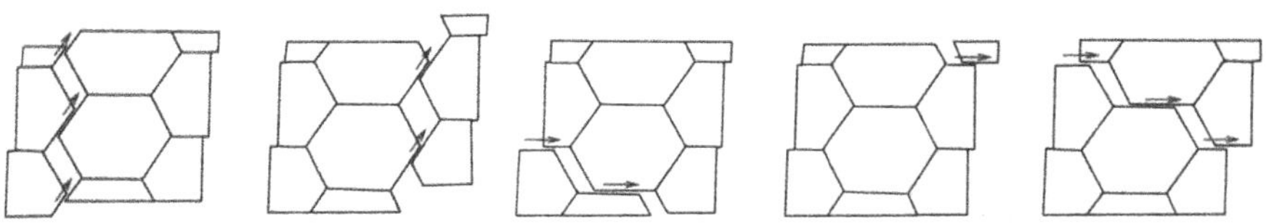

Figure 2.4. *Additional rigid modes associated with the overall status in Figure 2.3 (source: Alart 2007). For a color version of this figure, see www.iste.co.uk/lebon/mechanics.zip*

The operator $\tilde{G}_*$ is then involved in defining a coarse problem associated with the matrix $\tilde{G}_*^t \tilde{S} \tilde{G}_*$. It should be noted that this modification is not specific to friction but is generic for any problem involving non-symmetric matrices. Thus, we define the *balancing method* preconditioner, also known as Neumann–Neumann level-2, for symmetric problems that eliminate the distinction between standard and dual rigid modes. Such a preconditioner makes it possible to restore numerical extensibility as evidenced by the results summarized in Figure 2.5 that plot the evolution of the number of iterations of the GMRES solver as a function of the number of subdomains for two stresses of a sample of deformable grains. For a frictionless problem ($\mu = 0$), the standard Neumann–Neumann Level 2 preconditioner is optimal. It stops being so when friction is introduced ($\mu = 0.2$). In the case of shear, in order to find or approximate optimality, it is necessary to introduce how non-symmetry is addressed, as we have previously explained. The almost optimal behavior of a "more mechanically inspired" preconditioner should be mentioned; described as "specific", it consists of using the matrix preconditioner without friction. The results are even more conclusive with a roller shutter problem (the model is introduced in Ach and Alart (2001) and Barboteu et al. (2001)) for which is provided, in Figure 2.6, the performance of the different preconditioners according to the coefficient of friction and the number of sub-domains. More comprehensive comments are available in Alart et al. (2001) and Barboteu et al. (2001).

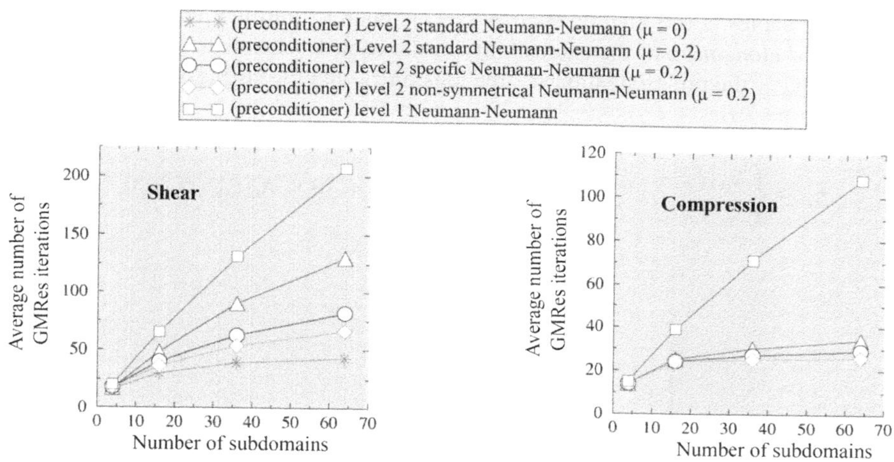

Figure 2.5. *Numerical scalability of the different preconditioners for two stresses with a sample of deformable grains (source: Alart et al. 2001). For a color version of this figure, see www.iste.co.uk/lebon/mechanics.zip*

To conclude, the *law non-smoothness* resulting from the frictional contact was treated within a finite element framework using a generalized Newton method. The techniques borrowed from high-performance computing were therefore confined to solving the linear problem obtained at each Newton iteration, taking into account the particular properties of the tangent matrices resulting from the mixed formulation. The *spatial non-smoothness*, particularly "exacerbated" in systems with *diffuse non-smoothness*, could be addressed through a decomposition of the domain in which a difference is made between (numerical) interfaces from subdomains and (physical) contact interfaces.

2.3. Diffuse non-smoothness in discrete structures: tensegrity

2.3.1. *Motivation*

This section discusses an intermediate situation between the equilibrium of continuous deformable structures in contact, discussed in section 2.2, and the dynamics of rigid body collections (typically granular systems) discussed in section 2.4. A tensegrity structure is a reticulated system comprising nodes, usually ball joints, connected by links or elements, with either smooth behavior such as bars or non-smooth such as cables. We are therefore no longer faced with a contact problem, but the unilateral behavior of cables can be expressed by complementarity relations similar to [2.2]. Therefore, if we consider along with deformable bars, extensible cables, a formulation such as [2.2] is obtained through the introduction of a *corrected elongation* of the extensibility of the cable α, denoted λ_α; it includes the elongation e_α, the pre-elongation e^0_α, the tension τ_α and the stiffness (in tension) k_α, that is, $\lambda_\alpha := -(e_\alpha + e^0_\alpha) + k^{-1}_\alpha \tau_\alpha$:

$$\tau_\alpha = \begin{cases} k_\alpha(e_\alpha + e^0_\alpha) & if\ e_\alpha + e^0_\alpha > 0 \\ 0 & if\ e_\alpha + e^0_\alpha \leq 0 \end{cases} \iff 0 \leq \lambda_\alpha \perp \tau_\alpha \geq 0 \qquad [2.15]$$

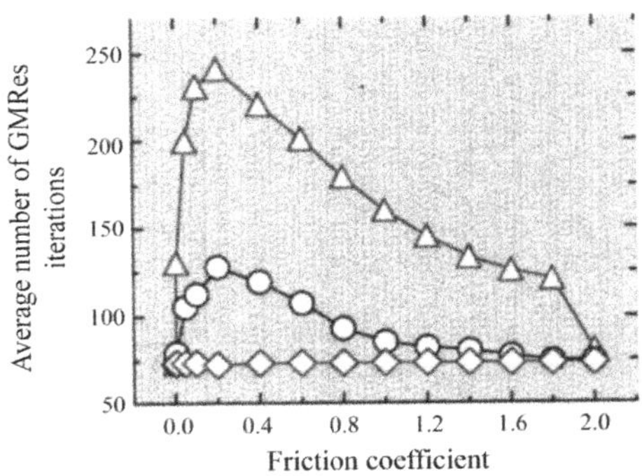

Figure 2.6. *Impact of the friction coefficient on the behavior of the solver for various Neumann–Neumann level 2 preconditioners (standard △, specific O, non-symmetric ◇) (source: Alart et al. 2001). For a color version of this figure, see www.iste.co.uk/lebon/mechanics.zip*

On the other hand, the ratio of the number of non-smooth relations to the number of degrees of freedom is comparable to that of a collection of rigid solids. This problem therefore corresponds to a system *with diffuse non-smoothness*. Despite this being a problem modeled from the outset as a discrete rather than a continuous system, the mathematical formulations of the equilibrium of such structures, consisting of deformable bars and extensible cables, are comparable to the formulations in section 2.2.

2.3.2. *Domain decomposition: micro-macro LATIN solver*

The decomposition strategy is identical to that developed in section 2.2 devoted to the Newton–Schur solver. It consists of distributing the elements, bars or cables, across the subdomains. Thereby, the non-smoothness of the cables remains localized in the sub-domains. On the other hand, the interfaces include duplicate nodes in each substructure for which equality relations of displacement or integral transmission of forces, namely smooth relations, can be written. In this sense, the interface is said to be *perfect* (see Figure 2.7). This strategy is dual to that developed by Ladevèze et al. (2002) with a LATIN approach or by Dureisseix and Farhat (2001) with a FETI approach, both concerning continuous problems.

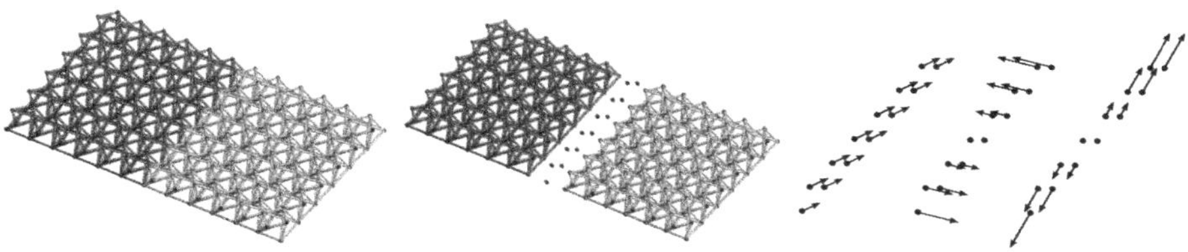

Figure 2.7. *Perfect substructuring-interface (nodes) – three macro displacement on the discrete interface: average translation, average rotation and average elongation (source: Alart and Dureisseix 2008)*

The approach was originally published in Nineb et al. (2007) and in Alart and Dureisseix (2008) for a multi-scale extensible version with a homogenized macro problem. This last version is briefly presented here without expanding the equations but with comments on what should be learned therefrom. The LATIN method initiated by Ladevèze (1996, 1999), extended to multiscale problems by Ladevèze and Dureisseix (1999) and specialized here to tensegrity systems, is an iterative solver which consists of two successive steps at each iteration, one referred to as local solving potentially nonlinear local equations, the other referred to as linear solving potentially global linear equations. In our example, the content of both steps is specified as follows.

The local step provides an approximate solution of the fields satisfying the following ingredients:

– a search direction per subdomain, as a linear relation linking tensions and elongations of elements, parameterized by a numerical stiffness l;

– the behavior law of each element, linear for bars, but non-regular for cables;

– a search direction per interface, as a linear relation linking displacements and inter-forces on interface nodes, parameterized by a numerical stiffness d;

– the law of perfect-linear interface behavior.

This step leads to linear, non-smooth problems whose solutions are explicit. Actually, under this formulation, the problem of local complementarity, by cable, can effectively be solved explicitly. This is the simplest choice. A more sophisticated version could consist of taking into account the equilibrium equations in order to define an LCP per subdomain coupling bars and cables. The LCP then has a reasonable size and can be solved using standard LCP solvers.

The global step provides a new approximate solution of the fields combining the following ingredients:

– a search direction per subdomain, as a linear relation linking tensions and elongations of elements, parameterized by a numerical stiffness l;

– balance and admissibility equations in subdomains and interfaces;

– additional admissibility constraints on macro fields at interfaces (resulting in a global coarse problem);

– a weakened search direction on the interface fields, consistent with the previous additional admissibility constraints, parameterized by a numerical stiffness d.

This step leads to a linear domain decomposition problem, similar to that created by the FETI method, consisting of solving linear systems by subdomain (factorable once and for all) and a coarse system enhanced by homogenized behavior through taking macro degrees of freedom into account. The macro displacements at the interfaces usually considered are average translations, average rotations, average elongations and shears, as shown in Figure 2.7. Alart and Dureisseix (2008) developed a study of the problem of *weak interfaces*, specific to discrete systems, but related to the topic of *corner modes* in continuous domain substructuring problems. The interesting result for the study of granular media that should be observed concerns the choice of numerical stiffness (l and d). The first parameter l is simply chosen as the characteristic stiffness of a bar, k_b, or a cable, k_c. The parameter d is chosen as a first approximation as the stiffness of springs at the interfaces which gives the same overall stiffness for a subdomain (without slack cables!), for a decomposition into identical subdomains. There is clearly an optimal value of this parameter, significantly smaller than the previous one, as shown in Table 2.2. This is not surprising, as the d parameter is estimated on a subdomain without slack cables, which is more rigid than a subdomain with between 10% and 21% of slack cables in the problem under consideration.

d / (10^6 N/m)	7.84	*3.92*	1.96	**0.98**	0.49	0.245
Number of iterations	68 (108)	39 (60)	23 (36)	17 (37)	20 (63)	29 (103)

Table 2.2. *3D tensegrity grid with 144 subdomains – number of iterations to achieve an error less than 0.1% (in parentheses: less than 0.01%) according to the numerical stiffness at the interface – first approximation value in italics, optimal value in bold (source: Alart and Dureisseix 2008)*

2.4. Granular dynamics

The framework for the development of the works on high-performance computing for granular dynamics is a discrete element method (and no longer that of finite elements in section 2.2) initiated by Moreau (1999, 2003) and Jean (1999), called *non-smooth contact dynamics* (NSCD). At the site of the LMGC90 platform[1], built based on this approach, various developments can be found, including in particular static equilibrium problems of aggregates associating NSCD and finite elements. Here, we focus on the dynamic behavior of collections of rigid bodies, without discretization of grains into deformable finite elements. We will therefore identify the similarities but also the differences with the techniques and results obtained in the previous sections. The NSCD approach is an exact non-regularized formulation of the contact dynamics but can also operate in discrete time, namely in an integration *time-stepping* based scheme. We will therefore pose the problem directly on a finite time step (although the use of differential measures in the NSCD formulation also allows expressing it at a given time). This problem can therefore be compared to that obtained with a loading step in a quasi-static simulation of a multicontact structure, for example, the formulation in [2.4].

2.4.1. *Velocity-impulse formulation*

As in the multi-contact problems of section 2.2 and the discrete structures of section 2.3, non-smoothness due to contact plays a major role for the behavior of the system and is the main challenge of numerical simulation and high-performance computing. Nonetheless, the formulation of contact interactions is adapted to the NSCD approach by linking not the displacements relative to contact reactions as in section 2.2, but the velocities relative to contact impulses. The *non-smoothness in time* is treated in this way, underlined by Moreau (1999), characterized by potential speed jumps in *collisions*, alongside *space and law non-regularities* already taken into account in the previous formulations. If d_α designates, as in section 2.2, the contact distance or gap at the contact of index α, the unilateral velocity–impulse contact law to be satisfied at time step dt can be expressed as:

$$\mathcal{R}(v_\alpha, r_\alpha) \; = \; 0 \Leftrightarrow \begin{cases} \text{If } d_\alpha > 0, \; r_\alpha = 0 \\ \text{If } d_\alpha \leq 0, \; 0 \leq v_\alpha \perp r_\alpha \geq 0 \end{cases} \qquad [2.16]$$

where v_α and r_α, respectively, designate the normal contact velocity and the normal contact impulse. When a friction law is to be taken into account for the tangential component, the previous law must be restricted to the normal components of the

1 See: https://git-xen.lmgc.univ-montp2.fr/lmgc90/lmgc90_user.

relative velocity and impulse and the friction law must be added, which all together constitute the contact law $\mathcal{R}(v_\alpha, r_\alpha) = 0$. It should be noted that Coulomb's law, for example, is naturally expressed in terms of tangential velocity and momentum in the form of complementarity relations. The need to control, prior to the verification of complementarity relations in velocity–impulse, the contact distance d_α via a prediction, at half step for example, can lead to referring to this formulation as a *prospective law*. The viability lemma of Moreau (1999) guarantees the satisfaction of the exact interpenetrability of bodies when the time step dt tends to 0.

The relative velocities v and impulses at contacts r are related to the velocities V of the bodies and the torque of forces of the contact R applied to them via the following admissibility conditions:

$$\begin{cases} v = H^T V + G^T V^* = H^T V + v^* \\ R = H r \end{cases} \qquad [2.17]$$

where linear operators H and G depend on local reference frames at the contacts and V^* represents the velocities imposed on certain bodies (such as the walls of a containment box, for instance). The problem to be solved at each time step includes the discrete equation of body dynamics and interaction relations at the contacts:

$$\begin{cases} MV - Hr = MV^i + R^d \\ \mathcal{R}(v_\alpha, r_\alpha) = 0, \alpha = 1, n_{contact} \end{cases} \qquad [2.18]$$

where M concatenates the inertial operators of all bodies, V^i and $V = V^f$ denote the initial (known) and final (unknown) velocities, R^d the torques of the external impulses. Finally, it proves useful, to interpret and optimize solver algorithms for high-performance computing, to favor local variables v and r by "condensing the dynamics at the contacts" by way of the *Delassus operator* $W := H^T M^{-1} H$:

$$\begin{cases} Wr - v = -v^d = -H^T(V^i + M^{-1} R^d) - v^* \\ \mathcal{R}(v, r) = 0 \end{cases} \qquad [2.19]$$

The second member of the reduced dynamics (v^d) represents the relative "free" velocities, that is, when there are no contacts.

Despite that formally the system [2.19] is very similar to the systems in [2.3] or [2.4], the differences are clearly visible. Equilibrium or dynamics are expressed in both cases – at least for [2.3] – by a linear equation, with nonlinearity (non-smoothness) being treated in local interaction relations. On the other hand, although the matrix K in [2.3], [2.4] or [2.5] can have a kernel relatively reduced to a few rigid modes of deformable bodies, the operator W presents a large kernel comprising all self-balancing contact impulse networks. This kernel is all the larger when friction is taken into account and with a high number of coordination (average number of contacts per body). This strong singularity of W has important consequences on the performance of solver algorithms. Non-smoothness is here no longer simply "diffuse", but "invasive", in the sense that, associated with a rigid, therefore coarse, modeling of solids, it leads to very poorly conditioned, even inconsistent problems. The problems of indeterminacy and inconsistency thus induced are not addressed here: we should refer to Alart (2014) and Alart and Renouf (2018).

2.4.2. *Parallelized and parallelizable solvers*

The generic solver of [2.19] is based on a Gauss–Seidel type solving technique of linear systems and is therefore called *nonlinear Gauss–Seidel* (NLGS). Algebraically, it consists of solving at each iteration k and each contact α the system of equations and inequalities of unknowns $(v_\alpha^{k+1}, r_\alpha^{k+1})$, that is, four unknown in two-dimensional discretization and six in three-dimensional discretization:

$$\begin{cases} W_{\alpha\alpha}\, r_\alpha^{k+1} - v_\alpha^{k+1} = v_\alpha^{free} - \sum_{\beta<\alpha} W_{\alpha\beta}\, r_\beta^{k+1} - \sum_{\beta<\alpha} W_{\alpha\beta}\, r_\beta^{k} \\ \mathcal{R}(v_\alpha^{k+1}, r_\alpha^{k+1}) = 0 \end{cases} \qquad [2.20]$$

Its convergence is difficult to demonstrate, except for a very particular situation and out of granular context (Jourdan et al. 1998). Its speed of convergence can be hampered both by the non-smoothness in the laws of interaction and the singularity of the operator W. Its slowness, inherent to stationary solving methods (Gauss–Seidel, Jacobi, successive relaxations), is compensated by a robustness confirmed by use.

2.4.2.1. *Multitasking (*multithreading*) parallelized NLGS*

For high-performance computing of large-sized systems, the optimization paths depend on the structure of the computer used, depending on whether it is a multiprocessor machine with shared or distributed memory. The easiest technique to implement is the distribution of tasks in shared memory or *multithreading*. It does not require distributing memory across multiple processors, but only distributing the computations of the loop of contacts of index α. Therefore, the set of contacts is broken down into n subsets of the same size, n being equal to the number of available processors, or *threads*. The inherent computational sequentiality in the Gauss–Seidel algorithm cannot be completely satisfied if, for example, a contact processed at the beginning of the list of one processor is connected with a contact at the end of the list of another processor, while the first has a higher index than the second. In fact, in a sequential processing, the first would have had the value of the impulse of the second, which will not be the case in *multithreading*. The sequence of iterations is therefore disrupted and the convergence solution modified. This remark does not apply to a Jacobi version of the algorithm, for which the impulses of the previous iteration are always used. However, numerical tests show that, since the convergence of the algorithm is slow, especially for dense systems, and therefore a large W kernel, this loss of sequentiality has only little impact on the number of convergent iterations. The first results of this approach were presented in Renouf et al. (2004).

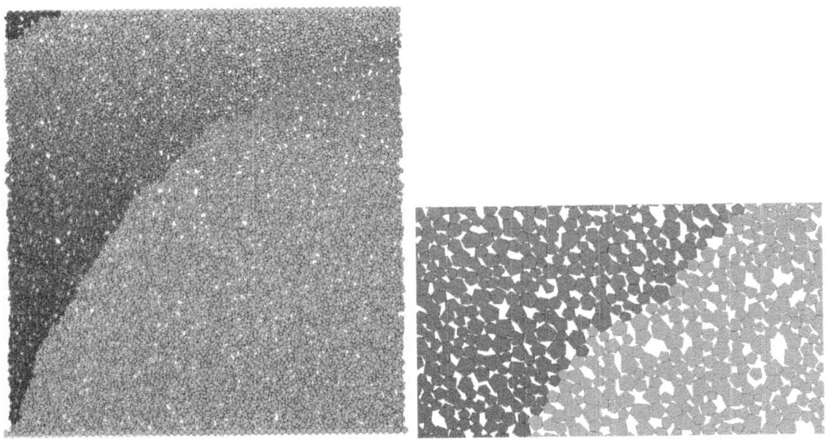

Figure 2.8. *Shear of 10,000 pentagonal particles generating around 39,800 contacts: velocity field and enlargement of the central area by E. Azéma. For a color version of this figure, see www.iste.co.uk/lebon/mechanics.zip*

The multiprocessor version available in LMGC90 is based on this technique. It is widely used by researchers, especially for the LMGC, for the statistical study of large systems in order to reduce computation time. This is, for example, the case with the ongoing study of Azéma on a simple shear of previously compacted pentagonal particles, as shown in Figure 2.8. The highly confined nature of this type of test leads to a very large number of iterations if accurate results with high tolerance thresholds are sought for. The conventional numerical behavior with this type of simulation is shown in Figure 2.9 and illustrates the CPU time improvement obtained by a very unintrusive multitasking approach and therefore is easy to implement. We should observe, however, gain saturation with an increasing number of processors. Many statistical studies on complex samples systematically make use of the multitasking version of LMGC90, for example, on the behavior of polydisperse three-dimensional assemblies in Cantor et al. (2018) or the mechanical strength and microstructure properties of a granular asteroid, modeled as a cohesive granular aggregate with self-gravitation in Azéma et al. (2018).

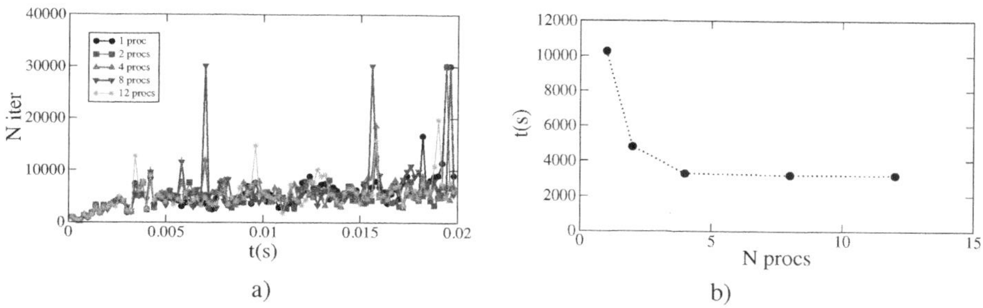

Figure 2.9. *Shear of 10,000 pentagonal particles, study according to the number of processors employed: (a) evolution of the number of iterations during the process; (b) CPU time by E. Azéma. For a color version of this figure, see www.iste.co.uk/lebon/mechanics.zip*

The parallel solver of LMGC90 proposes two alternative methods for constructing the second member of [2.19]: one consists, at the initialization of the algorithm, of computing and storing the matrices $W_{\alpha\beta}$, called local-level data storage (LDS), and the other consists of directly computing the second member (at the so-called local level of contacts) through exchanges between the local and global levels (that of the bodies) or LGE. The details are given in the thesis of Renouf (2004). The first method requires more memory and the second more computations. This distinction does not interfere with the behavior of the algorithm, but takes into account the memory capacities of the processors.

2.4.3. *Conjugate projected gradient solver*

Another way for high-performance computing is to adapt conjugate gradient-based solver algorithms to friction contact problems, which have the advantage of being efficient, when correctly preconditioned, and intrinsically parallelizable because they are based on scalar product and matrix-vector product calculations. This approach was developed in the thesis of Renouf (2004) and in Renouf and Alart (2004); the reader should refer thereto for details. The problem in [2.19] can be rewritten in terms of a constraint optimization problem of the single impulse variable r, more precisely in the frictional contact context, a *quasi-optimization* problem parameterized by the normal components of the impulses involved as both unknowns and stress parameters:

$$r \in \underbrace{argmin}_{\bar{r} \in \mathcal{C}(\mu r_n)} \frac{1}{2}\bar{r}.W\bar{r} + \bar{r} \ \ with \ \mathcal{C}(s) := \prod_{\alpha=1}^{n_{contact}} \mathbb{R}^+ \times \mathcal{B}(0, s_\alpha) \qquad [2.21]$$

where the set of constraints, itself depending on the solution r_n (hence the prefix *quasi*), is the Cartesian product of infinite half-cylinders in $\mathbb{R}^2$ or $\mathbb{R}^3$ with a section $\mathcal{B}(0, s_\alpha)$ the segment or disk of radius s_α. To process this parametric optimization within a single iteration loop, the conjugate projected gradient algorithm (*conjugate projected gradient* (CPG)) thus comprises at each iteration (a) a projection, before conjugation, of the gradient on a cylindrical local approximation of the Coulomb cone and (b) a correction of the iterate to replace it if necessary on the Coulomb cone (see details in Renouf and Alart (2004)). For two-dimensional modeling, this algorithm performs far better than NLGS as shown in Figures 2.10–2.12.

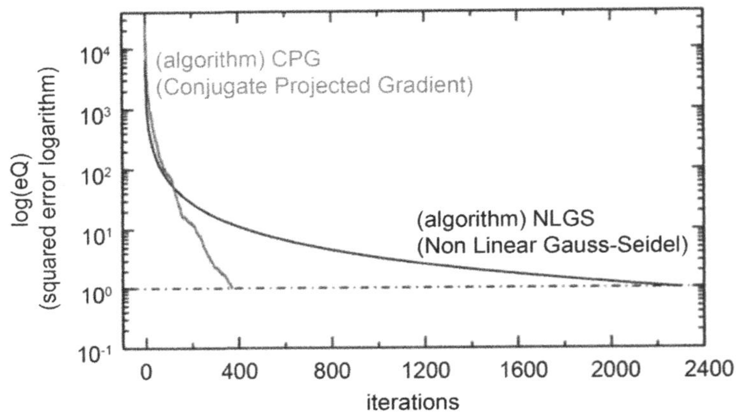

Figure 2.10. *Convergence of CPG and NLGS algorithms on a time step for a two-dimensional sample of 15,000 frictionless contacts (μ = 0). CPG is eight times faster than NLGS and convergence is regular because for zero friction, we are solving a real convex optimization problem (source: Renouf and Alart 2004). For a color version of this figure, see www.iste.co.uk/lebon/mechanics.zip*

The extension to three-dimensional modeling with isotropic friction no longer takes advantage of the polyhedral nature of the set of constraints. The notion of active constraint associated with polyhedral assemblies is no longer available and the gains are less important compared to NLGS than for two-dimensional models (see Renouf 2004). In conclusion, if the CPG algorithm is a performing alternative to the NLGS solver in two-dimensional modeling for aggregates with frictional contact, it cannot be used as a new generic solver, given that it is moreover difficult to extend to other laws of interactions other than frictional contact.

2.4.4. *Domain decomposition: FETI-NLGS solver*

While the multitasking version of NLGS achieves a distribution of the contacts across different processors, it does not represent a domain decomposition strategy

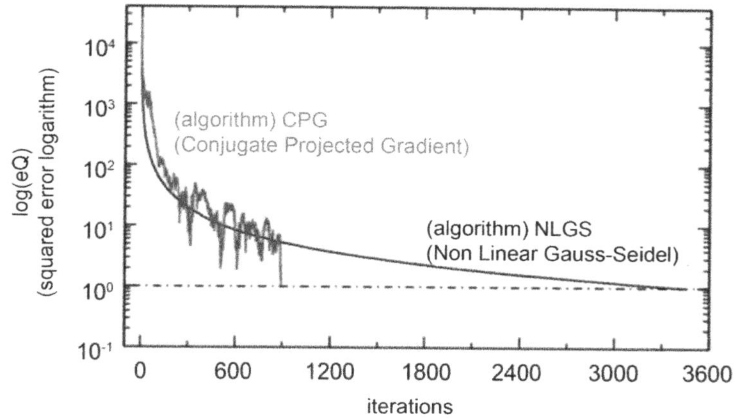

Figure 2.11. *Convergence of CPG and NLGS algorithms on a time step for a two-dimensional sample of 15,000 contacts with moderate friction (μ = 0.2). CPG is three times faster than NLGS but convergence is more erratic when friction is non-zero (source: Renouf and Alart 2004). For a color version of this figure, see www.iste.co.uk/lebon/mechanics.zip*

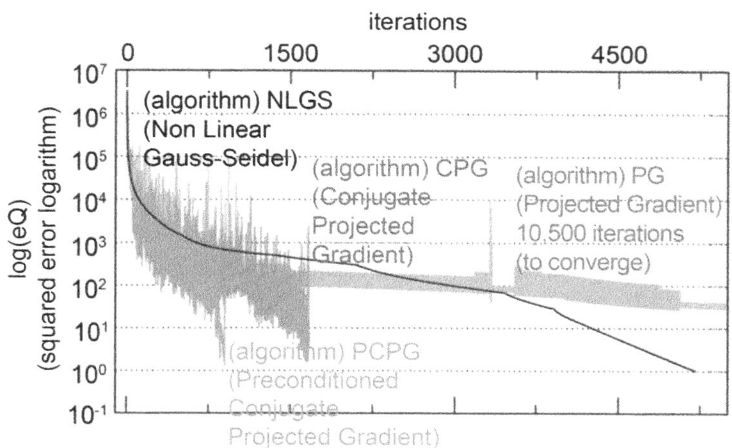

Figure 2.12. *Convergence of CPG, PCPG (CPG with diagonal preconditioning), PG (projected gradient without conjugation) and NLGS algorithms on a time step for a two-dimensional sample of 15,000 contacts with fairly high friction (μ = 0.4). PCPG is five times faster than NLGS, convergence is erratic and preconditioning is effective although rudimentary (source: Renouf and Alart 2004). For a color version of this figure, see www.iste.co.uk/lebon/mechanics.zip*

in the sense that contacts are not a priori geometrically but randomly distributed. Consequently, there are no subdomains or interfaces. A domain decomposition assumes a geometric distribution of bodies and/or contacts into subdomains. The notion of interface is less clear for discrete systems, and particularly in granular dynamics, than for continuous media whose meshes include entities of different topological dimensions, namely finite elements of maximal dimension, and then facets, edges or corners of decreasing dimension. Therefore, the first approach of this kind called *box method* and presented by Breitkopf and Jean (1999) does not involve new interface unknowns, but only exchanges between processors by way of the torques exercised on the interface grains. Visseq (2013) extended this approach into the *additive Schwarz method* in his thesis and a study thereof is proposed in Visseq et al. (2013).

In an algebraic version, unknowns and equations must be distributed across subdomains, but also across interfaces. If discrete systems do not introduce any topological distinction, we distinguish two levels and therefore two types of unknowns, one called global associated with bodies, the other local associated with contacts. Non-smoothness concerns contact relations while the equations of body dynamics are linear.

The space Ω occupied by the interacting bodies is decomposed following a decomposition grid with no overlapping set by the user: $\Omega = \Omega_1 \cup ... \cup\Omega_E \cup ...\Omega_{n_{sd}}$, shown in Figure 2.13.

The *primal partitioning* consists of distributing the bodies over the subdomains by assigning to a subdomain Ω_E all the bodies whose center of mass belongs thereto. An interaction between bodies is:

– internal, if the candidate and the antagonist belong to the same subdomain;

– at the interface, if the candidate and antagonist do not belong to the same subdomain.

Consequently, subdomains and interface are made up of interactions, as shown in Figure 2.14(a). In an algebraic substructure of the system of equations [2.19], the interface appears as a $(n+1)$th subdomain coupling the first n subdomains. The NLGS solver can therefore be directly applied, since the algorithm behaves as a multitasking version of section 2.4.2 with a special numbering for the contacts. The interest here is to be able to be implemented, not only on shared memory computers, but also with distributed memory. Applications of this approach are available in Hoang et al. (2011a, 2011b).

Dual partitioning consists of distributing interactions across subdomains, for example, by controlling the middle of the centers of mass of the candidate and antagonistic bodies. The interface then consists of bodies that have been schematically sliced in Figure 2.14(b) to illustrate that their mass will equally be distributed over adjacent subdomains. The non-smoothness of (contact for granular media) interactions being isolated in the subdomains, it is possible to define a linear interface problem. This strategy, initially implemented in Iceta's thesis and the publications extracted therefrom (see Iceta et al. 2010; Alart et al. 2012), consists of imposing a zero speed jump of the interface bodies using Lagrange multipliers. The linearity of the relation allows us to call it the *perfect* interface, similar to the numeric interfaces in the formulations of multicontact problems of section 2.2.3, and the "nodal" interfaces of the discrete tensegrity structures of section 2.3. The treatment of interface relations using Lagrange multipliers is very similar to the FETI method approach of Farhat and Roux (1991), developed in elastostatics. More precisely, the equations to be solved in [2.22] have a system based on [2.19] per subdomain E where a continuity equation of the velocities of the interface bodies is added:

$$\left\{\begin{array}{l} \left.\begin{array}{l}\tilde{W}_E r_E - v_E - \hat{A}^T_{\Gamma E} F_\Gamma = -v^d_E \\ \mathcal{R}(v_E, r_E) = 0\end{array}\right\} E = 1, \ldots, n_{sd} \\ \displaystyle\sum_{E=1}^{n_{sd}} A_{\Gamma E} V_E = 0 \end{array}\right. \qquad [2.22]$$

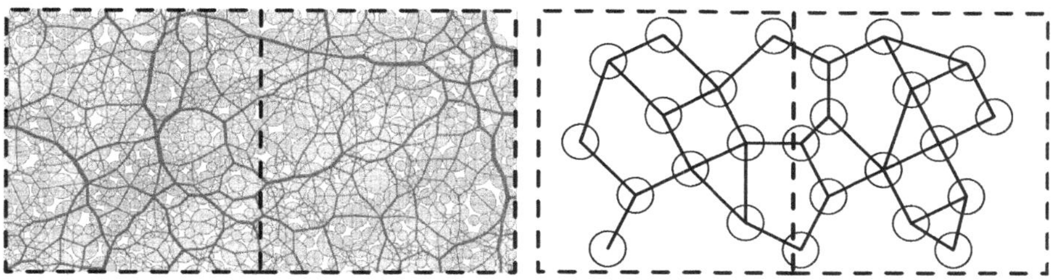

Figure 2.13. *"Box" distribution: application to a two-dimensional aggregate and schematic principle (source: Hoang et al. 2012)*

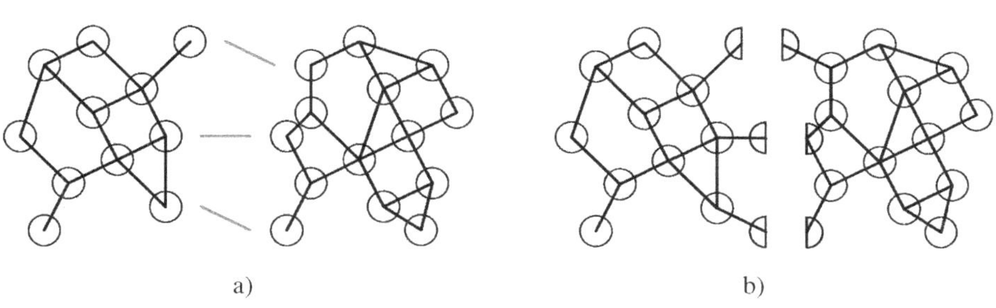

a) b)

Figure 2.14. *Primal (a) and dual partitioning (b)*

The Delassus operator $\tilde{W}_E$ takes into account, by virtue of the mass matrix $\tilde{M}_E$, the distribution of the masses of the interface bodies, $\tilde{W}_E = H_E^T \tilde{M}_E^{-1} H_E$, which also impacts the operator $\hat{A}_{\Gamma E}^T = H_E^T \tilde{M}_E^{-1} A_{\Gamma E}^T$ acting on the multiplicator F_Γ. This brings us back to a system [2.23] of unknowns (r_E, v_E, F_Γ), where X is the interface operator $X = \sum_E A_{\Gamma E} \tilde{M}_E^{-1} A_{\Gamma E}^T$:

$$\begin{cases} \left.\begin{array}{l} \tilde{W}_E r_E - v_E - \hat{A}_{\Gamma E}^T F_\Gamma = -v_E^d \\ \mathcal{R}(v_E, r_E) = 0 \end{array}\right\} E = 1, \ldots, n_{sd} \quad \text{a)} \\ X F_\Gamma - \displaystyle\sum_{E=1}^{n_{sd}} \hat{A}_{\Gamma E} r_E = \hat{f} = \sum_E A_{\Gamma E} \tilde{M}_E^{-1} R_E^d \quad \text{b)} \end{cases} \qquad [2.23]$$

The nonlinear solver associated with this formulation must take into account the singular nature of the operators $\tilde{W}_E$ presenting large kernels. It is therefore not possible to implement, as in section 2.2.3, a Newton method, and even a generalized one, solving at each iteration a linear system by an FETI-type linear solver. Exploiting the fact that non-smoothness is isolated in the subdomains, the strategy called FETI-NLGS consists, from the FETI-type formulation of [2.23], of the following two steps:

1) solve, possibly incompletely, using the NLGS algorithm the first two equations in [2.23a] per subdomain;

2) solve the third equation in [2.23b], which is linear and restricted to the interface.

When the interface bodies connect only two subdomains (multiplicity equal to 2), the operator X is blockwise diagonal and each block is easily invertible. The solution to the second step is then simple and inexpensive. If we overlook an adaptation presented in Visseq et al. (2012) and Visseq (2013), the problem can be reduced to this situation for bodies of multiplicity greater than 2; this will be spoken of wedge grains. A formulation extended to collections of deformable bodies is proposed in Visseq et al. (2012) using a corotational approach, without the numerical developments having been carried out.

It is also demonstrated that the optimal version in terms of number of iterations involves performing only one iteration of the NLGS algorithm per subdomain of the first step. This is because the second step, characterized by a diagonal operator, does not transfer long-distance information across subdomains. There is therefore no real advantage in iterating several times in each subdomain. The number of iterations of the FETI-NLGS algorithm is therefore quite stable, depending on the number of subdomains, as shown in Table 2.3, which summarizes some results from a two-dimensional biaxial compaction test shown in Figure 2.15.

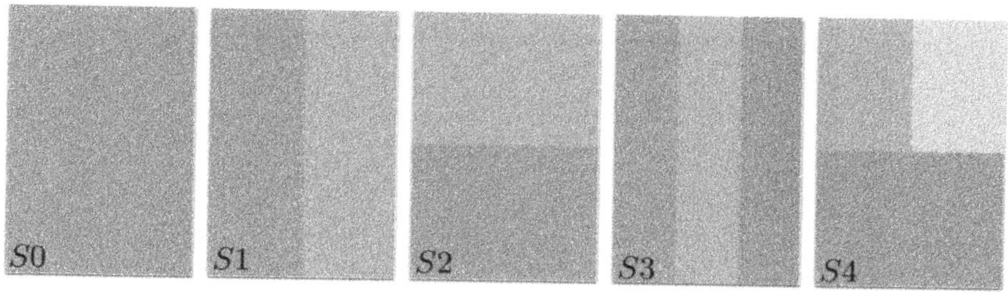

Figure 2.15. *Different decompositions at their initial state for a biaxial test sample of 12,000 particles (source: Visseq et al. 2012). For a color version of this figure, see www.iste.co.uk/lebon/mechanics.zip*

	$S0$	$S1$	$S2$	$S3$	$S4$
Number of FETI-NLGS iterations	$164\ 10^5$	$164\ 10^5$	$164\ 10^5$	$166\ 10^5$	$166\ 10^5$
Number of interface fields	0	117	104	207	227
Number of $W_{\alpha\beta}$	$418\ 10^3$	$380\ 10^3$	$341\ 10^3$	$366\ 10^3$	$303\ 10^3$

Table 2.3. *Total number of FETI-NLGS iterations, average number of interface bodies and extra-diagonal blocks of the Delassus operator ($W_{\alpha\beta}$) on 250 10^3 time steps for S0, S1, S2, S3, and S4. The decomposition into subdomains significantly reduces the number of matrices $W_{\alpha\beta}$ to store (source: Visseq et al. 2012; Visseq 2013)*

Although scalability in terms of iterations is more or less guaranteed, it remains to evaluate the additional cost of message exchanges between processors inherent in the MPI-type shared memory structure. This additional cost is no longer dependent on the mathematical formulation of the decomposition, but on the computer implementation. Two communication schemes have been implemented. The first, called centralized (*Centralized Communication Scheme* for CCS), amounts to dedicating a *master* processor to solving the interface problem; it must therefore receive data from processors dedicated to *slave* subdomains. The second, called decentralized (*Decentralized Communication Scheme* for DCS), enables each processor or subdomain to solve the interface problem that is assigned to it, associated with a partition of the unit for reattachment. The DCS scheme is more efficient and above all leads to obtaining strong scalability, namely, a quasi-independence of computational costs per processor vis-à-vis the number of processors on the same problem (fixed number of global contacts). Indeed, in this configuration, the size of the interface increases with the number of processors, while the size of the problem per subdomain decreases. Table 2.4 provides a comparison of the two diagrams with an isotropic compaction of a sample of 16,000 spheres.

N_{sd}	Partitioning grid ($n_x \times n_y \times n_z$)	CCS	DCS
2	$1 \times 1 \times 2$	6%	2%
4	$1 \times 2 \times 2$	25%	3%
8	$2 \times 2 \times 2$	47%	9%
12	$2 \times 2 \times 3$	57%	10%
16	$2 \times 2 \times 4$	69%	16%

Table 2.4. *Comparison of the percentage of time spent in message exchanges for centralized (CCS) and decentralized (DCS) versions of the FETI-NLGS approach; isotropic compression of a sample of 16,000 spheres (source: Visseq 2013; Visseq et al. 2013)*

Despite the fact that the FETI-NLGS method applied to the system of [2.23] proves to be scalable and therefore allows performing computations in parallel, maintaining the number of convergent iterations comparable to sequential computation, this number of iterations highly increases with the number of contacts. The diagonal interface problem [2.23b] is too poor to couple the various interfaces of the same subdomain and thereby accelerate convergence by providing a coarse macroscopic solution. The interface problem can be improved by introducing additional intermediate variables per subdomain, $\widetilde{r}_E$ and $\widetilde{v}_E$, satisfying specific admissibility requirements, that is, which may be different from relations [2.17] originating from potential contacts determined by the contact detection phase ($G_E \neq H_E$):

$$\widetilde{R}_E = G_E \widetilde{r}_E \text{ and } \widetilde{v}_E = G_E^T \widetilde{V}_E \quad [2.24]$$

and by giving the subdomains a linear tangent behavior, using of the numerical parameter ℓ_E:

$$(\widetilde{r}_E - r_E) + \ell_E(\widetilde{v}_E - v_E) = 0 \quad [2.25]$$

The enhanced interface problem, replacing [2.23b], therefore has the following expression:

$$X_\ell F_\Gamma = \sum_{E=1}^{n_{sd}} A_{\Gamma E} \widetilde{M}_{\ell,E}^{-1} \left[R_E^d + G_E \left(r_E + \ell_E v_E \right) \right] \quad [2.26]$$

where:

$$X_\ell = \sum_{E=1}^{n_{sd}} A_{\Gamma E} \widetilde{M}_{\ell,E}^{-1} A_{\Gamma E}^T, \;\; \widetilde{M}_{\ell,E} = \widetilde{M}_E + \ell_E K_E, \;\; K_E = G_E G_E^T \quad [2.27]$$

The linear interface problem in [2.26] is no longer blockwise diagonal, but amounts to solving a global linear problem on the reticulated structure domain. It can be solved using a conventional FETI-based domain decomposition linear solver (see Farhat and Roux 1991). A first version of such a strategy was studied, from an essentially theoretical point of view, in Alart et al. (2012) by taking $G_E = H_E$ as the compatibility operator in [2.24]. In Visseq (2013) and Visseq et al. (2014), other variants were studied, leading to adopt as compatibility operator G_E the one restricted to the normal components of active contacts only. However, the complete solution of the two steps listed above is computationally costly, especially since enhancing the interface problem does not accelerate convergence. As a matter of fact, elementary examples, studied in Alart et al. (2012), show that an optimal ℓ_E parameter, that is, accelerating convergence, will be difficult to determine, unlike what could be done at the end of section 2.3 on tensegrity systems. An alternative no longer consists of focusing on an exact solution of all relations in a time step, but to relax certain relations according to the scale. More precisely, the idea is to combine an explicit solution of the (linear) interface problem at the subdomain level (macro level), based on the network of active contacts defined at the beginning of the step, with an implicit contact-wise solution of non-regular problems in each subdomain (micro level). This strategy suggests that the interface efforts involved in the overall medium strength evolve more slowly than the local impulses governed by non-smooth dynamics. The work of Radjai et al. (1998) on the bimodality of the contact network also highlights that the strong network, involved in mechanical strength, is dominated by normal impulses in contacts that engage little tangential slip. This also consolidates the choice to limit the compatibility operator G_E to the normal components of active contacts. The results presented in Visseq et al. (2014) significantly depend on the more or less dynamic nature of the simulated process. Therefore, errors resulting from incompletely solving the relations can accumulate, correct themselves, or migrate into the domain. Figures 2.16 and 2.17 provide an analysis on the simulation of a granular flow of 750 particles on an inclined plane with periodic conditions (fast dynamics with frequent rearrangements), decomposed into two subdomains.

2.5. Conclusion

The research summarized in this chapter and conducted over 20 years pertain to high-performance computing for contact mechanics. The initial position, and subsequently maintained, was to use high-performance computing to deal with large problems, particularly involving a large number of contacts. Despite the inherent non-smoothness of contact interactions, *the strategy consisted of extending the high-performance computing techniques developed in the linear domain.*

This strategy has proven effective with the structural problems discussed in section 2.2. Parallel computational methods for large linear systems have been adapted to solve linear systems obtained at each iteration of a Newton algorithm. The non-smoothness resulting from contact was managed by a mixed formulation and a generalized Newton algorithm. The EBE preconditioner and the primal Schur domain decomposition method, as techniques developed for parallel computing, have been successfully extended to the resulting linear systems, requiring adaptations to the mixed nature of the formulation and to the non-symmetry induced by dry friction. Therefore, for domain decomposition, numerical scalability by enhancing a coarse problem has been recovered. This confirms a result from another domain. In nonlinear elasticity, De Roeck et al. (1992); De Roeck (1993) demonstrate that it is more efficient to insert a domain decomposition method at each iteration of a nonlinear solver (Newton type) than to perform a decomposition before implementing a nonlinear solver.

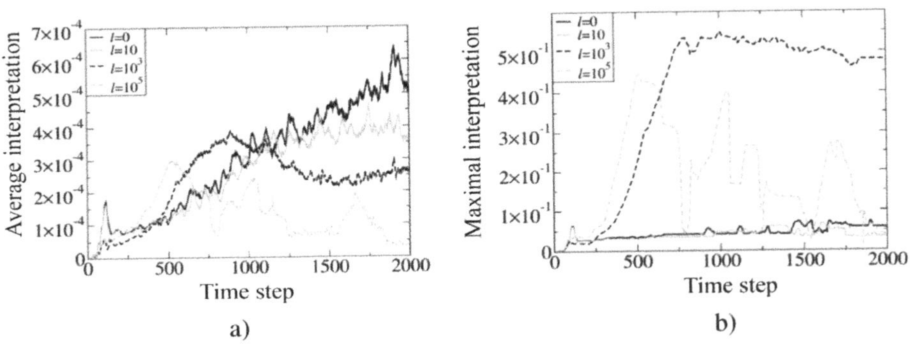

Figure 2.16. *Granular flow: evolution of the average (a) and maximum (b) interpenetration according to the parameter ℓ (source: Visseq et al. 2014). In most cases, strong numerical coefficients ℓ make it possible to find substantial but acceptable interpenetrations*

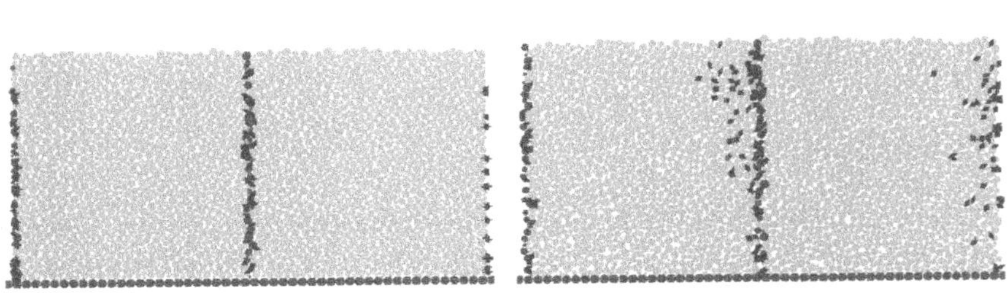

Figure 2.17. *Granular flow: configuration of interpenetrations at the beginning and end of the simulation; red lines indicate interpenetrations beyond 90% of maximal interpenetration (source: Visseq et al. 2014). Interpenetrations are initiated at interfaces because the solution is explicit, and then migrate to domains. For a color version of this figure, see www.iste.co.uk/lebon/mechanics.zip*

This strategy is still effective for computing the equilibrium of tensegrity systems with particularly diffuse non-smoothness throughout the structure. Indeed, to paraphrase a quote from RB Fuller[2], these systems are characterized by "islands of smoothness in an ocean of non-smoothness". The use of a LATIN-based method, instead of a Newton algorithm, requires enhancing the linear global step of a homogenized problem. However, the optimal choice of numerical stiffnesses, l and d in section 2.3, is a priori difficult because it depends on the solution itself.

This strategy reaches its limits in granular dynamics, as detailed in section 2.4. To obtain a system of relations comparable to the two previous cases, a velocity–impulse formulation should be adopted, but we are no longer really in the presence of islands of regularity. The multitasking version of the NLGS solver results in interesting time improvements on a few processors, but its efficiency decreases with a large number of processors. Several domain decomposition strategies have been tested. However, due to an "invasive" non-smoothness, the algorithms developed invert the strategy developed in section 2.2, by inserting a nonlinear solver, here NLGS, in a domain decomposition method, here FETI. The FETI–NLGS algorithm, with a single NLGS iteration per subdomain and an interface problem inexpensive for being diagonal, makes it possible to find a total number of global iterations comparable to the single-processor NLGS solver. Nonetheless, it is difficult to enhance the interface problem in order to accelerate convergence and/or implement a relevant multi-scale approach. In other words, a large non-smooth problem that must satisfy accurate steric exclusions between rigid bodies does not seem to be able to be correctly predicted by a linear problem, condensed at the interfaces. It is because non-smooth local corrections strongly perturb the overall dynamics crossing interfaces. The most effective alternatives to the strategies developed are difficult to devise. We should certainly consider not only a multi-scale approach in space, but also in time, commit the temporal integration of non-smooth systems, discriminate the issues arising from indeterminacy and the possible inconsistency of the system of relations to be solved.

2.6. References

Ach, K. and Alart, P. (2001). Numerical simulation of the behaviour of a multi-jointed structure. *Philosophical Transactions of the Royal Society of London. Series A: Mathematical, Physical and Engineering Sciences*, 359(1789), 2557–2573.

Alart, P. (1997). Méthode de Newton généralisée en mécanique du contact. *Journal de Mathématiques Pures et Appliquées*, 76(1), 83–108.

Alart, P. (2007). *Contact on Multiprocessor Environment: From Multicontact Problems to Multiscale Approaches*. Springer, Vienna.

---

2 Islands of compression in an ocean of tensions.

Alart, P. (2014). How to overcome indetermination and interpenetration in granular systems via nonsmooth contact dynamics. An exploratory investigation. *Computer Methods in Applied Mechanics and Engineering*, 270, 37–56.

Alart, P. and Curnier, A. (1991). A mixed formulation for frictional contact problems prone to Newton like solution methods. *Computer Methods in Applied Mechanics and Engineering*, 92(3), 253–375.

Alart, P. and Dureisseix, D. (2008). A scalable multiscale LATIN method adapted to nonsmooth discrete media. *Computer Methods in Applied Mechanics and Engineering*, 197(5), 319–331.

Alart, P. and Renouf, M. (2018). On inconsistency in frictional granular systems. *Computational Particle Mechanics*, 5(2), 161–174.

Alart, P., Barboteu, M., Lebon, F. (1997). Solution of frictional contact problems by an EBE preconditioner. *Computational Mechanics*, 20, 370–378.

Alart, P., Barboteu, M., Le Tallec, P., Vidrascu, M. (2000). Méthode de Schwarz additive avec solveur grossier pour problèmes non symétriques. *Comptes Rendus de l'Académie des Sciences. Series I, Mathématiques*, 331(5), 399–404.

Alart, P., Barboteu, M., Le Tallec, P., Vidrascu, M. (2001). Additive Schwarz method for nonsymmetric problems: Application to frictional multicontact problems. In *Thirteenth International Conference on Domain Decomposition Methods — DD13*. CIMNE, Barcelona/Lyon.

Alart, P., Barboteu, M., Gril, J. (2004). A numerical modelling of non linear 2D-frictional multicontact problems: Application to post-buckling in cellular media. *Computational Mechanics*, 34(4), 298–309.

Alart, P., Iceta, D., Dureisseix, D. (2012). A nonlinear domain decomposition formulation with application to granular dynamics. *Computer Methods in Applied Mechanics and Engineering*, 205/208, 59–67.

Azéma, E., Sánchez, P., Scheeres, D.J. (2018). Scaling behavior of cohesive self-gravitating aggregates. *Phys. Rev. E*, 98, 030901.

Barboteu, M., Alart, P., Lebon, F. (1998). A modified element-by-element preconditioner for elastostatics. *ASME. J. Appl. Mech.*, 65(2), 531–533.

Barboteu, M., Alart, P., Vidrascu, M. (2001). A domain decomposition strategy for nonclassical frictional multi-contact problems. *Computer Methods in Applied Mechanics and Engineering*, 190, 4785–4803.

Breitkopf, P. and Jean, M. (1999). Modélisation parallèle des matériaux granulaires. In *4e Colloque National en Calcul des Structures, CSMA*, Giens.

Cantor, D., Azéma, E., Sornay, P., Radjai, F. (2018). Rheology and structure of polydisperse three-dimensional packings of spheres. *Phys. Rev. E*, 98, 052910.

Clarke, F.H. (1990). *Optimization and Nonsmooth Analysis*, volume 5. SIAM, Philadelphia.

Curnier, A. and Alart, P. (1988). A generalized Newton method for contact problems with friction. *Journal of Theoretical and Applied Mechanics*, 7, 67–82.

De Roeck, Y.-H. (1993). Nonlinear elasticity solved by a domain decomposition method on a hypercube. *Applied Numerical Mathematics*, 12(5), 459 – 471.

De Roeck, Y.-H., Le Tallec, P., Vidrascu, M. (1992). A domain-decomposed solver for non linear elasticity. *Comp. Meth. Appl. Mech. Engrg.*, 99, 187–207.

Dureisseix, D. and Farhat, C. (2001). A numerically scalable domain decomposition method for the solution of frictionless contact problems. *International Journal for Numerical Methods in Engineering*, 50(12), 2643–2666.

Farhat, C. and Roux, F.-X. (1991). A method of finite element tearing and interconnecting and its parallel solution algorithm. *International Journal for Numerical Methods in Engineering*, 32(6), 1205–1227.

Hoang, T.M.P., Alart, P., Dureisseix, D., Saussine, G. (2011a). A domain decomposition method for granular dynamics using discrete elements and application to railway ballast. *Annals of Solid and Structural Mechanics*, 2(2/4), 87–98.

Hoang, T.M.P., Saussine, G., Dureisseix, D., Alart, P. (2011b). Behaviour of a portion of railway track under maintenance operation. In *9th World Congress on Railway Research (WCRR2011)*, Lille.

Hoang, T.M.P., Saussine, G., Dureisseix, D., Alart, P. (2012). Domain decomposition with discrete element simulations using shared-memory parallel computing for railways applications. *European Journal of Computational Mechanics*, 21(3–6), 242–253.

Hughes, T.J., Levit, I., Winget, J. (1983). An element-by-element solution algorithm for problems of structural and solid mechanics. *Computer Methods in Applied Mechanics and Engineering*, 36(2), 241–254.

Iceta, D., Dureisseix, D., Alart, P. (2010). Domain decomposition for granular dynamics: Scalability issue. In *5th European Conference on Computational Mechanics – ECCM2010*, Paris.

Jean, M. (1999). The non-smooth contact dynamics method. *Computer Methods in Applied Mechanics and Engineering*, 177, 235–257.

Jourdan, F., Alart, P., Jean, M. (1998). A Gauss-Seidel like algorithm to solve frictional contact problems. *Computer Methods in Applied Mechanics and Engineering*, 155(1/2), 31–47.

Ladevèze, P. (1996). *Mécanique non-linéaire des structures. Nouvelle approche et méthodes de calcul non incrémentales*. Hermès, Paris.

Ladevèze, P. (1999). *Nonlinear Computational Structural Mechanics: New Approaches and Non-incremental Methods of Calculation*. Springer, Verlag.

Ladevèze, P. and Dureisseix, D. (1999). Une nouvelle stratégie de calcul micro/macro en mécanique des structures. *Comptes-Rendus de l'Académie des Sciences*, 327(12), 1237–1244.

Ladevèze, P., Nouy, A., Loiseau, O. (2002). A multiscale computational approach for contact problems. *Computer Methods in Applied Mechanics and Engineering*, 191, 4869–4891.

Le Tallec, P. (1994). Domain-decomposition methods in computational mechanics. *Computational Mechanics Advances*, 1(2), 121–220.

Mandel, J. (1993). Balancing domain decomposition. *Communications on Applied Numerical Methods*, 9, 233–241.

Moreau, J.J. (1999). Numerical aspects of sweeping process. *Computer Methods in Applied Mechanics and Engineering*, 177, 329–349.

Moreau, J.J. (2003). Modélisation et simulation de matériaux granulaires. In *Actes du 35e Congrès National d'Analyse Numérique*, France.

Nineb, S., Alart, P., Dureisseix, D. (2007). Domain decomposition approach for nonsmooth discrete problems, example of a tensegrity structure. *Computers and Structures*, 85(9), 499–511.

Radjai, F., Wolf, D.E., Jean, M., Moreau, J.J. (1998). Bimodal character of stress transmission in granular packings. *Physical Review Letters*, 80(1), 61–64.

Raous, M., Chabrand, P., Lebon, F. (1988). Numerical methods for solving unilateral contact problem with friction. *Journal of Theoretical and Applied Mechanics*, 7, 111–128.

Renouf, M. (2004). Numerical optimisation and parallel computing applied to the simulation of 2D/3D discrete element. Thesis, Université Montpellier II, Montpellier.

Renouf, M. and Alart, P. (2004). Conjugate gradient type algorithms for frictional multi-contact problems: Applications to granular materials. *Computer Methods in Applied Mechanics and Engineering*, 194, 2019–2041.

Renouf, M., Dubois, F., Alart, P. (2004). A parallel version of non smooth contact dynamics algorithm applied to simulation of granular medium. *Journal of Computational and Applied Mathematics*, 168, 375–382.

Visseq, V. (2013). Calcul haute performance en dynamique des contacts via deux familles de décomposition de domaine. Thesis, Université Montpellier II, Montpellier.

Visseq, V., Martin, A., Iceta, D., Azema, E., Dureisseix, D., Alart, P. (2012). Dense granular dynamics analysis by a domain decomposition approach. *Computational Mechanics*, 49, 709–723. doi: 10.1007/s00466-012-0699-5.

Visseq, V., Alart, P., Dureisseix, D. (2013). High performance computing of discrete nonsmooth contact dynamics with domain decomposition. *International Journal for Numerical Methods in Engineering*, 96(9), 584–598.

Visseq, V., Alart, P., Dureisseix, D. (2014). Towards an augmented domain decomposition method for nonsmooth contact dynamics models. *Computational Particle Mechanics*, 1(1), 15–26.

Wriggers, P. (2012). Computational contact mechanics. *Computational Mechanics*, 49(6), 685–685.

3

Numerical Methods in Micromechanical Contact

Vladislav A. YASTREBOV
CNRS, Mines Paris PSL, France

3.1. Introduction

Mechanical contact and all associated mechanisms such as friction, wear, adhesion and mixed lubrication regimes represent critical mechanisms in many natural (such as small animal legs, joints, glaciers and faults) and industrial systems (tires/pavement, gears, bearings and piston/cylinder, for example). Very often, in the analysis of these systems, it is assumed that contact surfaces have nominal shapes smooth almost everywhere. However, in reality, all real surfaces, especially those found in nature, exhibit much more complex shapes than their nominal forms: in other words, we can say that these surfaces are rough. This roughness influences almost all interfacing phenomena (Bowden and Tabor 1986; Vakis et al. 2018): the stress state near the contact, the contact interface stiffness, wear, friction, adhesion, heat and electrical transfer as well as watertightness are all affected by the roughness of the surfaces in contact. In most cases, at the macroscopic scale (the scale of the nominal contact area and structure), the effect of roughness can be taken into account by phenomenological laws or those based on microscopic considerations (Greenwood 1966; Cooper et al. 1969; Zavarise et al. 1992), including Archard's law of wear (Archard and Hirst 1956), Coulomb's law of friction as well as other laws

For a color version of all figures in this chapter, see www.iste.co.uk/lebon/mechanics.zip.

Numerical Methods for Strong Nonlinearities in Mechanics,
coordinated by Jacques BESSON, Frédéric LEBON and Eric LORENTZ. © ISTE Ltd 2024.

such as "rate and state" friction (Rice and Tse 1986; Dieterich 1992; Dieterich and Kilgore 1994; Rice 2006). However, micromechanical models used in the past were based on fairly simplistic roughness and mechanical behavior models. Often, these limitations came from the need to make strong enough assumptions to obtain results because the contact problem formulated for complex geometry cannot be solved analytically due to its high nonlinearity. Now, considering advanced numerical methods and today's computing power, a large field of applications in micromechanical contact is becoming highly accessible to researchers and engineers. This chapter discusses advances in numerical methods and our increasing understanding of small-scale contact physics.

3.1.1. *Plan*

This chapter is organized as follows. First, the contact problem and its micromechanical nature are presented in section 3.2. The finite element method (FEM) is briefly discussed in section 3.3 with emphasis on features relevant to small-scale contact. The boundary element method is not presented in detail in this chapter, so an interested reader should refer to a recent review by Bemporad and Paggi (2015). Nevertheless, many applications of this method to small-scale contact problems are presented. Two main applications of micromechanical contact: the contact between rough edges and contact between rough surfaces are covered in sections 3.4 and 3.5, respectively. Some problems for the future are formulated in section 3.6.

3.2. Contact micromechanical problem

3.2.1. *Surface geometry: mathematical description*

In this section, we formulate the micromechanical problem of contact. This will form the subject of the following sections. We consider the contact between two simply related bodies[1] (open sets $\Omega^1(t), \Omega^2(t)$) with one or more surfaces exhibiting "roughness"[2]. The configuration is defined at each time t, in the following this explicit notation of this time dependency will be omitted. The mechanical behavior of materials (homogeneous or heterogeneous) can be linear or nonlinear, while the contact model can be considered without or with friction/adhesion. In the general case, in the initial state, the two potential contact surfaces are described by vectors:

1 The simple connectedness condition of bodies in contact makes it possible to somewhat simplify the mathematical formulation, especially in terms of the application of parametric space in physical space.

2 "Roughness" refers here to any surface shape.

$$\begin{cases} \boldsymbol{r}^1(\xi^1,\eta^1) \in \partial\Omega^1 \\ \boldsymbol{r}^2(\xi^2,\eta^2) \in \partial\Omega^2 \end{cases} \quad \boldsymbol{r}^1, \boldsymbol{r}^2 \in \mathbb{R}^3$$

where $\partial\Omega^i$ defines the closure of the set, $\boldsymbol{r}^1, \boldsymbol{r}^2$ are at least surjective mappings[3] of parameters $\{\xi^i, \eta^i\} \in \mathcal{X}^i \subset \mathbb{R}^2$, $i = 1, 2$ in two-surfaces $\partial\Omega^i$ emerged in the three-dimensional physical space $\mathbb{R}^3$:

$$\boldsymbol{r}^i : \mathcal{X}^i \subset \mathbb{R}^2 \to \partial\Omega^i \subset \mathbb{R}^3$$

The mapping $\boldsymbol{r}^i$ is continuous. Let us assume that $\mathcal{X}^i$ is simply connected and convex. Hereafter, we merely consider the problem of three-dimensional contact; the indexes $1, 2$ in superscript denote both surfaces. In principle, the formulation can be adapted to self-contact problems, where different parts of the same surface come into contact.

Contact can be achieved by way of different types of loading (mechanical, thermal, magnetic). This formulation of the contact problem is rather general and can be adapted to macroscopic and microscopic contact. However, we will consider that the geometry of both surfaces is very complex, unlike most macroscopic applications where the surfaces studied remain quite simple.

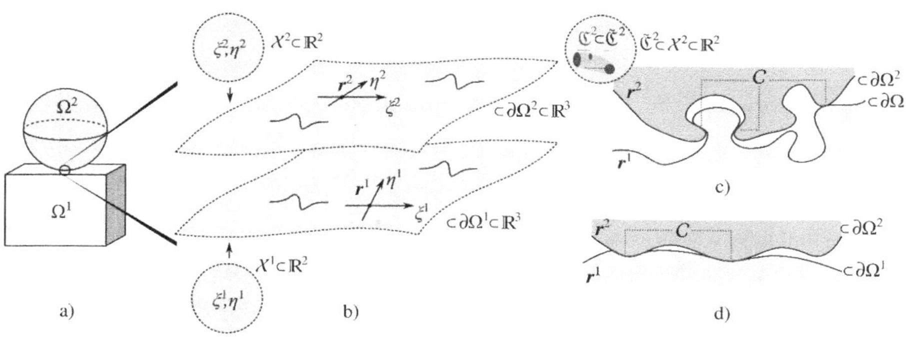

Figure 3.1. *Contact problem between two bodies on the surface: (a) the macroscopic scale; (b) magnification of surfaces parts potentially in contact (parameterization is indicated); (c) example of the non-simplifiable contact problem; (d) example of surfaces that allow an explicit representation at the given time*

Let us introduce the contact area $\mathcal{C}$ (with no self-contact) in the current configuration as a set of points $\boldsymbol{r}^c$ such that:

$$\mathcal{C} = \left\{ \boldsymbol{r}^c \in \mathbb{R}^3 \quad \middle| \quad \exists \boldsymbol{r}^1 \in \partial\Omega^1, \boldsymbol{r}^2 \in \partial\Omega^2 : \boldsymbol{r}^c = \boldsymbol{r}^1 = \boldsymbol{r}^2 \right\}$$

3 When there is no self-contact, these maps can be considered to be bijective.

Its images are denoted as:

$$\boldsymbol{r}^i : \mathfrak{C}^i \subset \mathcal{X} \rightarrow C \subset \partial\Omega^i$$

Let the set $\tilde{\mathfrak{C}}^i \supset \mathfrak{C}^i$ also be introduced, such that $\tilde{\mathfrak{C}}^i$ is a simply connected and convex two-manifold of minimal size[4]. It should be noted that the sets $\mathfrak{C}$ and C are not necessarily connected sets, nor are they manifolds, and are not necessarily compact either. So that geometric quantities of different sizes can be compared, we define the linear size $T(\bullet)$ of an entity $\mathcal{A}$:

$$T(\mathcal{A}) = \max_{\boldsymbol{r},\boldsymbol{r}' \in \mathcal{A}} \|\boldsymbol{r} - \boldsymbol{r}'\|$$

where $\|\boldsymbol{a}\|$ defines a vector Euclidean norm. Let $\mathfrak{C}_0^i$ denote the maximal extension of the contact area in the parametric space $\mathfrak{C}_0^i \subset \mathcal{X}^i$, that is, $\forall t : \mathfrak{C}(t) \subseteq \mathfrak{C}_0^i \subseteq \mathcal{X}^i$. Naturally, $\mathfrak{C}_0^i$ is problem dependent and cannot always be guessed.

The surface has an *explicit form* in the area of interest. If at every time instant $t \in \mathcal{T}$ there exists an orthonormed basis $\boldsymbol{e}_x, \boldsymbol{e}_y, \boldsymbol{e}_z \in \mathbb{R}^3$, $\boldsymbol{e}_\alpha \cdot \boldsymbol{e}_\beta = \delta_{\alpha\beta}$, $\alpha, \beta \in \{x, y, z\}$ such that the surfaces in which the potential contact is located can be represented as functions $z^i(x, y)$, where $z^i = \boldsymbol{r}^i \cdot \boldsymbol{e}_z$, $x = \boldsymbol{r}^i \cdot \boldsymbol{e}_x$, $y = \boldsymbol{r}^i \cdot \boldsymbol{e}_y$:

$$\forall t \in \mathcal{T}, \exists \boldsymbol{e}_x, \boldsymbol{e}_y, \boldsymbol{e}_z \in \mathbb{R}^3 \text{ s.t. } \forall \boldsymbol{r}^i(\tilde{\mathfrak{C}}_0^i) : \exists!\ z^i(x, y) \text{ s.t. } \boldsymbol{r}^i = \boldsymbol{e}_z z^i(x, y) + \boldsymbol{e}_x x + \boldsymbol{e}_y y$$

where $\tilde{\mathfrak{C}}_0^i \supset \mathfrak{C}_0^i$ is, as introduced earlier, a simply connected and convex two-manifold of minimal size that includes $\mathfrak{C}_0^i$. More broadly, we will allow ourselves to work with surfaces that have a *semi-explicit* representation, which will be defined below. First, let the set $\mathbb{X} \in \mathbb{R}^2$ be defined such that it consists of all points: $\{\boldsymbol{r}^i(\tilde{\mathfrak{C}}_0^i) \cdot \boldsymbol{e}_x,\ \boldsymbol{r}^i(\tilde{\mathfrak{C}}_0^i) \cdot \boldsymbol{e}_y\}$. A semi-explicit surface is defined as a surface for which the set of points $\{x, y\} \in \mathbb{X}_n \subset \mathbb{X}$ for which $z^i(x, y)$ is not always unique and of zero measure. If the two surfaces do not have an explicit or semi-explicit shape as defined previously in the neighborhood of the potential contact area, the associated contact problem will be referred to as *unsimplifiable*. An example of a unsimplifiable problem is shown in Figure 3.1(c).

From this very general formulation, we could propose some particularly appealing sub-cases. When the linear dimension of the contact area is very small compared to the size of the bodies in contact $T(C) \ll T(\Omega_1), T(\Omega_2)$ and if we are

4 The convexity condition of $\mathcal{X}^i$ that was requested earlier comes from the need to ensure that: $\tilde{\mathfrak{C}}^i \subset \mathcal{X}^i$.

exclusively focusing on the state of strain and stress in the contact area, fields can be thought to be semi-infinite (see, for example, the theory of Hertz (1881)). If also, there is in the area of interest an existing explicit representation of the surfaces, and the maximum gradient is infinitesimal, that is, $\max |\nabla z(x, y)| \ll 1$, fundamental Boussinesq, Cerutti and Flamant solutions can be used. It is important to emphasize that these solutions are not usable when a deformable surface does not have an explicit representation in the vicinity of the contact zone.

3.2.2. *Surface geometry: examples and discussions*

"Roughness", or more generally the geometry, of natural and industrial surfaces, can take very varied forms. In Figure 3.2, some pictures of the various surfaces are assembled: (a) beach wrinkles formed by the interaction of sand particles with wind or current; (b) pebbles of different sizes; (c) the granite fracture surface; (d) the texture of a painted wall; (e) the creep fracture surface of a titanium superalloy; (f) "elephant skin" on rocks in the forest of Fontainebleau; (g) the surface of the sea; (h) the surface of a fibrous material; (i) peeling paint; (j) pebble pavement; (k) dry soil; (l) asphalt; (m) bark; (n) finger skin; (o) drops of water on car paint; and (p) the fracture surface (lower part) on a rock in the forest of Fontainebleau. The geometries presented here are all complex and show a random character. However, except for examples (c) and (g), the surface geometries in the pictures rather represents the mesoscopic scale which additionally contains a small-scale roughness, which will be discussed in a little more detail in section 3.2.3.

In addition, this geometry can be considered resolution dependent. This assertion is all the more true because at the atomic scale matter has no continuity, the latter is only a model. Nonetheless, this is a model that remains valid from the nanoscale and the one that will be employed in the context of this presentation. But even at the validity scales of the continuous medium model, surface geometry can be described at different levels of accuracy. Consider an example of a beach wrinkle (see Figure 3.2(a) where a "continuous" shape, which is nevertheless made up of sand particles which form at their scale a very complex "surface", similar to that of Figure 3.2(b) can be distinguished. In addition, the surface of the sand particles is also rough as in Figure 3.2(c). For this example, at least three scales can be distinguished: wrinkle, particles and roughness. Clearly, these three scales cannot be simultaneously treated in order to understand the interaction of the wheels of a rover on Mars with the ground of the Red Planet. Besides the difference in the scales of shape and roughness, there are many examples of surface geometries that distinguish different scales: turned surfaces, abrasion-polished surfaces and architected surfaces (Yoon et al. 2006; Autumn 2007; Costa and Hutchings 2007). We can note that although these scales are quite distinct at first glance, in some cases there is no separation between these scales in terms of mechanics (Greenwood and Tripp 1967; Yastrebov 2019).

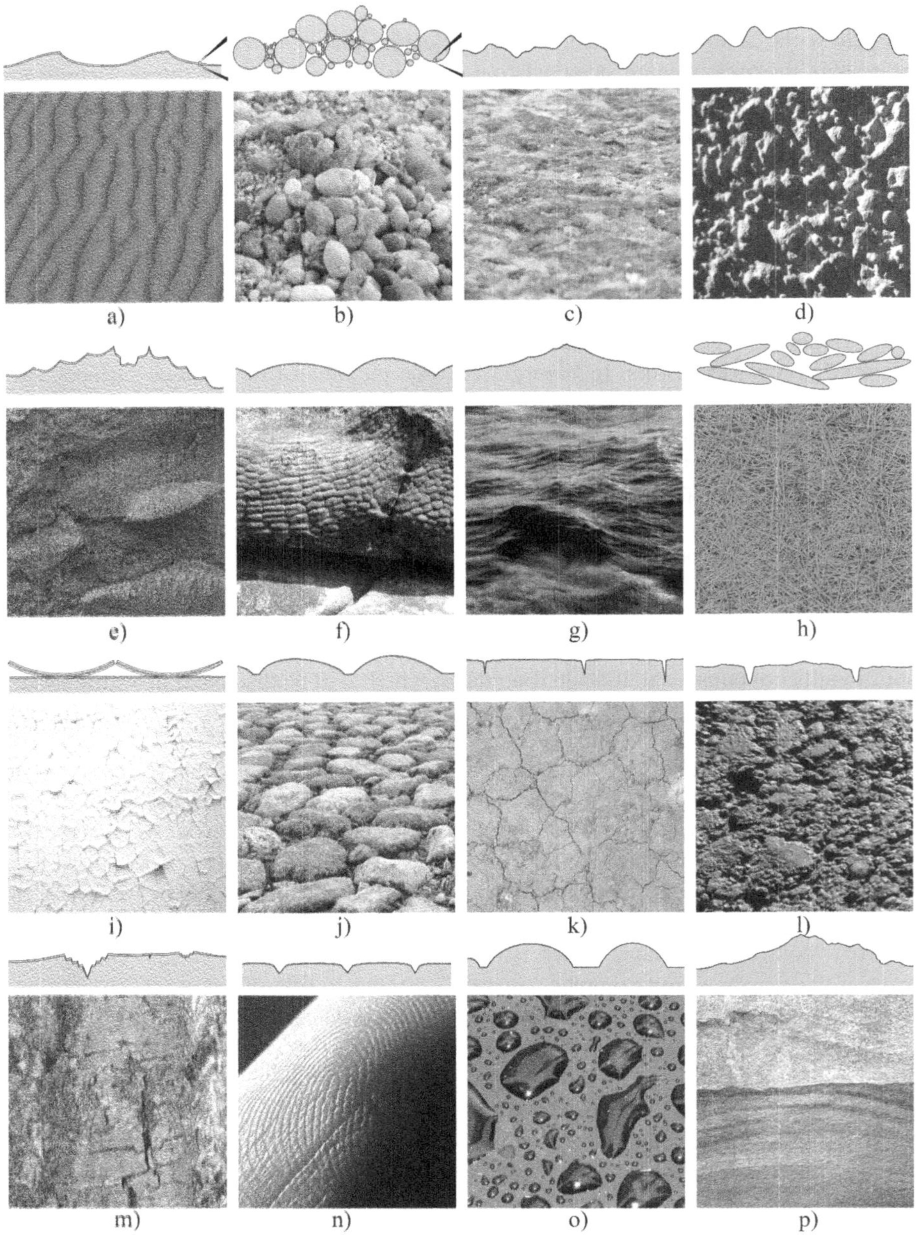

Figure 3.2. *Different geometries of natural surfaces*

Often, in contact problems, only the "shape" of the parts, as it was designed, is taken into consideration, but inevitably this shape hides aspects of the roughness that is the subject of this section. Form can often be defined as deterministic, especially in engineering applications while roughness is stochastic and random. However, it can also be addressed deterministically. Otherwise, the aspect of roughness can be taken into account by means of phenomenological relations that can be obtained, among other things, by deterministic calculations of the contact of rough surfaces or through statistical models.

3.2.3. *Roughness models*

The objective here is not to present in detail all aspects of rough surfaces; all of this information can be found in engineering (Thomas 1999; Whitehouse 2010), mathematical (Adler and Taylor 2009) or physics texts (Mandelbrot 1983; Meakin 1998). Nevertheless, it would be important to propose some references on modeling the geometry of rough surfaces. First, a model for a rough surface can be obtained by direct surface measurements, but since the resolution of all instruments is limited, the reconstruction of a final surface can be considered as modeling. In addition to direct measurements, different methods for modeling rough surfaces can be found. Often, these are methods that are based on a fractal characteristic of most natural and industrial surfaces (Russ 1994; Renard et al. 2013). A good review of these methods for addressing the mechanics of contact of rough surfaces was given in Zahouani et al. (1998). The simplest method is the *mid-point random displacement* (Russ 1994; Meakin 1998) which, in a hierarchical way, subdivides the grid of points at each iteration (a segment in two, a face in four) and in doing so, it disturbs the vertical coordinate of each new point in a random manner. To reach the fractal aspect, the standard deviation of these perturbations decreases with each new iteration. The rate of this decay depends on the fractal dimension of the modeled surface. On the other hand, the resulting surface requires smoothing (interpolation) in order to obtain a well-posed mechanical problem, that is, a problem for which mesh refinement leads to the convergence of the numerical solution. Another fairly common method makes use of the Weierstrass–Mandelbrot function generalized to problems with 2D surfaces by Majumdar and Tien (1990). This method does however not easily obtain isotropic surfaces (Zahouani et al. 1998). A powerful and flexible method is the spectral filtering method for white noise (Hu and Tonder 1992), which makes it possible to reproduce on average a self-affining spectrum with a defined Hurst exponent while presenting a certain level of stochasticity. In addition, this filtering can be used to introduce low- and high-frequency cuts, which results in obtaining smooth surfaces at the scale of discretization and the Gaussian distribution of heights (Yastrebov et al. 2015). Recently, a method for modeling surfaces with controlled height distribution was proposed in Pérez-Ràfols and Almqvist (2019).

3.2.4. *Contact formalization*

In this section, we will formalize contact conditions which can be summarized (with no friction and adhesion) by the following three conditions:

1) No penetration, that is, no intersection between solids $\Omega^1 \cap \Omega^2 = \emptyset$.

2) No adhesion $\boldsymbol{n} \cdot \boldsymbol{\sigma} \cdot \boldsymbol{n} \leq 0$, that is, the normal stress is not positive.

3) No shear at the contact interface, that is, the tangential component cancels out $(\boldsymbol{I} - \boldsymbol{n} \otimes \boldsymbol{n}) \cdot \boldsymbol{\sigma} = 0$, where $\boldsymbol{\sigma}$ is the Cauchy stress tensor, $\boldsymbol{n}$ is the outgoing normal and $\boldsymbol{I}$ is an identity tensor. Condition (1) can be reformulated as a kinematic inequality as follows: none of the points of a surface $\partial\Omega^1$ can penetrate the volume Ω^2, nor can penetrate under the surface $\partial\Omega^2$, which should therefore be orientable. The surface indexes 1 and 2 are arbitrary and can be exchanged. This condition can be reformulated even more easily as a relation between each surface point and the nearest points $\boldsymbol{r}_p$ of another surface:

$$\forall \boldsymbol{r}^1 \in \partial\Omega^1, \exists \boldsymbol{r}_p^2 \in \partial\Omega^2 \text{ s.t. } \forall \boldsymbol{r}^2 \in \partial\Omega^2, \; \|\boldsymbol{r}^1 - \boldsymbol{r}_p^2\| \leq \|\boldsymbol{r}^1 - \boldsymbol{r}^2\| \qquad [3.1]$$

Moreover, if the surface $\partial\Omega^2$ is sufficiently smooth, it can be said that for these points:

$$(\boldsymbol{r}^1 - \boldsymbol{r}_p^2) \cdot \boldsymbol{n}(\boldsymbol{r}_p^2) \geq 0$$

where $\boldsymbol{n}(\boldsymbol{r}_p^2)$ is the outgoing normal in the current configuration of point $\boldsymbol{r}_p^2$. If the surface, here $\partial\Omega^2$, is C^1-smooth, a normal vector can be uniquely defined at each point as:

$$\boldsymbol{n}(\boldsymbol{r}^2) = \frac{\frac{\partial \boldsymbol{r}^2}{\partial \xi^2} \times \frac{\partial \boldsymbol{r}^2}{\partial \eta^2}}{\left\| \frac{\partial \boldsymbol{r}^2}{\partial \xi^2} \times \frac{\partial \boldsymbol{r}^2}{\partial \eta^2} \right\|}$$

The parameterization of the surface ξ^2, η^2 must actually be chosen in such a way that the normal thus defined be outgoing. If these partial derivatives do not exist, the normal vector $\boldsymbol{n}$ cannot be uniquely defined. On the other hand, its derivatives can be defined using the notions of subdifferentials, and then defined as follows: the normal should be regarded as the set of normal vectors originating from the subdifferential of the surface at the point of interest (Moreau 1966; Heegaard and Curnier 1996; Pietrzak 1997; Yastrebov 2013). To simplify equation [3.1], the following notation is introduced:

$$g_n = g_n(\boldsymbol{r}^1, \partial\Omega^2) = (\boldsymbol{r}^1 - \boldsymbol{r}_p^2) \cdot \boldsymbol{n}(\boldsymbol{r}_p^2) \qquad [3.2]$$

where g_n is called the normal *gap*, whose absolute value defines the minimal distance between the point $\boldsymbol{r}^1$ and the surface $\partial\Omega^2$ if the latter is at least C^1-smooth (at least locally). The sign of g_n determines whether there is local penetration ($g_n < 0$) or separation ($g_n > 0$). We should bear in mind that in the general case, $g_n(\boldsymbol{r}^1, \partial\Omega^2)$ is not necessarily equal to $g_n(\boldsymbol{r}_p^2, \partial\Omega^2)$, where $\boldsymbol{r}_p^2$ is the closest point to $\boldsymbol{r}^1$ even if $\partial\Omega^1$ is C^1-smooth. With the definition of the normal *gap*, the contact conditions can be formulated as follows:

$$g_n \geq 0, \; \sigma_n \leq 0, \; g_n\sigma_n = 0 \qquad [3.3]$$

where $\sigma_n = \boldsymbol{n} \cdot \boldsymbol{\sigma} \cdot \boldsymbol{n}$ is the normal contact stress or the contact pressure negative $\sigma_n = -p$. Conditions [3.3] are called Hertz–Signorini–Moreau conditions (Wriggers 2006) or KKT (Karush–Kuhn–Tucker) in the convex analysis literature.

To give an example, consider a contact problem between a rough surface $\partial\Omega^1$ defined by $z(t, x, y)$ and an elastic half-space Ω^2 with the boundary $z = 0$. With no deformation, at each time instant, the gap is $g_n(\boldsymbol{r}^1, \partial\Omega^2) = z(t, x, y)$. However, in the current configuration, the gap is more complex because of the deformation of the half-space. It should be noted here, in the case considered and of a surface $z(x, y)$ nominally flat, that is, for any surface $S \subset \mathbb{R}^2$ large enough $\langle z \rangle = 0$, that when there are no external forces, $\forall\delta < \infty : \; z(t, x, y) = z(x, y) - \delta(t)$, the contact area remains zero.

This conclusion, which may seem surprising at first glance, is related to the Cheseaux–Olbers paradox (Harrison 1990): nonzero contact forces distributed over the surface of the elastic half-space with nonzero density create a nonzero average pressure over the entire infinite surface that results in infinite vertical displacement. Hence, if the displacement is finite $\delta < \infty$, the average pressure must be zero, so are the forces and the area of contact. A cleaner formulation would be to fix the rough surface $z(t, x, y) = z(x, y)$ and apply external pressure at infinity $-\sigma_{zz} = p_0$. In this example, the fraction of the contact area will be nonzero, regardless of the pressure applied.

NOTE.– The conditions [3.3] are strictly valid if and only if there is no fluid around the solids in contact. In real-life situations, a nonzero pressure fluid often surrounds bodies. To this end, to establish contact, the contact pressure must be greater than the fluid pressure everywhere at the edge of the contact zones $\forall \boldsymbol{r} \in \partial\mathcal{C} : \sigma_n \leq -p_f$, where p_f is the local fluid pressure. This correction is particularly relevant when contact pressure is comparable to fluid pressure (Shvarts and Yastrebov 2018a, 2018b).

NOTE.– At the macroscopic scale, it is sometimes assumed that the contact interface has a certain stiffness (Shi and Polycarpou 2005; Campañá et al. 2011; Pohrt and Popov 2012). This additional stiffness originates from the deformation of the roughness of both surfaces. At the macroscopic scale, if the surfaces are considered as averaged surfaces of the real surfaces (when there are no fine details), the actual contact can be established at a nonzero distance between averaged surfaces. On the other hand, the complementary stiffness must become infinite when the contact surface saturates and when in the contact area, the real contact area is equal to the contact area between average surfaces. This consideration justified the use of a barrier method with nonlinear parameters that captured the evolution of contact stiffness. Alternatively, a penalty method can be used to capture this complementary contact stiffness (Wriggers and Zavarise 1993; Andersson and Kropp 2008). However, in this case, the geometry of the surfaces should not be the average line of the actual surfaces but rather the surface that contains them and that exhibits the highest roughnesses. These techniques will not be physically justified in the context of small-scale contact because it is assumed that there is no longer any underlying roughness.

3.2.5. *Laws of friction*

There is a wide variety of friction laws which in their majority are nevertheless only relevant at the macroscopic scale. For example, Coulomb's law of friction according to which the tangential force $\boldsymbol{T}$ cannot exceed a threshold proportional to the normal force $\boldsymbol{N}$, that is, $\|\boldsymbol{T}\| \leq \mu\|\boldsymbol{N}\|$, can be considered as the consequence of the linear evolution of the local real contact area with the pressure $a \sim p$. According to the adhesive theory of contact (Rabinowicz 1965; Bowden and Tabor 1986; Straffelini 2001), the maximal resistance of asperities in contact to shear is given by a constant τ_c, then the maximal tangential force $T \leq \int a\tau_c dA \sim \int p\tau_c dA = \|\boldsymbol{N}\|\tau_c$. In sliding, in addition to dissipative losses on the surface, there exists dissipation present in the volume as it occurs with elastomers with viscoelastic behavior, metals with visco-elastoplastic behavior or the dissipation can originate from the formation of damage in any materials (micro cracks). At the macroscopic level, all of these energy losses are included in the law of friction. In addition, the heat produced near the contact interface alters the behavior of the material (thermal conductivity, Young's modulus, etc.) and can even cause phase change (glass transition, austenitization, melting, evaporation, etc.). All of these microscopic phenomena give rise to macroscopic friction laws (Vakis et al. 2018), which can be used for complex systems. The use of these laws at the roughness scale is confusing and cannot properly reproduce macroscopic friction laws. For this purpose, at the roughness scale, it is relevant to use Tresca-type friction laws, in other words, laws that assume that the local tangential resistance does not depend on the local pressure (Krim 1996, 2002; Mo et al. 2009) and which can be formulated as follows:

$$\begin{cases} \|\boldsymbol{v}_t\| = 0, \text{ if } \|\boldsymbol{\sigma}_t\| < \tau_c \text{ adhesion} \\ \|\boldsymbol{v}_t\| > 0, \text{ if } \|\boldsymbol{\sigma}_t\| = \tau_c \text{ sliding} \end{cases} \qquad [3.4]$$

where $\boldsymbol{v}_t$ is the tangential sliding speed, and $\boldsymbol{\sigma}_t = (\boldsymbol{I} - \boldsymbol{n} \otimes \boldsymbol{n}) \cdot \boldsymbol{\sigma}$ is the stress vector stripped of its normal component, that is, the stress vector in the tangent plane. To enhance the friction model, the dependency of the temperature and velocity threshold τ_c can be considered originating from the atomic scale (Gnecco et al. 2000; Gnecco and Meyer 2015):

$$\tau_c(T, v_t) = \tau_c^0 + \alpha T \log(v_t/v_0)$$

where α [Pa K^{-1}] is a constant related to the atomic structure and v_0 a normalization rate. However, at high speeds, thermal activation should no longer be significant and resistance should saturate. In the interval of these velocities, the following law can be used (Sang et al. 2001):

$$\tau_c(T, v_t) = \begin{cases} \tau_c^e - \alpha T^{2/3} \log(\beta T/v_t)^{2/3}, \text{ si } v_t < \beta T \\ \tau_c^e, \text{ if } v_t \geq \beta T \end{cases}$$

where α [Pa K$^{-2/3}$] and β [m s^{-1} K^{-1}] are constants related to materials in contact.

The KKT conditions can thus be formulated as follows:

$$\|\boldsymbol{v}_t\| \geq 0, \ \|\boldsymbol{\sigma}_t\| - \tau_c \leq 0, \ (\|\boldsymbol{\sigma}_t\| - \tau_c)\,\|\boldsymbol{v}_t\| = 0 \qquad [3.5]$$

Unlike Coulomb's law, this friction law is associative, which greatly simplifies how associated friction contact problems are addressed (Michalowski and Mróz 1978; Curnier 1984) because there is no coupling between contact pressure and shear. In addition, in the more general framework, Coulomb friction can be approximated by "fixed point"-based methods that make use of the Tresca law at each iteration (Panagiotopoulos 1975; Gwinner 2013).

NOTE.– Similarly to the way how penalization methods and barriers lose their physical motivation in the framework of small-scale normal contact of roughness, the same thing happens when addressing friction with the penalizationmethod, which should be avoided or utilized with particular caution, since there is no physical basis.

3.3. Finite element method

The FEM is a powerful and versatile tool for solving various problems in mechanics and physics (Bathe 1996; Zienkiewicz and Taylor 2000a, 2000b). However, contact remains a particular problem for this method because it requires the description of interaction between solids that each have their own discretization into finite elements. In the general case where the original surfaces are not geometrically compatible and/or their discretizations are not compatible either, processing contact remained a sensitive issue until recently. In this section, we will consider some questions of general interest in order to apply the FEM to micromechanical contact problems. All implementation details of contact algorithms can be found in the following monographs and doctoral theses: Kikuchi and Oden (1988); Pietrzak (1997); Laursen (2002); Wriggers (2006); Konyukhov and Schweizerhof (2012); Popp (2012); Yastrebov (2013); Akula (2019); Shvarts (2019).

Before covering the particularities of using FEM with microcontacts, we recall the basics of this method and its utilization for contact. The FEM is based on the principle of virtual works which, when there is no contact and in the infinitesimal context, can be described in the following form[5]:

$$\int_{\Omega} \boldsymbol{\sigma} : \delta\boldsymbol{\varepsilon} dV = \int_{\Omega} \rho(\boldsymbol{b} - \ddot{\boldsymbol{u}}) \cdot \delta\boldsymbol{u} dV + \int_{\Gamma_f} \boldsymbol{t}_0 \cdot \delta\boldsymbol{u}\, dS \qquad [3.6]$$

where $\delta\boldsymbol{u}$ denotes virtual displacements (or test displacements) that cancel out at the edges where Dirichlet conditions are prescribed Γ_u, that is, $\delta\boldsymbol{u} = 0$ on Γ_u, $\boldsymbol{b}$ are density forces, $\ddot{\boldsymbol{u}}$ is an acceleration vector, ρ is the density and $\boldsymbol{t}_0$ is the surface intensity of the prescribed forces on the Neumann surface Γ_f. The displacement $\boldsymbol{u}$ must verify the Dirichlet conditions $\boldsymbol{u} = \boldsymbol{u}_0$ on Γ_u and be in a functional space such that integrals in [3.6] make sense; this also applies to virtual displacements.

The introduction of contact constraints transforms the variational equality [3.6] into a variational inequality:

$$\int_{\Omega} \boldsymbol{\sigma} : \delta\boldsymbol{\varepsilon} dV + \int_{\Gamma_c} \boldsymbol{\sigma}_t \cdot \delta\boldsymbol{g}_t dS \geq \int_{\Omega} \rho(\boldsymbol{b} - \ddot{\boldsymbol{u}}) \cdot \delta\boldsymbol{u} dV + \int_{\Gamma_f} \boldsymbol{t}_0 \cdot \delta\boldsymbol{u}\, dS \qquad [3.7]$$

5 This form is called the weak form in contrast to the strong form because the former is obtained by the integration of the strong form, which allows, among other things, the conditions on the regularity of the solution to be weakened.

with the dissipation term put in the second integral evaluated on one side of the contact surface which remains unknown. At the same time, the solution $\boldsymbol{u}$ and the virtual displacement $\delta\boldsymbol{u}$ must verify the contact conditions $g_n(\boldsymbol{u}) \geq 0$ and $g_n(\boldsymbol{u} + \delta\boldsymbol{u}) \geq 0$. For friction laws relative to small scales [3.4], the friction dissipation integral can be reformulated as:

$$\int_{\Gamma_c} \boldsymbol{\sigma}_t \cdot \delta\boldsymbol{g}_t dS = \int_{\Gamma_{sl}} \tau_c \|\delta\boldsymbol{g}_t\| dS$$

where Γ_{sl} determines the sliding area in which $\|\sigma_t\| = \tau_c$. This form of the dissipative term preserves non-differentiability (Duvaut and Lions 1972; Kikuchi and Oden 1988) but eliminates the dependency of pressure.

In order to address the contact problem, inequality [3.7] (which can be thought of as a constraint optimization problem) can be transformed into an unconstrained optimization problem by means of classical methods such as the penalty method, the barrier method, the Lagrange multiplier method, the augmented Lagrangian method or using other optimization methods (Bertsekas et al. 2003). These methods can be used to remove the contact constraints imposed on the choice of real and virtual displacements and bring them back into the functional itself. After discretization of the continuous finite element formulation, a system of nonlinear equations is obtained that is generally solved using the Newton method in the implicit formulation of finite elements. This procedure naturally requires the linearization of the weak form (to obtain the tangent matrix), which is cumbersome in terms of implementation in the general case (Pietrzak 1997; Popp 2012; Yastrebov 2013). However, automatic bypass methods can lead to simplifying this procedure (Korelc 1997; Wriggers 2008; Lengiewicz et al. 2011).

Before moving on to the discussion of the applications of the FEM to small-scale contact problems, it would be important to make a few observations about the method itself.

3.3.1. *Convergence, parameters and loading step*

The convergence of the Newton iterations in contact problems cannot always be ensured for the given change of boundary conditions. However, the user can define control parameters such as the penalty coefficient(s) or the increase coefficient(s) in the augmented Lagrangian method. It is preferable that these parameters are defined independently for normal contact and tangential contact. In addition, it is also preferable that these parameters be independently defined for each localization, that is, for each contact element because these parameters are related to the local structural stiffness which is not uniform and which changes to nonlinear material models. It is known that for the penalty method, the solution can be greatly impacted by to the choice of the penalty parameter.

It is therefore good practice to redo each calculation with the penalty twice as high, for example, which will make it possible to validate the solution. The practice of "naked eye" penetration control, in general, is never conclusive. The low penalization value often gives inaccurate results but ensures fairly rapid and robust convergence.

On the other hand, the high value of the penalization deteriorates the number of conditioning of the tangent matrix, which may be critical for iterative solvers and at the same time deteriorates convergence. For the augmented Lagrangian method, the choice of increase parameters does not influence the converged solution as much but how this solution (convergence itself) is reached strongly depends thereupon: a wrong choice of the increase parameter can easily result in an infinite convergence loop (Alart 1997; Yastrebov 2013). For optimal convergence, increase parameters can be adjusted on the fly during iterations (Bussetta et al. 2012; Sewerin and Papadopoulos 2017).

3.3.2. *Convergence of friction problems*

The convergence of quasi-static friction problems does not necessarily mean convergence towards the right solution. A striking example is the Hertzian contact between two cylinders or a plane and a 2D or 3D cylinder with Coulomb friction (see Figure 3.3). If the materials which the two solids are made of are different[6], introducing normal contact (without tangential forces) will produce shear stresses at the interface. On the other hand, the self-similar semi-analytic solution (Spence 1975; Johnson 1985) can be obtained if and only if the contact takes place in such a way that at each loading step very few elements come into contact.

It is however possible to reach the ultimate load in a single increment without convergence difficulty, but the solution obtained is different from the solution obtained for the same case achieved with 100 time steps (see Figures 3.3(d) and 3.3(e)) (Klarbring and Bjöourkman 1992; Christensen et al. 1998; Kravchuk 2008; Spinu and Frunza 2015; Shvarts 2019). The importance of automatically choosing the loading step presents a relevant problem in contact mechanics that has only received very little attention (Torstenfelt 1984). Moreover, the solution of friction problems is not always unique, even theoretically in quasi-static special cases (Schatzman 1978; Klarbring 1990a, 1990b; Ballard 1999; Ballard and Basseville 2005).

6 More precisely, this statement is correct if the combination of the following elastic parameters $(1 - 2\nu_i)(1 + \nu_i)/E_i$ is different for both materials (Johnson 1985).

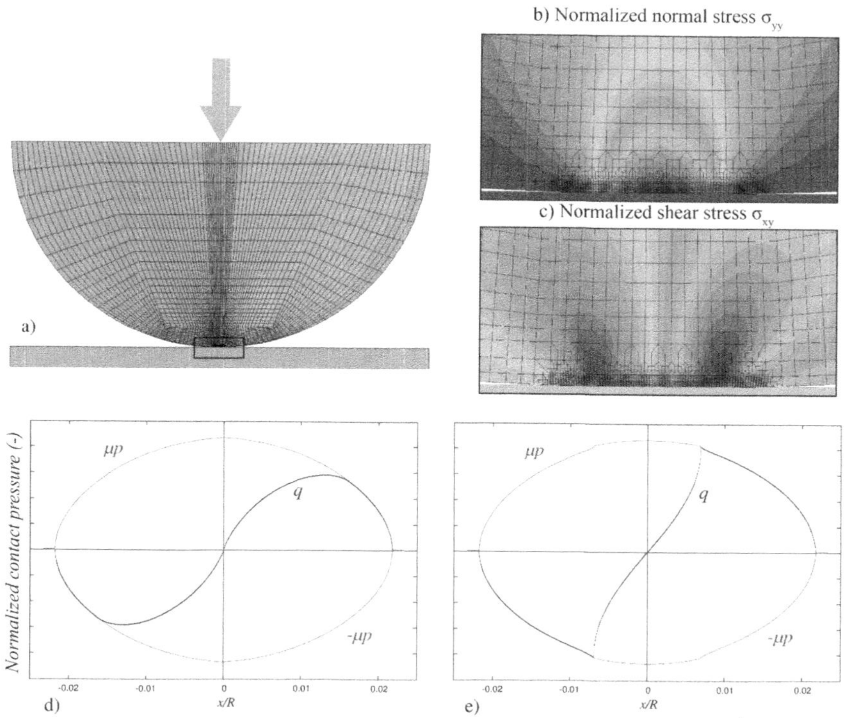

Figure 3.3. *Hertz problem with friction*

COMMENT ON FIGURE 3.3.– *Contact between an isotropic elastic half cylinder with a rigid plane under the displacement imposed on the upper surface of the cylinder (a). The magnified contact zone (b, c) demonstrates the distribution of normal stresses σ_{yy} and shear stress σ_{xy}, respectively, obtained with 100 loading steps with the imposed displacement following the parabolic form $u_y \sim i^2$, where i is the number of the time step. The stress distribution along the interface for this case is demonstrated in (d). For comparison, the distribution of the same forces for a solution obtained in a single loading step is demonstrated in (e), and this solution is erroneous. Figure is adapted from Shvarts (2019).*

The problem of elastodynamic friction is even worse in terms of solution existence and uniqueness. In the case of Coulomb friction between elastically different semi-spaces, the uniform sliding solution in the interface may be unstable. Even worse, the problem itself can be ill-posed. For example, in the case of friction between a rigid plane and an elastic half-plane, the problem remains well posed and there is an existing uniform slip solution for Coulomb friction coefficients of less than one. On the other hand, for a coefficient of friction greater than one, the slightest stress disturbance at the interface grows exponentially in time with the exponent proportional to the wavenumber of this perturbation (Renardy 1992; Martins et al. 1995). Since there is no relevant wavelength limit in classical continuous mechanics, the problem becomes ill-posed and the lack of convergence with the mesh of these problems may lead researchers to erroneous conclusions (Andrews and Ben-Zion 1997). In problems with two deformable half-planes, the stability of the solution depends on the contrast of their elastic properties and their densities (Adams 1995; Cochard and Rice 2000; Ranjith and Rice 2001): there are situations where the stable solution exists but only for friction values below a certain critical threshold, and there are cases where the solution is unstable or ill-posed, regardless of the coefficient of friction. These instabilities are related to the strong coupling between contact pressure and shear stress in the case of different materials. These effects give rise, for example, to brake squeal noise (Kinkaid et al. 2003; Massi et al. 2007) and slip localizations in geological faults that are responsible for earthquakes (Heaton 1990). For finite-sized or stratified systems, uniform slip can always be unstable (Brener et al. 2016; Mohammadi and Adams 2018), which manifests itself in slip spatio-temporal localization (Renard 1998; Adams 2000; Bui and Oueslati 2010), adhesion and even opening in some cases (Gerde and Marder 2001; Moirot et al. 2003; Yastrebov 2016). To meet this challenge, studies of friction dynamics must reject Coulomb-type laws where the shear stress is proportional to the contact pressure. This bond can be softened by a viscoelastic type relationship as can be seen in the Prakash–Clifton law of friction, which is based on an experimental observation (Prakash and Clifton 1993; Prakash 1995) of non-immediate friction adjustment in response to an instantaneous pressure change. This regularization of friction results in obtaining physical and converged results (Cochard and Rice 2000; Ranjith and Rice 2001; Kammer et al. 2014). An alternative is the use of nonlocal friction laws (Simoes and Martins 1998). However, for relevant friction laws at the roughness scale, friction dynamics must not represent the following defects due to the absence of normal-tangential stress coupling.

3.3.3. *Quadratic convergence*

Unlike many nonlinear problems in continuous media mechanics, contact problems in most cases do not quadratically converge with the Newton method. This lack of quadratic convergence is caused by the change in status of the contact elements between contact/noncontact and slip/adhesion. If the set of statuses is fixed, the convergence becomes quadratic when all of the terms of the tangent matrix have been included. In contact problems with the very rapid change of boundary conditions (e.g. initial penetration greater than element size), it is preferable not to include all terms in the tangent matrix, which can be added as soon as the convergence of statuses occurs (Wriggers 2006).

3.3.4. *Mesh and computation time*

The stress state considerably varies near the contact area which therefore requires fine meshes in these areas to ensure the accuracy of the digital solution. The contact pressure near the edge of the contact area within the framework of smooth surfaces changes as $p \sim t^{1/2}$, where t is the distance to the edge. Because of this fast decreasing contact pressure, the very fine mesh is necessary to be able to determine the boundary between contact and noncontact and to measure the contact area, which is important for a large number of micromechanical contact problems (Yastrebov et al. 2017a, 2017b). For flat punches, the pressure near the edge evolves as $p \sim t^{-1/2}$. In this case, it is easy to refine the mesh near this edge that is a priori known. On the other hand, for a sliding contact, the uniformly fine mesh would be necessary for this problem. Mesh refinement is also necessary if the coefficient of friction is not continuous along the contact interface (Ballard 2016) or when studying a mode II or III crack. In all of these cases, constraint singularities appear at the interface (Coker et al. 2005; Svetlizky and Fineberg 2014; Barras et al. 2014; Kammer et al. 2015; Svetlizky et al. 2016).

The simulation of the contact between nonlinear materials requires refined volume meshes near contact zones (Nigro et al. 2014). This is the case in studies of the behavior of isolated asperities that will be discussed in detail in section 3.4 where mesh refinement is very important for detecting the initiation of the plasticization process that begins in the volume when there is no surface friction (Johnson 1985; Hill et al. 1989; Mesarovic and Fleck 1999; Kogut and Etsion 2002), but also for obtaining a precise macroscopic force–displacement response that strongly depends on the fineness of the volume mesh. This requirement on mesh fineness incurs a bottleneck in computing rough surfaces in contact with the FEM even when using very advanced parallel methods deployed on high-performance computing clusters (Dostál et al. 2016), because it involves very high computation times. In addition, in order to conduct a statistically significant study of contact between rough surfaces, a very large number of calculations must be performed (Yastrebov et al. 2015; Rey et al. 2019). Nonetheless, the fine mesh can be restricted to the area near the contact interface only and remain coarse enough away from this area to simply capture the stiffness of the system or avoid edge effects. Some examples of such meshes are shown in Figure 3.4.

3.3.5. *Contact constraint*

In contact problems, the focus is often on the distribution of surface stress. However, this is not always easy to achieve because the constraint is often evaluated at the integration points located inside the elements. In order to have nodal values and use interpolation later, finite element software programs extrapolate the values found at the points of interest and averaging. The direct consequence of this procedure is that the contact stress obtained in such a way is not very accurate and may have nonzero values outside the contact area. To recover true stress values, the stresses must be found at the surface using directly $\boldsymbol{\sigma} = \boldsymbol{C} : (\nabla \boldsymbol{u} + \boldsymbol{u}\nabla)/2|_{\boldsymbol{x}\in\Gamma_c}$ in the elastic case, where $\boldsymbol{C}$ is the fourth-order elasticity modulus tensor, this calculation could be done during *post-processing*. For nonlinear materials involving state variables, the surface stress computation procedure is more costly and requires the addition of surface integration points at the very beginning of the computation and involves the integration of the material model locally (Simo and Hughes 1998). Contrary to what may appear to be the case, this discussion is as relevant for using the conventional penalization method as it is for the Lagrange multiplier method. It is true that in the latter it is more trivial to have values of the contact forces (forces [N]), which are represented by Lagrange multipliers in node-to-surface discretization. For this purpose, in order to determine stresses [N/m^2], it is necessary to associate an area with each surface node to normalize the value of the Lagrange multipliers. Although performing this procedure in 2D is trivial, this is not what happens in 3D. Another possibility to have a good estimation of surface stresses is to have a thin layer of the elements near the contact surface to minimize extrapolation error. In state-of-the-art methods, surface stresses can be directly obtained, these are surface-to-surface type discretization methods: mortar and Nitsche methods (Belgacem et al. 1999; Puso and Laursen 2004; Puso et al. 2008; Wriggers and Zavarise 2008; Popp et al. 2010; Temizer et al. 2012; Chouly and Hild 2013; Chouly et al. 2015).

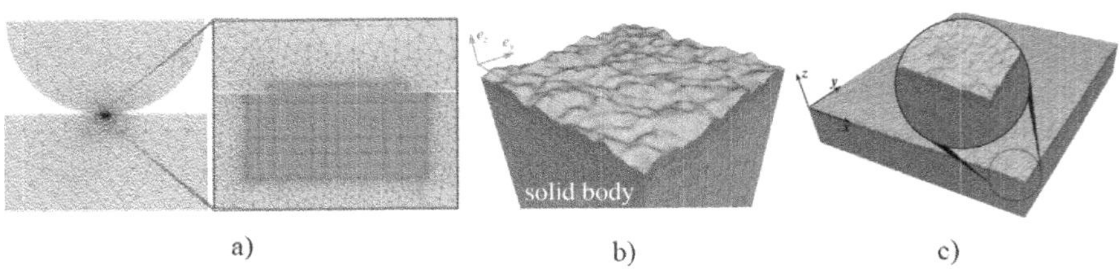

Figure 3.4. *Finite element meshes for (a) 2D fatigue fretting contact (source: Sun 2012); (b and c) rough contact with roughness (source: Shvarts 2019)*

It is important to notice that node-to-surface methods cannot transfer uniform surface forces between non-compliant meshes, which represents a "patch-test" of interface problems (Taylor and Papadopoulos 1991) (see Figure 3.5). This effect is related to the under-integration of contact stresses at the interface from which the advanced analysis was derived to meet this challenge (Crisfield 2000; El-Abbasi and Bathe 2001; Tan 2003; Chen and Hisada 2006; Hartmann et al. 2009; Oliver et al. 2009; Zavarise and De Lorenzis 2009a, 2009b). Despite that parasitic oscillations due to subintegration might have very small amplitudes, they can lead to erroneous solutions if we are focusing on the surface stress state and if, for example, wear laws based on these stresses are used to update the geometry of worn surfaces (Lengiewicz and Stupkiewicz 2013; Basseville and Cailletaud 2015; Farah et al. 2016; Basseville et al. 2019). Surface-to-surface type methods can be used to avoid this defect and ensure an accurate stress state. However, if the contrast between materials is very high, then similar oscillations may appear in the contact even with these methods (Akula et al. 2019).

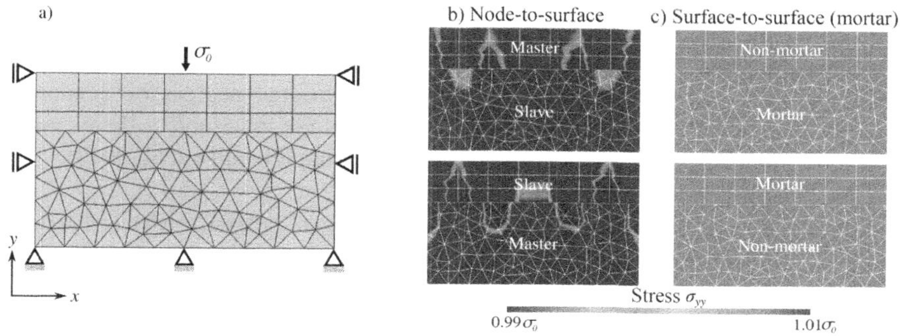

Figure 3.5. *Comparison of normal stresses σ_{yy} with (b) node-to-surface (NTS) and (c) surface-to-surface (STS) discretization for different choices of master-slave and mortar/non-mortar surfaces with NTS and STS, respectively (adapted from Akula 2019)*

3.3.6. *Surface regularity*

Naturally, finite element surfaces are of regularity class C^0 (only continuity), even if they are C^∞-smooth inside the regions of the elements. The incompatibility of gradients at the junction between elements is related to the compact nature of finite element shape functions. This irregularity in surface geometry poses problems for addressing contact because it induces oscillations in local stresses that can result in macroscopic oscillation of forces.

To avoid these problems, in applications of contact between bodies of very different stiffness (tire/road, hardness test indenter/substrate, touch screen/finger and shoe/floor), it is preferable to use rigid surfaces described by analytical functions or CAD models (Heege and Alart 1996): the simplest example would be a rigid plane, which is widely used in micromechanical contact for isolated asperities as well as for rough surfaces (Mesarovic and Fleck 1999; Kogut and Etsion 2002; Pei et al. 2005; Yastrebov et al. 2011). However, the iso-geometric analysis (IGA) method, which takes advantage of smooth and at the same time deformable surfaces, makes it possible to forget the issues of surface regularity.

However, the formulation of the contact in the context of the IGA is not direct because the too regular nature of the solutions does not allow the representation of the contact pressure, which reaches a zero with vertical slope. To this end, different techniques have been developed and can be found in the review articles by Temizer et al. (2011); De Lorenzis et al. (2014).

In the context of classical finite elements, smoothing techniques can be utilized: this can either exclusively concern the normal vector field (Yang et al. 2005; Popp et al. 2010), or directly surface geometry (Wriggers and Krstulović-Opara 2000; Wriggers et al. 2001; Belytschko et al. 2002; Puso and Laursen 2002; Chamoret et al. 2004; Munoz 2008). In addition, the FEM with additional degrees of freedom of rotation, which ensures the regularity of the normal at the junctions of the elements, can be considered as an alternative solution to smoothing (Batailly et al. 2013).

UNINTERRUPTED FRACTAL SURFACE.– A contact problem between rough surfaces with a high-frequency uninterrupted self-affine spectrum (see, for example, Hyun et al. 2004; Pei et al. 2005) presents a problem poorly posed from a numerical mechanics point of view. Since the shortest surface wavelength is limited by the size of the element, convergence cannot be achieved by refining the mesh, which of course also poses the problem of interpreting the computational results.

3.4. Application I: study of an isolated asperity

3.4.1. *Elastic asperity*

In the continuous and smooth description[7] of "rough" surfaces, "asperities" can always be isolated, namely, the elements of the simple geometry that can be locally approximated by elliptical paraboloids:

$$z(x, y) = \frac{x^2}{2R_x} + \frac{y^2}{2R_y} + z_0$$

in an orthonormed basis $\{x, y\}$ with two principal radii of curvature R_x, R_y such that $R_x R_y > 0$. However, crystal (minerals) surfaces can be described by different elementary shapes such as steps, ridges and pyramids. It should be noted that apart from paraboloids, which are essential for the study of weak contact, another quadric – hyperbolic paraboloid or saddle point ($R_x Ry < 0$) – is essential for the study of critical junctions in runoff problems between rough surfaces in contact (Plouraboué et al. 2004; Dapp et al. 2012; Dapp and Müser 2016; Shvarts and Yastrebov 2018a) but also for contact beyond the infinitesimal area hypothesis (Johnson et al. 1985; Yastrebov et al. 2014).

7 If the surface is smooth, it is automatically *non-fractal* because fractals are not differentiable manifolds: elements of these surfaces cannot be described by simple and differentiable shapes.

The study of contact between rough edges represents an important branch of the study of microscopic contacts. First of all, it should be noted that this type of contact is directly related to the indentation study, especially if the asperities are described at the scale where the classical theories of continuous media remain valid. On the other hand, at a small scale, where the purely local characterization of the deformation ceases to be valid, the link between the contact of asperities and micro- or nano-indentation should rather be considered. The objective of this section is to present a review of studies of these elementary contacts carried out with the finite element or boundary method. At the same time, in relevant cases, we will highlight numerical innovations that have been used.

The problem of purely linear elastic contact between elliptical paraboloids represents the first study of deformable contact which was carried out by Hertz (1881). In the simple case of revolution surfaces, the contact radius a, the gap δ and the contact pressure distribution are found:

$$a = \left(\frac{3NR^*}{4E^*}\right)^{1/3} \qquad [3.8]$$

$$\delta = \frac{a^2}{R^*} = \left(\frac{9N^2}{16R^*E^{*2}}\right)^{1/3} \qquad [3.9]$$

$$p(x,y) = \frac{3P}{2\pi a^2}\sqrt{1 - \frac{x^2+y^2}{a^2}} \qquad [3.10]$$

where N is the normal force, $R^* = R_1R_2/(R_1 + R_2)$ is the effective radius of curvature with R_i being the radius of curvature of the surface $i = \{1,2\}$, $E^* = E_1E_2/[(1-\nu_2^2)E_1 + (1-\nu_1^2)E_2]$ is the effective modulus of elasticity with ν_i, E_i is the Poisson coefficient and Young's modulus of the field $i = \{1,2\}$. In the vicinity of the vertex, the parabolic surface $z = r^2/(2R)$ is equivalent to a spherical surface $z = R - \sqrt{R^2 - r^2}$ up to $o(r^2/R^2)$. To this end, the Hertz theory formulated for paraboloids works just as well for spheres if $a/R \ll 1$.

Now, Hertzian contact is the conventional test for validating numerical contact processing techniques. However, these validation tests are often limited to validating the method for two-dimensional surfaces of revolution $R_x = R_y$ with respect to the axial symmetry of the problem. The simplest validation can be made for the contact between a rigid plane $R_1 = \infty$, $E_1 = \infty$ and a deformable paraboloid (sphere or its half) $R_2 = R^*$, $E_2 = E^*(1-\nu_2^2)$. On the other hand, an axisymmetric representation may be satisfying only in the case of normal contact (with no tangential loading).

CALCULATION OF THE CONTACT AREA.– Since the computation of the contact area or contact radius is based on the calculation of the number of contact areas/elements, the numerical evaluation of the contact area is not continuous: the contact radius grows by steps of comparable size to the discretization of contact surfaces $\Delta a \lessapprox h$, which therefore determines the numerical error of the contact radius. In the general case, as this error is located at the perimeter P of the contact region of area A, the error on the contact area depends on the ratio $E_r \sim Ph/A$ that for the revolution contact is reduced to $E_r \sim 2h/a$, which explains the need for very fine meshes near the contact boundary to correctly capture the contact area. For rough surfaces with the minimal wavelength λ_s, the curvature of the asperities $R \sim \lambda_s^2/(4\pi^2\Delta)$, where Δ is the amplitude related to this wavelength. Given that $a \sim \sqrt{R^*\delta}$, the computation error of the contact area evolves as $E_r \sim \frac{4\pi h}{\lambda_s}\sqrt{\frac{\Delta}{\delta}}$ for each asperity in contact whose number can be very large (Yastrebov et al. 2017a; Shvarts 2019).

Apart from this simple case which is exclusively used to validate numerical methods, the contact interaction can be enhanced with friction and/or adhesion. In the presence of friction, the analytical solution exists only for the following three cases: (1) normal and tangential contact between identical materials (see footnote 6, in section 3.3.2) or noncompressible materials; (2) normal contact with infinite friction (Mossakovskii 1954, 1963; Abramian et al. 1966; Spence 1968); and (3) stationary slip asperity (Goodman and Hamilton 1966; Hamilton 1983). These three cases can be used to validate the implementation of friction contact algorithms. The most used and richest case of the three is notably the case of contact between two similar materials that was derived in Cattaneo (1938); Mindlin (1949). This test was used in Laursen and Simo (1993); Yang et al. (2005); Gitterle et al. (2010); De Lorenzis et al. (2011); Wei et al. (2016); Akula et al. (2019) to name some. Another friction test was used in Alart and Curnier (1991); Yastrebov (2013) based on a semi-analytical solution of Klang (1979). The test consists of a contact between a cylinder and a circular hole in a plane that have quasi-conformal geometries. Nevertheless, the two numerical solutions (Alart and Curnier 1991; Yastrebov 2013), which agree, do not converge to the analytical solution in terms of shear stress distribution. A more consistent numerical solution was obtained in Pietrzak and Curnier (1999) for a slightly different geometry than that used in the two attempts mentioned above.

3.4.1.1. *Normal and tangential contact separability*

In the general case, the contact problem can be divided into two decoupled problems (normal contact and tangential contact problems) if and only if the elastic constants of two bodies in contact are the same (see footnote 6, section 3.3.2) and when the edges or interfaces are far enough from the contact area. Otherwise, the two problems remain coupled and cannot be rigorously solved separately. However, still with the same material, numerical schemes need to be used with great caution to obtain correct results even for normal loading; see examples of erroneous solutions (Andersson 1981, appl. 1; Shyu et al. 1989). The coupling between tangential forces $\tau(x)$/normal $p(x)$ applied to $x \in (-b, a)$ and the associated displacements v, u can be explicitly visualized by the shape of the fundamental Flamant and Cerutti solution for an elastic half-plane (Johnson 1985):

$$\left.\frac{\partial u_i}{\partial x}\right|_{y=0} = -\frac{(1-2\nu_i)(1+\nu_i)}{E_i}p(x) - \frac{2(1-\nu_i^2)}{\pi E_i}\int_{-b}^{a}\frac{\tau(s)}{x-s}\,ds \qquad [3.11]$$

$$\left.\frac{\partial v_i}{\partial x}\right|_{y=0} = \frac{(1-2\nu_i)(1+\nu_i)}{E_i}\tau(x) - \frac{2(1-\nu_i^2)}{\pi E_i}\int_{-b}^{a}\frac{p(s)}{x-s}\,ds \qquad [3.12]$$

where $i = 1, 2$ designates the values associated with the ith body.

A semi-analytic solution for a problem involving a rigid indentor based on the consideration of self-similarity was obtained by Spence (1975). It should however be noted that self-similarity is lost on unloading (Stingl et al. 2013; Kim and Jang 2014). This problem, generalized to the case of two deformable solids (Nowell et al. 1988), represents a reference test for the implementation of friction contact algorithms (Jinn 1989; Klarbring and Bjöourkman 1992; Christensen et al. 1998; Guyot et al. 2000; Gallego et al. 2010; Akula et al. 2019). This normal contact with no coupling between normal and tangential tractions was considered in Hills and Sackfield (1987). However, we need to bear in mind that this decoupling is exclusively feasible with the boundary element method and cannot be used with finite elements where coupling is inherent. We can note that examples of false numerical solutions of this problem are recurringly found in the literature; as examples see Jing and Liao (1990); Lee (1994); Kosior et al. (1999); Elkilani (2003); Li and Berger (2003); Chen and Wang (2008) and see a discussion on the subject in Kwak and Lee (1988). The self-similarity of the solution can be used to speed up numerical computations by reducing the calculation to a single loading step with additional conditions to be verified (Storåkers and Elaguine 2005; Jelagin and Larsson 2012).

In addition to being used as a test case for numerical algorithms, this problem of normal contact between different bodies is also important for several applications, especially to understand failure modes (wear and fatigue crack initiation) of micro- and macroscopic cyclic normal contact.

Another test that involves compression and torsion of the contact area with finite friction between a rigid plane and a deformable sphere was used in Chaudhary and Bathe (1986) based on an analytical solution (Hetenyi and McDonald 1958). This problem involves a normal loading phase with no friction that is followed by a torsion phase with finite friction. This hypothesis is non-physical because the influence of history is precisely what makes the friction problem so complex and it is not taken into account in this solution. Another numerical test for frictional contact between a deformable body and a rigid indenter is the axisymmetric indentation of an elastic half-space by a cylindrical punch whose solution has been shown to be similar to parabolic indentation (Spence 1975): this problem was used to test the implementation of augmented Lagrangian in Torstenfelt (1984); Pietrzak and Curnier (1999); Yastrebov (2013). An automation of the choice of loading step for friction problems was proposed in Torstenfelt (1984).

This section will be concluded by highlighting the lack of analytical and numerical solutions for normal and tangential contact between paraboloids made up of similar or different materials with a Tresca-type friction law that is more relevant at a small scale of asperities than Coulomb's law.

3.4.2. *Elastoplastic asperity*

The elastoplastic behavior of the material in the vicinity of contact zones is highly relevant for many engineering applications that involve metallic materials. From the force–displacement curves of a hard indentor (diamond or tungsten carbide) on a metallic material or based on the impression, several properties of the material can be found (Tabor 1951; Oliver and Pharr 1992; Fischer-Cripps 2011; Herrmann et al. 2011). The relevance of this type of indentor/substrate interaction is obvious for microscopic contact: at a small scale, asperities of the harder ("less plastic") material can be seen as indentors for the asperities of the other material. Understanding the normal and tangential interaction of parabolic surfaces in the elastoplastic regime is necessary for understanding the microscopic contact of metals and alloys. In this section, we briefly review the numerical calculations of elastoplastic contact either between a rigid parabolic/spherical indenter and a deformable substrate, or the opposite, or between both deformable bodies.

Many studies dedicated to elastoplastic contact focus exclusively on the loading phase. However, the unloading phase as well as repetitive contact (cyclic loading) are very important to consider for asperity contact. Figure 3.6 shows the results of a finite element calculation of the elastoplastic behavior of a spherical body under cyclic loading where each new cycle has a higher loading amplitude than the previous cycle. The zero cycle is purely elastic and follows the Hertz solution. For cycles with a higher force (branch 1), the contact pressure saturates with the hardness of the material and the derivative of the contact area A with respect to the compressive force $\partial A/\partial F$ remains constant, that is, the contact area evolves in an affine manner with the force $A = A_0 + F/H$, where H indicates the hardness of the material. The constant A_0 is often neglected because it can be very small for developed plastic contact. Unloading (branch 2), in this case, is completely elastic, but does not follow the Hertz solution $A \sim F^{2/3}$ because the residual curvature is not constant at all points as it should be for the Hertz solution to be valid. The consecutive reloading, branch 3, again follows branch 2 (since we still are in the frictionless elastic regime), but as soon as the load exceeds the maximum load of the previous cycle, the solution qualitatively changes and again follows the saturation curve (branch 4) and so on: elastic unloading (5); elastic reloading (6), saturation under load (7). It is therefore clear with this example that the contact area cannot be predicted for a given load because the loading history is involved as in all nonconservative evolving problems.

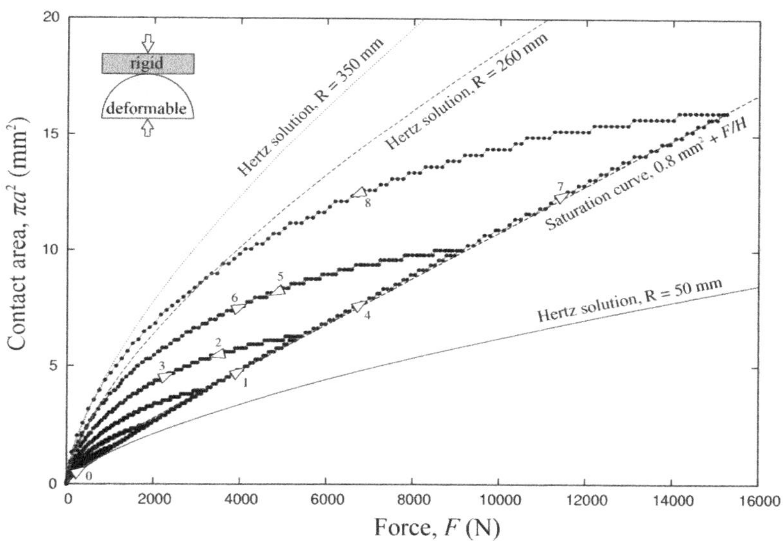

Figure 3.6. *Cyclic loading of an elastoplastic "asperity" of the initial radius R = 50 mm in contact with a rigid plane*

Many studies on elastoplastic indentation have been performed and include Komvopoulos (1989); Kral et al. (1993); Song and Komvopoulos (2013) as well as Kogut and Etsion (2002, 2003a); Etsion et al. (2005); Kadin et al. (2006); Brizmer et al. (2007). The elastoplastic contact in the presence of coating was studied in Sun et al. (1995); Song et al. (2012) using the FEM.

In an idealized case of normal contact between parabolic surfaces and with no friction and adhesion, the Hertz theory makes it possible to predict the beginning of plastic flow. To obtain the start point located on the axis of symmetry, the von Mises stress σ_{eq} for example, simply has to be differentiated with respect to the normalized depth $z' = z/a$ where a and the contact radius. Using the following equations:

$$\frac{\sigma_z(z)}{p_0} = -\frac{1}{1+z'^2} \qquad [3.13]$$

$$\frac{\sigma_r(z)}{p_0} = \frac{\sigma_\theta(z)}{p_0} = (1+\nu)\left(z' \arctan(1/z') - 1\right) + \frac{1}{2(1+z'^2)} \qquad [3.14]$$

and the von Mises stress is given by:

$$\sigma_{\text{eq}} = |\sigma_z - \sigma_r| = p_0 m(\nu, z') \qquad [3.15]$$

with:

$$m(\nu, z') = \left|(1+\nu)\left(1 - z' \arctan(1/z')\right) - \frac{3}{2(1+z'^2)}\right|$$

It can then easily be shown that the von Mises stress reaches the maximal value at a distance z'^* which verifies $\partial\sigma_{\text{eq}}/\partial z' = 0$. The last equation is reduced to the following equation:

$$(1+\nu)\left(\arctan(1/z'^*) - z'/(1+z'^{*2})\right) = 3z'/\left(1+z'^{*2}\right)^2 \qquad [3.16]$$

The solution to this equation can be approximated by an affine function (see Figure 3.7(b.4)):

$$z'^* \approx 0.3819375 + 0.33187\nu \qquad [3.17]$$

Then, the function $m(\nu, z'^*)$ can be approximated by another affine function (see Figure 3.7(b.3)):

$$\max(\sigma_{\text{eq}}/p_0) = m(\nu, z'^*) \approx 0.7696422 - 0.4738901915\nu \qquad [3.18]$$

Finally, if the yield strength is given by σ_Y, then maximal pressure at the center of the contact p_0^* needed to initiate plasticization is given by p_0^* and the depth is given by z^*, respectively:

$$p_0^* = \frac{\sigma_Y}{m(\nu, z'^*)}, \quad z^* = z'^* a^* = \frac{\pi z'^* p_0^* R^*}{2E^*} = \frac{\pi z'^* \sigma_Y R^*}{2E^* m(\nu, z'^*)}$$

where R^* is the effective radius of curvature, and E^* is the effective modulus of elasticity. From these equations, the onset of the normalized depth of plastic flow can be found approximately using [3.18] and [3.17] (Figure 3.7(b.2)):

$$\frac{z^*}{R^*} \approx \frac{1,199892 + 1,042600\nu}{1,53928444 - 0,947780\nu} \frac{\sigma_y}{E^*} \qquad [3.19]$$

If assuming that the indenter is rigid, it follows that the effective modulus of elasticity depends only on the Poisson coefficient of the deformable solid $E^* = E/(1-\nu^2)$, which gives us the following approximate form for the normalized depth of the onset of plasticity (Figure 3.7(b.1)):

$$\frac{z^*}{R^*} \approx \frac{(1.199892 + 1.042600\nu)(1-\nu^2)}{1.53928444 - 0.947780\nu} \frac{\sigma_y}{E} \qquad [3.20]$$

This analytical estimate is used to build a finite element mesh adapted to elastoplastic calculations, in other words, to have the depth of the fine mesh that would be adapted to the calculation. It should noted that the onset of plasticity cannot be easily detected during the indentation test because if the plastic area remains very small, then it does not affect the loading curve.

Box 3.1. *Onset of plasticity*

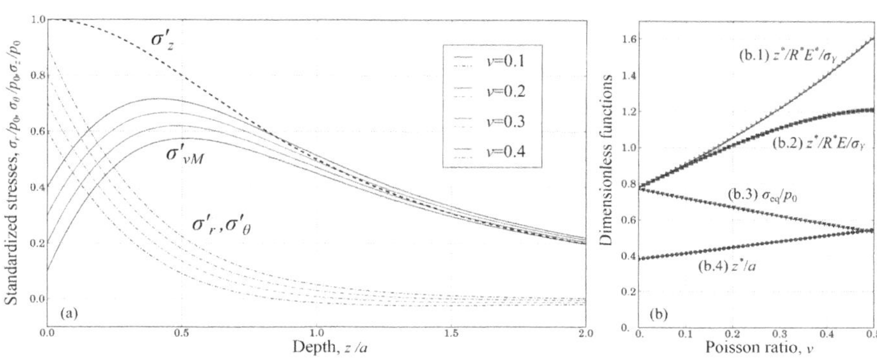

Figure 3.7. *Stress under Hertzian contact*

COMMENT ON FIGURE 3.7.– *(a) Stress σ_θ, σ_r and σ_z along the axis of symmetry; (b) dimensionless functions which represent: (b.1) the normalized depth of plastic flow* *onset $z^*/R^*E^*/\sigma_Y$, (b.2) the normalized depth with the rigid indenter $z/R^*E/\sigma_Y$, (b.3) the maximal normalized von Mises stress σ_{eq}/p_0, (b.4) the normalized depth z^*/a, the solution to equation [3.16]; The solid lines in (b) represent approximate solutions [3.19], [3.20], [3.18], [3.17], respectively.*

Of course, with small-scale roughness, the microstructure of the metal must be taken into account because the von Mises plasticity model is not very relevant at this scale. Nevertheless, hardness and force–displacement curves are not very affected by CFC crystal plasticity (namely crystallographic orientation) because many slip systems are activated (Casals and Forest 2009; Durand 2012; Sabnis et al. 2013). This is however not the case for the HCP network (Casals and Forest 2009). However, the morphology of the plastic area is very sensitive to the crystallographic orientation of the indented crystal. In addition to crystal plasticity anisotropy, elastic anisotropy must be taken into account at this scale. Since even the Boussinesq-based solution cannot be analytically obtained in the case of general anisotropy, the solution to the contact problem between anisotropic bodies is a difficult problem. It was, however, solved by Willis (1966, 1967) through the use of Fourier transforms, where the problem is reduced to the numerical evaluation of contour integrals. The two-dimensional case is simpler and there are analytical methods to address it (Lekhnitskii 1981).

The direct calculation of the depth of the onset of plasticity for very small curvature radii yield values comparable to interatomic distances. Consider polycrystalline gold which has yield strength $\sigma_Y = 140$ MPa, modulus of elasticity $E = 79$ GPa, Poisson coefficient $\nu = 0.4$, CFC crystalline parameter $a = 0.4$ nm. The onset depth can be estimated based on equation [3.20] at $z^* \approx 0.0021R^*$. When a gold surface is indented by a harder asperity of micrometric radius, the depth z^* is nanometric, for $R^* = 1$ μm, $z^* \approx 2.1$ nm $= 5.25a$, which cannot be properly described in the context of continuous media mechanics and isotropic plasticity. At this scale, the discrete nature of the crystal lattice must be taken into account, as in the following multi-scale studies (Fivel et al. 1998; Chang et al. 2010). Under the high pressure, at this small scale, the pressure saturates with the hardness of the material, which in turn depends on the scale (either the depth of the indentation for non-smooth indenters (Nix and Gao 1998; Qiu et al. 2001; Feng and Nix 2004), or the radius of curvature for spherical indenters (Swadener et al. 2002; Gao et al. 2015)). Through geometric (accommodation of shape by geometrically necessary dislocations) and phenomenological considerations (based on the saturation of dislocations), simplified laws for contact hardness H were deduced. The simplest distribution for a conical indenter takes the following form Nix and Gao (1998):

$$\frac{H}{H_0} = \sqrt{1 + \frac{d_0}{d}} \qquad [3.21]$$

where the macroscopic hardness is H_0, d is the indentation depth and d_0 is an internal distance characterizing the material. A more elaborate law has been proposed in

Qiu et al. (2001) and Feng and Nix (2004), which takes into account the high frictional stress that is characteristic of body-centered cubic (CC) crystals:

$$\frac{H - H_P}{H_0 - H_P} = \sqrt{1 + \frac{1}{(1 + \beta \exp(-d/d_1))^3} \frac{d_0}{d}} \qquad [3.22]$$

where $H_P = 3\sqrt{3}\,\tau_P$ and where τ_P is the Peierls stress, $1 + \beta \exp(-d/d_1) = r/a(d)$ represents the ratio between the radius of extension of the plastic zone with respect to the contact radius $a(d)$, which is a sinking function d, H_0 is the macroscopic hardness, and d_1, d_0 are characteristics of the material.

In the case of a spherical/parabolic indenter, the depth of sinking no longer modifies the material hardness, but the radius of curvature of the indentor R does. The simple model and, unlike conical indentation, without adjustable parameters, can be proposed in this case (Swadener et al. 2002):

$$\frac{H}{H_0} = \sqrt{1 + \frac{R_0}{fR}} \qquad [3.23]$$

where $R_0 = \bar{r}/(\rho_s b)$, and $\bar{r}$ is the Nye factor[8], ρ_s is the density of statistical dislocations in the deformed region, b is the Burgers vector, and the factor f is the ratio of the imprint curvature radius to the indentor radius that can be taken approximately $f = 1.17$ (Swadener et al. 2002). For gold, $b = 0.288$ nm, the statistical density of dislocations in the strongly deformed region under the indenter can be estimated as $\rho_s \approx 10^{16}$ m^{-2}, and for the Nye factor $\bar{r} = 2$, we get $R_0 \approx 0.69$ μm, which agrees with Kim et al. (2018). For macroscopic hardness $H_0 = 250$ MPa, we obtain the following hardness:

$$H = 250 \text{ MPa} \sqrt{1 + \frac{0.59\ \mu\text{m}}{R}}$$

Clearly, this equation cannot be used in finite element calculations, but it can be used in calculations made with multi-asperity models (Kogut and Etsion 2003a; Ciavarella et al. 2006; Afferrante et al. 2012; Yastrebov 2019) or with boundary element codes involving the phenomenological consideration of plasticity by contact pressure saturation (Sahlin et al. 2010; Manoylov et al. 2013). To simulate the size effect with finite elements, generalized models of the continuous medium

8 The Nye dimensionless factor determines the fraction of dislocations that must be created to accommodate the plastic strain relative to the number of geometrically necessary dislocations.

(Forest 1998; Maugin and Metrikine 2010; Altenbach et al. 2011; Eringen 1999) should be used, as in Zhang et al. (2005); Zisis et al. (2014); Song et al. (2017); Lewandowski and Stupkiewicz (2018).

With friction, the von Mises stress can be quite high at the surface and it can therefore create another source of plasticity growth that will aggregate at the plastic core developed below the surface. The effect of friction on indentation was discussed in several works (Tabor 1951; Mata and Alcala 2004), even if the discussion is limited to Coulomb's law which can be imprecise in indentation[9] and is not valid at a small scale. In addition, very high pressures are frequent in other applications such as, for example, indentation, machining (Özel 2006) or metal forming and rolling (Wilson and Sheu 1988; Ma et al. 2010; Hol et al. 2012). In some studies, the effect of friction is included in the analysis by considering two borderline cases: frictionless and with infinite friction, that is, purely adherent contact (Mesarovic and Fleck 1999). Nonlocal friction was studied in Jelagin and Larsson (2013).

There is a very substantial literature on indentation and the interaction of isolated asperities. The contact between different elastoplastic materials was studied in Eriten et al. (2012); Olsson and Larsson (2016). In addition to the properties of normal contact with and without friction, contact with adhesion plays an important role at small scales especially for noble metals that oxidize less than other metals. This is also the case for wear studies, where surface oxides are removed by wear making the interface more adhesive. Studies on adhesive contact between elastoplastic rough asperities can be found in Olsson and Larsson (2013). Even more important for the understanding of interface physics are studies of tangential motion between elastic or elastoplastic asperities starting with work by Green (1954); Greenwood and Tabor (1955) on the deformation of asperities in plasticine, which was followed by many studies (Cocks 1966; Challen and Oxley 1979; Black et al. 1993), among other numerical studies (such as for example, Tangena and Wijnhoven 1985; Faulkner and Arnell 2000; Kogut and Etsion 2003b; Jackson et al. 2007; Eriten et al. 2010; Mulvihill et al. 2011). However, in many studies, asperities are simulated as half-spheres, which reduces the relevance of quantitative results for true roughness; the same remark concerns the use of the Coulomb-type law of friction which, as has already been mentioned several times, is not relevant for contact between asperities and which must be replaced by Tresca friction.

The interest of all studies of asperities or isolated indentors is to be able to feed mesoscopic models that are based on the notion of asperities and their interaction (Yastrebov et al. 2011; Afferrante et al. 2012). In addition, the results obtained for isolated asperities can be used for the construction or validation of

9 The invalidity of Coulomb's law for very high pressure is explained by the fact that the contact area saturates and the maximum resistance also does so accordingly.

phenomenological models (Hulikal et al. 2015). In the study by Yastrebov et al. (2011), numerous large-strain elastoplastic calculations on isolated asperities resulted in constructing a response surface of characteristic asperities found on experimental rough surfaces, which was then used in a multi-asperity model with an elastic but short-range interaction[10]. The scale effect in this type of multi-scale calculation was briefly analyzed in Yastrebov (2019) assuming that the hardness depends on the radius of curvature. However, it must be realized that although the interaction between asperities can be taken into account (Ciavarella et al. 2006; Paggi and Ciavarella 2010; Yastrebov et al. 2011; Yastrebov 2019), these models based on the notion of asperities can predict the evolution of the actual area of contact only under low pressure when the contact areas remain associated with isolated asperities. The consideration of junction of contact areas associated with neighboring asperities cannot be rigorously achieved in these models and remains a very important growth mechanism of the real contact area (Eid and Adams 2007; Greenwood 2007; Yastrebov et al. 2014).

3.5. Application II: rough surface contact

As explained above, the study of contact between isolated asperities has its limitations and cannot be directly used to build mesoscopic models capable of dealing with the problem of contact between rough surfaces for arbitrary loading. This type of problem can therefore be directly addressed by simulating the contact between two deformable rough surfaces or a deformable rough surface and a rigid plane or a rigid rough surface and a deformable half-space. The requirements for tackling this problem are (1) the construction of one or more rough surfaces and (2) the solution of normal, and possibly tangential, contact between these surfaces. As already mentioned, at the discretization scale, the surfaces must be smooth to allow convergence to a solution with mesh refinement. This involves the implementation of a smoothing procedure for real topographic measurements, which is not very difficult because these are often carried out on regular grids. The smoothing can be done by Béziers surfaces (Pietrzak 1997; Yastrebov et al. 2011), NURBS (Laursen 2002), or even by Shanon interpolation[11] (Hyun and Robbins 2007; An et al. 2019). This issue was discussed in detail in Thompson and Thompson (2010).

---

10 There are configuration cases where plasticity is localized near the asperities in contact, and the interaction between asperities can be considered elastic.

11 Here, Shanon interpolation refer to Whittaker–Kotel'nikov–Shannon interpolation (Marks 2012) applicable to limited-band signals that interpolates Fourier series that can be evaluated at all points. This interpolation is applicable to two- and three-dimensional signals. The fact that it uses limited band indeed concurs with the need for small-scale smooth surfaces which is required for the convergence of mechanical computation.

For part (2) simulations, the challenges related to the capability of supercomputers to make this type of computation arises immediately. If we take as a reference mesh the one utilized for modeling the contact between two spherical asperities (Mulvihill et al. 2011) with $\approx 190,000$ tetrahedral and hexahedral elements for two quarters of the spheres, it then implies that for simulating a hundred asperities, it would be necessary to solve a system with a number of degrees of freedom $N_{dof} \approx 3 \times 100 \times 190,000 \times 2 \times 1.5 = 171 \times 10^6$, where the factor 3 takes into account the number of degrees of freedom (DOF) per node, the factor 100 is the number of asperities, the factor 2 is used for simulating the half-spheres of the asperities instead of the quarters, and the factor 1.5 to take into account the coarse mesh of the underlying volume and the spacing between asperities. A solution to the problem associated with so many DOFs seems rather realistic for high-performance computers capable of modern performances and for the level of parallelism of computational codes[12] (Dostál et al. 2019). On the other hand, contact calculations between asperities of such magnitude do not yet exist. Moreover, it should be noted that mesh optimization and remeshing procedures will be very significant to perform this type of calculation.

The history of contact calculations between rough surfaces begins with the first calculations that were made with boundary elements combined with a finite element layer on the surface: the two (Francis 1982, 1983b) and three-dimensional methods (Francis 1983a) called surface finite elements[13]. Also two-dimensional, a similar technique was used for slip analysis for an indentor on a surface with fingerprints (Webster and Sayles 1986). Often, this kind of computation is carried out with the boundary element method or similar which does not require volumetric discretization, but is normally limited to linear, isotropic and homogeneous elastic materials because it is based on the fundamental Boussinesq–Cerutti solution. As examples, we can cite the pioneering research of Lai and Cheng (1985); Seabra and Berthe (1987); Sainsot et al. (1990). One of the first results that took plasticity into account, which was done by saturating the contact pressure to $1.6\sigma_Y$[14], in 3D, was presented in West and Sayles (1987). The importance of solving topographic measurements for addressing contact was highlighted in Myshkin et al. (1998). The study of rough contact between coated solids presents an important problem of tribology, one of the first simulations was presented in Cole and Sayles (1992), other studies can be found in Nogi and Kato (1997); Peng (2001). It should be pointed out that in Nogi and Kato (1997), rough contact between a coated substrate and an indentor was studied in 3D using a spectral technique (see Stanley and Kato 1997 also for spectral methods). In Borri-Brunetto et al. (1998), interesting considerations

12 See: https://www.lnm.mw.tum.de/en/research/research-fields/highperformance-computing.

13 This is probably the first article in which three-dimensional simulations were presented.

14 Strangely, the authors took a value equal to $1.6\sigma_Y$ for hardness which corresponds to the plasticity initiation pressure even though it was already well known that plasticity onset pressure and pressure saturation (hardness) values correspond to the different pressures.

and ideas are presented as well as surface models and boundary element method-based analysis. The entire volume, from which this publication is taken, is of interest for roughness and contact studies. One of the earliest treatments of rough contact with friction was carried out by Kalker et al. (1997). Moreover, it is important to note that most boundary element methods for the simulation of rough contact are based on variational principles formulated by Kalker and Van Randen (1972) and Kalker (1977). A fairly detailed study on elastoplastic contact was carried out in Lee and Ren (1996) using a boundary element method (Ren and Lee 1994) to which pressure saturation was added. Moreover, the comment made by Etsion on this last article and the response of the authors reiterate the important question of the relevance of the study of elastic contact. Indeed, as already demonstrated by Greenwood and Williamson (1966), the plasticity index decreases due to the irreversible plastic deformation that flattens the surface and which, after some time, is deformed following an elastic regime: this is often the case with engineering systems which all operate in cyclic loading. In this light, the study of elastoplastic contact is only relevant for the *running-in* regime in which plastic deformation can take place.

A two- and three-dimensional contact analysis method was developed by Lubrecht and Ioannides (1991) based on conventional boundary elements but endowed with a multi-level multi-integration procedure or *multi-level multi-summation*) (Brandt and Lubrecht 1990) making it possible to significantly reduce the cost of convolution evaluation, moreover the authors use the associated multigrid method for also improving the solver. The applications of these methods can be found in many works (Venner and Lubrecht 1996; Polonsky and Keer 1999; Medina and Dini 2014). It should mentioned that the MLMI method with the *Full Multi-Grid* solver did not converge for sufficiently rough surfaces, and by replacing this solver with a conjugate gradient much more stable and faster convergence was ensured (Polonsky and Keer 1999); the use of the gradient conjugated with the FFT method was presented in Ai and Sawamiphakdi (1999). The couplings of the conjugate gradient with MLMI/MLMS or FFT were compared in Polonsky and Keer (2000). Strangely, Xiaolan Ai was thanked by the authors but his article on this type of coupling (see below) was not cited. In Gao et al. (2000), the authors performed one of the first thermo-mechanical simulations of contact heating at the roughness scale due to localized friction on the contact areas. A review of numerical methods for the treatment of rough contact was done in Sayles (1996). Another study that compared different methods (boundary elements) was retaken in Allwood (2005). The most recent review was carried out by Bemporad and Paggi (2015), where the authors compared several methods: (i) a greedy solver algorithm without stress (conjugate gradient, Cholesky, Gauss–Seidel), for example (Borri-Brunetto et al. 1999); (ii) the restricted conjugate gradient method (Polonsky and Keer 1999); (iii) *alternating direction method of multipliers* (Boyd et al. 2011); and (iv) the nonnegative least squares method (Lawson and Hanson 1995, p. 161). In addition, the authors stress the importance of the initial estimate of contact (*warm start*) for achieving very fast convergence.

In the scientific landscape of the last 15 years, simulations of rough contact by the FEM have begun to emerge (Hyun et al. 2004; Pei et al. 2005; Hyun and Robbins 2007; Yastrebov et al. 2011; Song et al. 2017; An et al. 2019). This method can be used in particular to address problems of nonlinear and heterogeneous materials. The methods of boundary elements are also still being developed by addressing contact with heterogeneous (Koumi et al. 2015), viscoelastic (Carbone and Putignano 2013; Putignano et al. 2015; van Dokkum and Nicola 2019) and elastoplastic bodies (Nelias et al. 2006; Pérez-Ràfols et al. 2016; Frérot et al. 2019) with adhesion (Carbone and Mangialardi 2008; Dapp and Müser 2015; Pohrt and Popov 2015; Müser et al. 2017; Popov et al. 2017; Rey et al. 2017) and friction (Paggi et al. 2014; Pohrt and Li 2014) also utilizing a remeshing procedure (Putignano et al. 2012). These new algorithms are still built based on the variational principles set up by Kalker (1977). The emergence of models based on vibration damping for a system of surface "atoms" interacting through an elastic operator of boundary elements allowed achieving unbelievably fine discretizations (Campañá and Müser 2006; Prodanov et al. 2014; Müser et al. 2017). In Prodanov et al. (2014), the authors present simulations made on a grid comprising $2^{34} \approx 17.18 \times 10^9$ (!) discretization points. However, this technique is not based on variational principles, in other words, there is no underlying minimization problem.

PURELY GEOMETRIC CONTACT MODEL.– In Pei et al. (2005), the authors demonstrated the inconsistency of the geometric subtraction model between two rough surfaces to determine the actual contact area, which was used in Sayles and Thomas (1978); Majumdar and Bhushan (1991) and Majumdar and Tien (1991), as well as other models based in particular on the notion of fractal. The geometric subtraction model (which is similar to the Winkler substrate model) is not accurate for elastoplastic contact, let alone elastic contact (a good demonstration of this fact can be found in Dapp et al. (2012)). On the other hand, the model based on the redistribution of the plastic volume (Pullen and Williamson 1972; Nayak 1973) makes more sense despite not taking the elastic interaction into account.

2D SIMULATION OF ROUGH CONTACT.– In the Flamant solution (2D), the stress decreases as $1/r$ and the displacement changes as $\ln(r)$ where r is the distance to the point of application of the force, which creates problems for defining the displacement as it diverges to infinity. In Boussinesq's solution (3D), the asymptotes are $1/r^2$ and $1/r$, respectively, for stress and displacement. As a result, the solutions to both problems are quite different. At the geometry level, the peaks that appear as asperities on a profile are not vertexes (true asperities) that may come into contact, the statistic of the two is not the same, although one can be deduced from the other (Nayak 1971). This has direct consequences on the mechanics of contact between profiles and surfaces. A direct correction of the 2D results for estimating the contact area as $A = \pi/4 \sum_i l_i^2$ where l_i are separate contact lengths, which, for example, were used in the 2D analysis by Mitchell et al. (2013). Despite that it leads to recovering the linear (or quasi-linear) growth of the contact area with the applied pressure, it cannot be considered as a precise solution. In addition, sealing and lubrication problems in mixed mode cannot be addressed in 2D because a single point of contact blocks the flow of a fluid, which therefore requires an artificial correction that is difficult to properly achieve. 2D and 3D computations were compared by Francis (1983a) to determine what is the level of anisotropy at which the 2D model can be representative of a sinusoidal surface. However, although the level of anisotropy required is not very large, it is unlikely that small-scale roughness be very anisotropic. In conclusion, 2D contact (for profiles) cannot be used for estimating the contact of rough surfaces in 3D.

THE MICROSTRUCTURAL ASPECT OF THE CONTACT.– In addition to surface geometry, the underlying microstructure plays an important part in micromechanical contact. We propose here some references (this list is largely incomprehensive) where this aspect is taken into account in the analysis of contact problems. One of the first microstructural models for *fretting fatigue* for studying microstructure on a regular 2D grid was introduced in Goh et al. (2003). More realistic and 3D microstructures for the same problem were studied in Dick and Cailletaud (2006); Dick et al. (2008). This work was followed by many others (McCarthy et al. 2014; Nigro et al. 2014; Lindroos et al. 2015; Tkalich et al. 2017; Lindroos et al. 2018). Crystal plasticity with the size effect (gradient plasticity) was taken into account in Ashton et al. (2018), which was motivated by observing and modeling the relationship between grain size and *fretting fatigue* crack onset (Ashton et al. 2017). Studies including microstructure as well as roughness are extremely rare (see Durand 2012).

3.6. Conclusion

Great progress has been made in the numerical simulation of microscopic contact with finite element and boundary element methods. However, a few shortcomings that should be addressed can be identified:

– Most studies of contact between asperities employ the Amontons–Coulomb law of friction which is not suitable for small scale. In the future, it would therefore be interesting to redo all of the studies using Tresca's law, which is more relevant for small-scale roughness.

– Another avenue for the future should be to study elastic and elastoplastic contact between *two* rough surfaces. It would be particularly important to better understand the transition between adhesion and slip that will make the link between static friction and roughness and material parameters. Taking into account the microstructure with its anisotropy and crystal plasticity will be important to draw quantitative results. In addition, it would be interesting to study the viscoelastic behavior that will allow quantifying the state of contact according to time and velocity (for low velocities), which will then make it possible to study dynamic friction. Furthermore, the viscoelastic effect will allow for better understanding the sliding (or flow) of glaciers on the rocky bed as well as accurately formulating the laws of friction that govern it.

– For higher speeds, the viscous effects will no longer be sufficient to understand small-scale frictional sliding. Since friction dissipation is highly localized, the heat produced is such that locally a fusion of the material occurs. Taking this phenomenon into account in highly coupled thermo-mechanical simulation will enable our understanding of this complex phenomenon to be improved.

– Much progress has been made on the simulation of fluid/solid interaction in the contact interface (Lubrecht and Venner 1999; Sahlin et al. 2010; Dapp et al. 2012; Pérez-Ràfols et al. 2016; Stupkiewicz et al. 2016; Temizer et Stupkiewicz 2016; Shvarts and Yastrebov 2018a, 2018b; Ager et al. 2019; Alp çakal et al. 2019; Shvarts 2019; Vlădescu et al. 2019). On the other hand, the strongly coupled finite element model that will allow roughness-scale lubrication in mixed regime to be simulated is still missing (Vakis et al. 2018).

3.7. References

Abramian, B., Arutiunian, N.K., Babloian, A. (1966). On symmetric pressure of a circular stamp on an elastic half-space in the presence of adhesion. *Journal of Applied Mathematics and Mechanics*, 30(1), 173–179.

Adams, G.G. (1995). Self-excited oscillations of two elastic half-spaces sliding with a constant coefficient of friction. *Journal of Applied Mechanics*, 62(4), 867–872.

Adams, G.G. (2000). Radiation of body waves induced by the sliding of an elastic half-space against a rigid surface. *Journal of Applied Mechanics*, 67(1), 1–5.

Adler, R.J. and Taylor, J.E. (2009). *Random Fields and Geometry*. Springer, New York.

Afferrante, L., Carbone, G., Demelio, G. (2012). Interacting and coalescing hertzian asperities: A new multiasperity contact model. *Wear*, 278, 28–33.

Ager, C., Schott, B., Vuong, A.-T., Popp, A., Wall, W.A. (2019). A consistent approach for fluid-structure-contact interaction based on a porous flow model for rough surface contact. *International Journal for Numerical Methods in Engineering*, 119(13), 1345–1378.

Ai, X. and Sawamiphakdi, K. (1999). Solving elastic contact between rough surfaces as an unconstrained strain energy minimization by using CGM and FFT techniques. *Journal of Tribology*, 121(4), 639–647.

Akula, B.R. (2019). Extended mortar method for contact and mesh-tying applications. PhD Thesis, PSL Research University, MINES ParisTech, Paris.

Akula, B.R., Vignollet, J., Yastrebov, V.A. (2019). Mortex method for contact along real and embedded surfaces: Coupling X-FEM with the mortar method. *arXiv:1902.04000*, 60440597.

Alart, P. (1997). Méthode de Newton généralisée en mécanique du contact. *Journal de mathématiques pures et appliqués*, 76, 83–108.

Alart, P. and Curnier, A. (1991). A mixed formulation for frictional contact problems prone to newton like solution methods. *Computer Methods in Applied Mechanics and Engineering*, 92(3), 353–375.

Allwood, J. (2005). Survey and performance assessment of solution methods for elastic rough contact problems. *Journal of Tribology*, 127(1), 10–23.

Altenbach, H., Maugin, G.A., Erofeev, V. (2011). *Mechanics of Generalized Continua*. Springer, Berlin, Heidelberg.

An, B., Wang, X., Xu, Y., Jackson, R.L. (2019). Deterministic elastic-plastic modelling of rough surface contact including spectral interpolation and comparison to theoretical model. *Tribology International*, 135, 246–258.

Andersson, T. (1981). The boundary element method applied to two-dimensional contact problems with friction. In *Boundary Element Methods: Proceedings of the Third International Seminar, Irvine, California, July 1981*, Brebia, C.A. (ed.). Springer, Berlin, Heidelberg, 239–258.

Andersson, P. and Kropp, W. (2008). Time domain contact model for tyre/road interaction including nonlinear contact stiffness due to small-scale roughness. *Journal of Sound and Vibration*, 318(1/2), 296–312.

Andrews, D.J. and Ben-Zion, Y. (1997). Wrinkle-like slip pulse on a fault between different material. *Journal of Geophysical Research: Solid Earth*, 102(B1), 553–571.

Archard, J. and Hirst, W. (1956). The wear of metals under unlubricated conditions. *Proc. R. Soc. Lond. A*, 236(1206), 397–410.

Ashton, P., Harte, A.M., Leen, S.B. (2017). Statistical grain size effects in fretting crack initiation. *Tribology International*, 108, 75–86.

Ashton, P., Harte, A., Leen, S. (2018). A strain-gradient, crystal plasticity model for microstructure-sensitive fretting crack initiation in ferritic-pearlitic steel for flexible marine risers. *International Journal of Fatigue*, 111, 81–92.

Autumn, K. (2007). Gecko adhesion: Structure, function, and applications. *Mrs Bulletin*, 32(6), 473–478.

Ballard, P. (1999). A counter-example to uniqueness in quasi-static elastic contact problems with small friction. *International Journal of Engineering Science*, 37(2), 163–178.

Ballard, P. (2016). Steady sliding frictional contact problem for a 2D elastic half-space with a discontinuous friction coefficient and related stress singularities. *Journal of the Mechanics and Physics of Solids*, 97, 225–259.

Ballard, P. and Basseville, S. (2005). Existence and uniqueness for dynamical unilateral contact with coulomb friction: A model problem. *ESAIM: Mathematical Modelling and Numerical Analysis*, 39(1), 59–77.

Barras, F., Kammer, D.S., Geubelle, P.H., Molinari, J.-F. (2014). A study of frictional contact in dynamic fracture along bimaterial interfaces. *International Journal of Fracture*, 189(2), 149–162.

Basseville, S. and Cailletaud, G. (2015). An evaluation of the competition between wear and crack initiation in fretting conditions for Ti–6Al–4V alloy. *Wear*, 328, 443–455.

Basseville, S., Niass, M., Missoum-Benziane, D., Leroux, J., Cailletaud, G. (2019). Effect of fretting wear on crack initiation for cylinder-plate and punch-plane tests. *Wear*, 420, 133–148.

Batailly, A., Magnain, B., Chevaugeon, N. (2013). A comparative study between two smoothing strategies for the simulation of contact with large sliding. *Computational Mechanics*, 51(5), 581–601.

Bathe, K.-J. (1996). *Finite Element Procedures*. Prentice Hall, Upper Saddle River, NJ.

Belgacem, F.B., Hild, P., Laborde, P. (1999). Extension of the mortar finite element method to a variational inequality modeling unilateral contact. *Mathematical Models and Methods in Applied Sciences*, 9(2), 287–303.

Belytschko, T., Daniel, W., Ventura, G. (2002). A monolithic smoothing-gap algorithm for contact-impact based on the signed distance function. *International Journal for Numerical Methods in Engineering*, 55(1), 101–125.

Bemporad, A. and Paggi, M. (2015). Optimization algorithms for the solution of the frictionless normal contact between rough surfaces. *International Journal of Solids and Structures*, 69, 94–105.

Bertsekas, D., Nedic, A., Ozdaglar, A. (2003). *Convex Analysis and Optimization*. Athena Scientific, Belmont.

Black, A., Kopalinsky, E., Oxley, P. (1993). Asperity deformation models for explaining the mechanisms involved in metallic sliding friction and wear – A review. *Proceedings of the Institution of Mechanical Engineers, Part C: Journal of Mechanical Engineering Science*, 207(5), 335–353.

Borri-Brunetto, M., Carpinteri, A., Chiaia, B. (1998). Lacunarity of the contact domain between elastic bodies with rough boundaries. In *PROBAMAT – 21st Century: Probabilities and Materials*, Frantziskonis, G.N. (ed.). Springer, Dordrecht. doi: 10.1007/978-94-011-5216-7_3.

Borri-Brunetto, M., Carpinteri, A., Chiaia, B. (1999). Scaling phenomena due to fractal contact in concrete and rock fractures. In *Fracture Scaling*, Bažant, Z.P. and Rajapakse, Y.D.S. (eds). Springer, Dordrecht. doi: 10.1007/978-94-011-4659-3_12.

Bowden, F.P. and Tabor, D. (1986). *The Friction and Lubrication of Solids*. Oxford University Press, Oxford.

Boyd, S., Parikh, N., Chu, E., Peleato, B., Eckstein, J. (2011). Distributed optimization and statistical learning via the alternating direction method of multipliers. *Foundations and Trends in Machine Learning*, 3(1), 1–122.

Brandt, A. and Lubrecht, A. (1990). Multilevel matrix multiplication and fast solution of integral equations. *Journal of Computational Physics*, 90(2), 348–370.

Brener, E.A., Weikamp, M., Spatschek, R., Bar-Sinai, Y., Bouchbinder, E. (2016). Dynamic instabilities of frictional sliding at a bimaterial interface. *Journal of the Mechanics and Physics of Solids*, 89, 149–173.

Brizmer, V., Kligerman, Y., Etsion, I. (2007). Elastic-plastic spherical contact under combined normal and tangential loading in full stick. *Tribology Letters*, 25(1), 61–70.

Bui, H.D. and Oueslati, A. (2010). On the stick-slip waves under unilateral contact and coulomb friction. *Annals of Solid and Structural Mechanics*, 1(3/4), 159–172.

Bussetta, P., Marceau, D., Ponthot, J.-P. (2012). The adapted augmented Lagrangian method: A new method for the resolution of the mechanical frictional contact problem. *Computational Mechanics*, 49(2), 259–275.

Çakal, B.A., Temizer, I., Terada, K., Kato, J. (2019). Microscopic design and optimization of hydrodynamically lubricated dissipative interfaces. *International Journal for Numerical Methods in Engineering*, 120(2), 153–178.

Campañá, C. and Müser, M.H. (2006). Practical Green's function approach to the simulation of elastic semi-infinite solids. *Physical Review B*, 74(7), 075420.

Campañá, C., Persson, B., Müser, M. (2011). Transverse and normal interfacial stiffness of solids with randomly rough surfaces. *Journal of Physics: Condensed Matter*, 23(8), 085001.

Carbone, G. and Mangialardi, L. (2008). Analysis of the adhesive contact of confined layers by using a Green's function approach. *Journal of the Mechanics and Physics of Solids*, 56(2), 684–706.

Carbone, G. and Putignano, C. (2013). A novel methodology to predict sliding and rolling friction of viscoelastic materials: Theory and experiments. *Journal of the Mechanics and Physics of Solids*, 61(8), 1822–1834.

Casals, O. and Forest, S. (2009). Finite element crystal plasticity analysis of spherical indentation in bulk single crystals and coatings. *Computational Materials Science*, 45(3), 774–782.

Cattaneo, C. (1938). Sul contatto de due corpi elastici: Distribuzione locale deglisforzi. *Rendiconti dell'Accademia nazionale dei Lincei*, 27, 342–349.

Challen, J. and Oxley, P. (1979). An explanation of the different regimes of friction and wear using asperity deformation models. *Wear*, 53(2), 229–243.

Chamoret, D., Saillard, P., Rassineux, A., Bergheau, J.-M. (2004). New smoothing procedures in contact mechanics. *Journal of Computational and Applied Mathematics*, 168, 107–116.

Chang, H.-J., Fivel, M., Rodney, D., Verdier, M. (2010). Multiscale modelling of indentation in fcc metals: From atomic to continuum. *Comptes Rendus Physique*, 11(3/4), 285–292.

Chaudhary, A.B. and Bathe, K.-J. (1986). A solution method for static and dynamic analysis of three-dimensional contact problems with friction. *Computers & Structures*, 24(6), 855–873.

Chen, X. and Hisada, T. (2006). Development of a finite element contact analysis algorithm to pass the patch test. *JSME International Journal Series A Solid Mechanics and Material Engineering*, 49(4), 483–491.

Chen, W.W. and Wang, Q.J. (2008). A numerical model for the point contact of dissimilar materials considering tangential tractions. *Mechanics of Materials*, 40(11), 936–948.

Chouly, F. and Hild, P. (2013). A nitsche-based method for unilateral contact problems: Numerical analysis. *SIAM Journal on Numerical Analysis*, 51(2), 1295–1307.

Chouly, F., Hild, P., Renard, Y. (2015). Symmetric and non-symmetric variants of Nitsche's method for contact problems in elasticity: Theory and numerical experiments. *Mathematics of Computation*, 84(293), 1089–1112.

Christensen, P., Klarbring, A., Pang, J.-S., Strömberg, N. (1998). Formulation and comparison of algorithms for frictional contact problems. *International Journal for Numerical Methods in Engineering*, 42(1), 145–173.

Ciavarella, M., Delfine, V., Demelio, G. (2006). A "re-vitalized" Greenwood and Williamson model of elastic contact between fractal surfaces. *Journal of the Mechanics and Physics of Solids*, 54(12), 2569–2591.

Cochard, A. and Rice, J. (2000). Fault rupture between dissimilar materials: Ill-posedness, regularization, and slip-pulse response. *Journal of Geophysical Research: Solid Earth*, 105(B11), 25891–25907.

Cocks, M. (1966). Shearing of junctions between metal surfaces. *Wear*, 9(4), 320–328.

Coker, D., Lykotrafitis, G., Needleman, A., Rosakis, A. (2005). Frictional sliding modes along an interface between identical elastic plates subject to shear impact loading. *Journal of the Mechanics and Physics of Solids*, 53(4), 884–922.

Cole, S. and Sayles, R. (1992). A numerical model for the contact of layered elastic bodies with real rough surfaces. *Journal of Tribology*, 114(2), 334–340.

Cooper, M., Mikic, B., Yovanovich, M. (1969). Thermal contact conductance. *International Journal of Heat and Mass Transfer*, 12(3), 279–300.

Costa, H. and Hutchings, I. (2007). Hydrodynamic lubrication of textured steel surfaces under reciprocating sliding conditions. *Tribology International*, 40(8), 1227–1238.

Crisfield, M. (2000). Re-visiting the contact patch test. *International Journal for Numerical Methods in Engineering*, 48(3), 435–449.

Curnier, A. (1984). A theory of friction. *International Journal of Solids and Structures*, 20, 637–647.

Dapp, W.B. and Müser, M.H. (2015). Contact mechanics of and reynolds flow through saddle points: On the coalescence of contact patches and the leakage rate through near-critical constrictions. *EPL (Europhysics Letters)*, 109(4), 44001.

Dapp, W.B. and Müser, M.H. (2016). Fluid leakage near the percolation threshold. *Scientific Reports*, 6, 19513. doi: 10.1038/srep19513.

Dapp, W.B., Lücke, A., Persson, B.N., Müser, M.H. (2012). Self-affine elastic contacts: Percolation and leakage. *Physical Review Letters*, 108(24), 244301.

De Lorenzis, L., Temizer, I., Wriggers, P., Zavarise, G. (2011). A large deformation frictional contact formulation using NURBS-based isogeometric analysis. *International Journal for Numerical Methods in Engineering*, 87(13), 1278–1300.

De Lorenzis, L., Wriggers, P., Hughes, T.J. (2014). Isogeometric contact: A review. *GAMM-Mitteilungen*, 37(1), 85–123.

Dick, T. and Cailletaud, G. (2006). Fretting modelling with a crystal plasticity model of Ti6Al4V. *Computational Materials Science*, 38(1), 113–125.

Dick, T., Basseville, S., Cailletaud, G. (2008). Fatigue modelling in fretting contact with a crystal plasticity model. *Computational Materials Science*, 43(1), 36–42.

Dieterich, J.H. (1992). Earthquake nucleation on faults with rate-and state-dependent strength. *Tectonophysics*, 211(1/4), 115–134.

Dieterich, J.H. and Kilgore, B.D. (1994). Direct observation of frictional contacts: New insights for state-dependent properties. *Pure and Applied Geophysics*, 143(1/3), 283–302.

van Dokkum, J.S. and Nicola, L. (2019). Green's function molecular dynamics including viscoelasticity. *Modelling and Simulation in Materials Science and Engineering*, 27, 075006. doi: 10.1088/1361-651X/ab3031.

Dostál, Z., Kozubek, T., Sadowská, M., Vondrák, V. (2016). *Scalable Algorithms for Contact Problems*. Springer, Cham.

Dostál, Z., Vlach, O., Brzobohatá, T. (2019). Scalable TFETI based algorithm with adaptive augmentation for contact problems with variationally consistent discretization of contact conditions. *Finite Elements in Analysis and Design*, 156, 34–43.

Durand, J. (2012). Approche multi-échelles des problèmes de contact et d'étanchéité. PhD Thesis, MINES ParisTech, Paris [Online]. Available at: https://pastel.archives-ouvertes.fr/pastel-00820173.

Duvaut, G. and Lions, J.L. (1972). *Les inéquations en mécanique et en physique*. Dunod, Paris.

Eid, H. and Adams, G.G. (2007). An elastic-plastic finite element analysis of interacting asperities in contact with a rigid flat. *Journal of Physics D: Applied Physics*, 40(23), 7432.

El-Abbasi, N. and Bathe, K.-J. (2001). Stability and patch test performance of contact discretizations and a new solution algorithm. *Computers & Structures*, 79(16), 1473–1486.

Elkilani, Y.S. (2003). A hybrid elasticity and finite element method for three-dimensional contact problems with friction. PhD Thesis, University of Cincinnati, Cincinnati.

Eringen, A.C. (1999). *Microcontinuum Field Theories: I. Foundations and Solids*. Springer, New York.

Eriten, M., Polycarpou, A., Bergman, L. (2010). Physics-based modeling for partial slip behavior of spherical contacts. *International Journal of Solids and Structures*, 47(18/19), 2554–2567.

Eriten, M., Petlicki, D., Polycarpou, A., Bergman, L. (2012). Influence of friction and adhesion on the onset of plasticity during normal loading of spherical contacts. *Mechanics of Materials*, 48, 26–42.

Etsion, I., Kligerman, Y., Kadin, Y. (2005). Unloading of an elastic-plastic loaded spherical contact. *International Journal of Solids and Structures*, 42(13), 3716–3729.

Farah, P., Gitterle, M., Wall, W.A., Popp, A. (2016). Computational wear and contact modeling for fretting analysis with isogeometric dual mortar methods. *Key Engineering Materials*, 681, 1–18. doi: 10.4028/www.scientific.net/kem.681.1.

Faulkner, A. and Arnell, R. (2000). The development of a finite element model to simulate the sliding interaction between two, three-dimensional, elastoplastic, hemispherical asperities. *Wear*, 242(1/2), 114–122.

Feng, G. and Nix, W.D. (2004). Indentation size effect in MgO. *Scripta Materialia*, 51(6), 599–603.

Fischer-Cripps, A.C. (2011). Nanoindentation testing. In *Nanoindentation*. Springer, New York.

Fivel, M., Robertson, C., Canova, G., Boulanger, L. (1998). Three-dimensional modeling of indent-induced plastic zone at a mesoscale. *Acta Materialia*, 46(17), 6183–6194.

Forest, S. (1998). Mechanics of generalized continua: Construction by homogenizaton. *Le journal de physique IV*, 8(PR4), Pr4-39–Pr4-48.

Francis, H. (1982). A finite surface element model for plane-strain elastic contact. *Wear*, 76(2), 221–245.

Francis, H. (1983a). The accuracy of plane strain models for the elastic contact of three-dimensional rough surfaces. *Wear*, 85(2), 239–256.

Francis, H. (1983b). A finite surface element model for two concentric nearly circular rings in partial contact. *Computers & Structures*, 17(2), 169–176.

Frérot, L., Bonnet, M., Molinari, J.-F., Anciaux, G. (2019). A Fourier-accelerated volume integral method for elastoplastic contact. *Computer Methods in Applied Mechanics and Engineering*, 351, 951–976.

Gallego, L., Nelias, D., Deyber, S. (2010). A fast and efficient contact algorithm for fretting problems applied to fretting modes i, ii and iii. *Wear*, 268(1/2), 208–222.

Gao, J., Lee, S.C., Ai, X., Nixon, H. (2000). An FFT-based transient flash temperature model for general three-dimensional rough surface contacts. *Transactions – ASME Journal of Tribology*, 122(3), 519–523.

Gao, Y., Larson, B.C., Lee, J., Nicola, L., Tischler, J., Pharr, G.M. (2015). Lattice rotation patterns and strain gradient effects in face-centered-cubic single crystals under spherical indentation. *Journal of Applied Mechanics*, 82(6), 061007.

Gerde, E. and Marder, M. (2001). Friction and fracture. *Nature*, 413(6853), 285.

Gitterle, M., Popp, A., Gee, M.W., Wall, W.A. (2010). Finite deformation frictional mortar contact using a semi-smooth Newton method with consistent linearization. *International Journal for Numerical Methods in Engineering*, 84(5), 543–571.

Gnecco, E. and Meyer, E. (2015). *Fundamentals of Friction and Wear on the Nanoscale*. Springer, Cham.

Gnecco, E., Bennewitz, R., Gyalog, T., Loppacher, C., Bammerlin, M., Meyer, E., Güntherodt, H.-J. (2000). Velocity dependence of atomic friction. *Physical Review Letters*, 84(6), 1172.

Goh, C.-H., Neu, R.W., McDowell, D.L. (2003). Crystallographic plasticity in fretting of TI–6AL–4V. *International Journal of Plasticity*, 19(10), 1627–1650.

Goodman, L. and Hamilton, G. (1966). The stress field created by a circular sliding contact. *ASME J. Appl. Mech.*, 33(2), 371–376.

Green, A. (1954). The plastic yielding of metal junctions due to combined shear and pressure. *Journal of the Mechanics and Physics of Solids*, 2(3), 197–211.

Greenwood, J.A. (1966). Constriction resistance and the real area of contact. *British Journal of Applied Physics*, 17(12), 1621.

Greenwood, J.A. (2007). A note on Nayak's third paper. *Wear*, 262(1/2), 225–227.

Greenwood, J.A. and Tabor, D. (1955). Deformation properties of friction junctions. *Proceedings of the Physical Society. Section B*, 68(9), 609.

Greenwood, J.A. and Tripp, J.H. (1967). The elastic contact of rough spheres. *Journal of Applied Mechanics*, 34(1), 153–159.

Greenwood, J.A. and Williamson, J.B.P. (1966). Contact of nominally flat surfaces. *P. Roy. Soc. Lond. A Mat.*, 295, 300–319.

Guyot, N., Kosior, F., Maurice, G. (2000). Coupling of finite elements and boundary elements methods for study of the frictional contact problem. *Computer Methods in Applied Mechanics and Engineering*, 181(1/3), 147–159.

Gwinner, J. (2013). hp-FEM convergence for unilateral contact problems with Tresca friction in plane linear elastostatics. *Journal of Computational and Applied Mathematics*, 254, 175–184.

Hamilton, G. (1983). Explicit equations for the stresses beneath a sliding spherical contact. *Proceedings of the Institution of Mechanical Engineers, Part C: Journal of Mechanical Engineering Science*, 197(1), 53–59.

Harrison, E. (1990). Olbers' paradox in recent times. In *Modern Cosmology in Retrospect*, Bertotti, B., Balbinot, R., Bergia, S., Messina, A. (eds). Cambridge University Press, Cambridge.

Hartmann, S., Oliver, J., Cante, J., Weyler, R., Hernández, J. (2009). A contact domain method for large deformation frictional contact problems. Part 2: Numerical aspects. *Computer Methods in Applied Mechanics and Engineering*, 198, 2607–2631.

Heaton, T.H. (1990). Evidence for and implications of self-healing pulses of slip in earthquake rupture. *Physics of the Earth and Planetary Interiors*, 64(1), 1–20.

Heegaard, J.-H. and Curnier, A. (1996). Geometric properties of 2D and 3D unilateral large slip contact operators. *Computer Methods in Applied Mechanics and Engineering*, 131, 263–286.

Heege, A. and Alart, P. (1996). A frictional contact element for strongly curved contact problems. *International Journal for Numerical Methods in Engineering*, 39(1), 165–184.

Herrmann, K. (ed.) (2011). *Hardness Testing: Principles and Applications*. ASM International, Almere.

Hertz, H. (1881). On the contact of elastic solids. *Z. Reine Angew. Mathematik*, 92, 156–171.

Hetenyi, M. and McDonald, P. (1958). Contact stresses under combined pressure and twist. *Journal of Applied Mechanics*, 25, 396–401.

Hill, R., Storåkers, B., Zdunek, A. (1989). A theoretical study of the brinell hardness test. *Proc. R. Soc. Lond. A*, 423(1865), 301–330.

Hills, D. and Sackfield, A. (1987). The stress field induced by normal contact between dissimilar spheres. *Journal of Applied Mechanics*, 54(1), 8–14.

Hol, J., Alfaro, M.C., de Rooij, M.B., Meinders, T. (2012). Advanced friction modeling for sheet metal forming. *Wear*, 286, 66–78.

Hu, Y.Z. and Tonder, K. (1992). Simulation of 3-D random rough surface by 2-D digital filter and fourier analysis. *Int. J. Mach. Tool Manu.*, 32, 83–90.

Hulikal, S., Bhattacharya, K., Lapusta, N. (2015). Collective behavior of viscoelastic asperities as a model for static and kinetic friction. *Journal of the Mechanics and Physics of Solids*, 76, 144–161.

Hyun, S. and Robbins, M.O. (2007). Elastic contact between rough surfaces: Effect of roughness at large and small wavelengths. *Tribology International*, 40(10/12), 1413–1422.

Hyun, S., Pei, L., Molinari, J.F., Robbins, M.O. (2004). Finite-element analysis of contact between elastic self-affine surfaces. *Phys. Rev. E.*, 70(2), 026117.

Jackson, R.L., Duvvuru, R.S., Meghani, H., Mahajan, M. (2007). An analysis of elasto-plastic sliding spherical asperity interaction. *Wear*, 262(1/2), 210–219.

Jelagin, D. and Larsson, P.-L. (2012). On indenter boundary effects at elastic contact. *Journal of Mechanics of Materials and Structures*, 7(2), 165–182.

Jelagin, D. and Larsson, P.-L. (2013). Nonlocal frictional effects at indentation of elastic materials. *Tribology Letters*, 51(3), 397–407.

Jing, H.-S. and Liao, M.-L. (1990). An improved finite element scheme for elastic contact problems with friction. *Computers & Structures*, 35(5), 571–578.

Jinn, J.-T. (1989). Finite element analysis of elastic contact problems with friction. PhD Thesis, Ohio State University, Columbus.

Johnson, K.L. (1985). *Contact Mechanics*. Cambridge University Press, Cambridge.

Johnson, K.L., Greenwood, J., Higginson, J. (1985). The contact of elastic regular wavy surfaces. *Int. J. Mech. Sci.*, 27(6), 383 –396.

Kadin, Y., Kligerman, Y., Etsion, I. (2006). Multiple loading-unloading of an elastic-plastic spherical contact. *International Journal of Solids and Structures*, 43(22/23), 7119–7127.

Kalker, J. (1977). Variational principles of contact elastostatics. *IMA Journal of Applied Mathematics*, 20(2), 199–219.

Kalker, J. and Van Randen, Y. (1972). A minimum principle for frictionless elastic contact with application to non-Hertzian half-space contact problems. *Journal of Engineering Mathematics*, 6(2), 193–206.

Kalker, J., Dekking, F., Vollebregt, E. (1997). Simulation of rough, elastic contacts. *Journal of Applied Mechanics*, 64(2), 361–368.

Kammer, D.S., Yastrebov, V., Anciaux, G., Molinari, J. (2014). The existence of a critical length scale in regularised friction. *Journal of the Mechanics and Physics of Solids*, 63, 40–50.

Kammer, D.S., Radiguet, M., Ampuero, J.-P., Molinari, J.-F. (2015). Linear elastic fracture mechanics predicts the propagation distance of frictional slip. *Tribology Letters*, 57(3), 23.

Kikuchi, N. and Oden, J. (1988). *Contact Problems in Elasticity: A Study of Variational Inequalities and Finite Element Methods*. SIAM, Philadelphia.

Kim, J.H. and Jang, Y.H. (2014). Frictional hertzian contact problems under cyclic loading using static reduction. *International Journal of Solids and Structures*, 51(1), 252–258.

Kim, Y.-C., Gwak, E.-J., Ahn, S.-M., Kang, N.-R., Han, H.N., Jang, J.-I., Kim, J.-Y. (2018). Indentation size effect for spherical nanoindentation on nanoporous gold. *Scripta Materialia*, 143, 10–14.

Kinkaid, N., O'Reilly, O.M., Papadopoulos, P. (2003). Automotive disc brake squeal. *Journal of Sound and Vibration*, 267(1), 105–166.

Klang, M. (1979). On interior contact under friction between cylindrical elastic bodies. PhD Thesis, Linköping University, Linköping.

Klarbring, A. (1990a). Derivation and analysis of rate boundary-value problems of frictional contact. *European Journal of Mechanics. A Solids*, 9(1), 53–85.

Klarbring, A. (1990b). Examples of non-uniqueness and non-existence of solutions to quasistatic contact problems with friction. *Archive of Applied Mechanics*, 60(8), 529–541.

Klarbring, A. and Bjöourkman, G. (1992). Solution of large displacement contact problems with friction using Newton's method for generalized equations. *International Journal for Numerical Methods in Engineering*, 34(1), 249–269.

Kogut, L. and Etsion, I. (2002). Elastic-plastic contact analysis of a sphere and a rigid flat. *Journal of Applied Mechanics*, 69(5), 657–662.

Kogut, L. and Etsion, I. (2003a). A finite element based elastic-plastic model for the contact of rough surfaces. *Tribology Transactions*, 46(3), 383–390.

Kogut, L. and Etsion, I. (2003b). A semi-analytical solution for the sliding inception of a spherical contact. *Journal of Tribology*, 125(3), 499–506.

Komvopoulos, K. (1989). Elastic-plastic finite element analysis of indented layered media. *Journal of Tribology*, 111(3), 430–439.

Konyukhov, A. and Schweizerhof, K. (2012). *Computational Contact Mechanics: Geometrically Exact Theory for Arbitrary Shaped Bodies*. Springer, Berlin, Heidelberg.

Korelc, J. (1997). Automatic generation of finite-element code by simultaneous optimization of expressions. *Theoretical Computer Science*, 187(1/2), 231–248.

Kosior, F., Guyot, N., Maurice, G. (1999). Analysis of frictional contact problem using boundary element method and domain decomposition method. *International Journal for Numerical Methods in Engineering*, 46(1), 65–82.

Koumi, K.E., Chaise, T., Nelias, D. (2015). Rolling contact of a rigid sphere/sliding of a spherical indenter upon a viscoelastic half-space containing an ellipsoidal inhomogeneity. *Journal of the Mechanics and Physics of Solids*, 80, 1–25.

Kral, E., Komvopoulos, K., Bogy, D. (1993). Elastic-plastic finite element analysis of repeated indentation of a half-space by a rigid sphere. *Journal of Applied Mechanics*, 60(4), 829–841.

Kravchuk, A. (2008). The solution of three-dimensional friction contact problems. *Journal of Applied Mathematics and Mechanics*, 72(3), 338–346.

Krim, J. (1996). Friction at the atomic scale. *Scientific American*, 275(4), 74–80.

Krim, J. (2002). Surface science and the atomic-scale origins of friction: What once was old is new again. *Surface Science*, 500(1/3), 741–758.

Kwak, B.M. and Lee, S.S. (1988). A complementarity problem formulation for two-dimensional frictional contact problems. *Computers & Structures*, 28(4), 469–480.

Lai, W. and Cheng, H. (1985). Computer simulation of elastic rough contacts. *ASLE Transactions*, 28(2), 172–180.

Laursen, T. (2002). *Computational Contact and Impact Mechanics: Fundamentals of Modeling Interfacial Phenomena in Nonlinear Finite Element Analysis*. Springer, Berlin, Heidelberg.

Laursen, T. and Simo, J. (1993). A continuum-based finite element formulation for the implicit solution of multibody, large deformation-frictional contact problems. *International Journal for Numerical Methods in Engineering*, 36(20), 3451–3485.

Lawson, C.L. and Hanson, R.J. (1995). *Solving Least Squares Problems*. SIAM, Philadelphia.

Lee, S.-S. (1994). A computational method for frictional contact problem using finite element method. *International Journal for Numerical Methods in Engineering*, 37(2), 217–228.

Lee, S.C. and Ren, N. (1996). Behavior of elastic-plastic rough surface contacts as affected by surface topography, load, and material hardness. *Tribology Transactions*, 39(1), 67–74.

Lekhnitskii, S. (1981). *Theory of Elasticity of an Anisotropic Elastic Body*, 2nd edition. Mir Publishers, Moscow.

Lengiewicz, J. and Stupkiewicz, S. (2013). Efficient model of evolution of wear in quasi-steady-state sliding contacts. *Wear*, 303(1/2), 611–621.

Lengiewicz, J., Korelc, J., Stupkiewicz, S. (2011). Automation of finite element formulations for large deformation contact problems. *International Journal for Numerical Methods in Engineering*, 85(10), 1252–1279.

Lewandowski, M. and Stupkiewicz, S. (2018). Size effects in wedge indentation predicted by a gradient-enhanced crystal-plasticity model. *International Journal of Plasticity*, 109, 54–78.

Li, J. and Berger, E. (2003). A semi-analytical approach to three-dimensional normal contact problems with friction. *Computational Mechanics*, 30(4), 310–322.

Lindroos, M., Ratia, V., Apostol, M., Valtonen, K., Laukkanen, A., Molnar, W., Holmberg, K., Kuokkala, V.-T. (2015). The effect of impact conditions on the wear and deformation behavior of wear resistant steels. *Wear*, 328, 197–205.

Lindroos, M., Laukkanen, A., Cailletaud, G., Kuokkala, V.-T. (2018). Microstructure based modeling of the strain rate history effect in wear resistant hadfield steels. *Wear*, 396, 56–66.

Lubrecht, A. and Ioannides, E. (1991). A fast solution of the dry contact problem and the associated sub-surface stress field, using multilevel techniques. *Journal of Tribology*, 113(1), 128–133.

Lubrecht, A. and Venner, C.H. (1999). Elastohydrodynamic lubrication of rough surfaces. *Proceedings of the Institution of Mechanical Engineers, Part J: Journal of Engineering Tribology*, 213(5), 397–404.

Ma, X., De Rooij, M., Schipper, D. (2010). A load dependent friction model for fully plastic contact conditions. *Wear*, 269(11/12), 790–796.

Majumdar, A. and Bhushan, B. (1991). Fractal model of elastic-plastic contact between rough surfaces. *Journal of Tribology*, 113(1), 1–11.

Majumdar, A. and Tien, C. (1990). Fractal characterization and simulation of rough surfaces. *Wear*, 136(2), 313–327.

Majumdar, A. and Tien, C. (1991). Fractal network model for contact conductance. *Journal of Heat Transfer*, 113(3), 516–525.

Mandelbrot, B.B. (1983). *The Fractal Geometry of Nature*. WH Freeman, New York.

Manoylov, A., Bryant, M.J., Evans, H.P. (2013). Dry elasto-plastic contact of nominally flat surfaces. *Tribology International*, 65, 248–258.

Marks, R.J.I. (2012). *Introduction to Shannon Sampling and Interpolation Theory*. Springer, New York.

Martins, J., Guimarães, J., Faria, L. (1995). Dynamic surface solutions in linear elasticity and viscoelasticity with frictional boundary conditions. *Journal of Vibration and Acoustics*, 117(4), 445–451.

Massi, F., Baillet, L., Giannini, O., Sestieri, A. (2007). Brake squeal: Linear and nonlinear numerical approaches. *Mechanical Systems and Signal Processing*, 21(6), 2374–2393.

Mata, M. and Alcala, J. (2004). The role of friction on sharp indentation. *Journal of the Mechanics and Physics of Solids*, 52(1), 145–165.

Maugin, G.A. and Metrikine, A.V. (eds) (2010). *Mechanics of Generalized Continua: One Hundred Years After the Cosserats*. Springer, New York.

McCarthy, O., McGarry, J., Leen, S. (2014). The effect of grain orientation on fretting fatigue plasticity and life prediction. *Tribology International*, 76, 100–115.

Meakin, P. (1998). *Fractals, Scaling and Growth Far from Equilibrium*. Cambridge University Press, Cambridge.

Medina, S. and Dini, D. (2014). A numerical model for the deterministic analysis of adhesive rough contacts down to the nano-scale. *International Journal of Solids and Structures*, 51(14), 2620–2632.

Mesarovic, S.D. and Fleck, N.A. (1999). Spherical indentation of elastic-plastic solids. *Proceedings of the Royal Society of London A: Mathematical, Physical and Engineering Sciences*, 455, 2707–2728.

Michalowski, R. and Mróz, Z. (1978). Associated and non-associated sliding rules in contact friction problems. *Archives of Mechanics*, 30, 259–276.

Mindlin, R. (1949). Compliance of elastic bodies in contact. *J. Appl. Mech., ASME*, 16, 259–268.

Mitchell, E., Fialko, Y., Brown, K. (2013). Temperature dependence of frictional healing of westerly granite: Experimental observations and numerical simulations. *Geochemistry, Geophysics, Geosystems*, 14(3), 567–582.

Mo, Y., Turner, K.T., Szlufarska, I. (2009). Friction laws at the nanoscale. *Nature*, 457(7233), 1116.

Mohammadi, N.K. and Adams, G.G. (2018). Self-excited oscillations of a finite-thickness elastic layer sliding against a rigid surface with a constant coefficient of friction. *Journal of Applied Mechanics*, 85(2), 021005.

Moirot, F., Nguyen, Q.-S., Oueslati, A. (2003). An example of stick-slip and stick-slip-separation waves. *European Journal of Mechanics – A/Solids*, 22(1), 107–118.

Moreau, J.-J. (1966). Fonctionnelles convexes. *Séminaire Jean Leray*, 1966–1967(2), 1–108.

Mossakovskii, V. (1954). The fundamental mixed problem of the theory of elasticity for a half-space with a circular line separating the boundary conditions. *Prikl. Mat. Mekh*, 18(2), 187–196.

Mossakovskii, V. (1963). Compression of elastic bodies under conditions of adhesion (axisymmetric case). *Journal of Applied Mathematics and Mechanics*, 27(3), 630–643.

Mulvihill, D.M., Kartal, M.E., Nowell, D., Hills, D.A. (2011). An elastic-plastic asperity interaction model for sliding friction. *Tribology International*, 44(12), 1679–1694.

Munoz, J. (2008). Modelling unilateral frictionless contact using the null-space method and cubic b-spline interpolation. *Computer Methods in Applied Mechanics and Engineering*, 197(9/12), 979–993.

Müser, M.H., Dapp, W.B., Bugnicourt, R., Sainsot, P., Lesaffre, N., Lubrecht, T.A., Persson, B.N., Harris, K., Bennett, A., Schulze, K. et al. (2017). Meeting the contact-mechanics challenge. *Tribology Letters*, 65(4), 118.

Myshkin, N., Petrokovets, M., Chizhik, S. (1998). Simulation of real contact in tribology. *Tribology International*, 31(1/3), 79–86.

Nayak, P.R. (1971). Random process model of rough surfaces. *J. Lubr. Technol. (ASME)*, 93, 398–407.

Nayak, P.R. (1973). Random process model of rough surfaces in plastic contact. *Wear*, 26(3), 305–333.

Nelias, D., Boucly, V., Brunet, M. (2006). Elastic-plastic contact between rough surfaces: Proposal for a wear or running-in model. *Journal of Tribology*, 128(2), 236–244.

Nigro, C., Sun, L., Meriaux, J., Proudhon, H. (2014). Microstructural simulations of the initiation and propagation of short fretting cracks in a Ti–6al–4v contact. *Tribology International*, 74, 103–109.

Nix, W.D. and Gao, H. (1998). Indentation size effects in crystalline materials: A law for strain gradient plasticity. *Journal of the Mechanics and Physics of Solids*, 46(3), 411–425.

Nogi, T. and Kato, T. (1997). Influence of a hard surface layer on the limit of elastic contact – Part I: Analysis using a real surface model. *Journal of Tribology*, 119(3), 493–500.

Nowell, D., Hills, D., Sackfield, A. (1988). Contact of dissimilar elastic cylinders under normal and tangential loading. *Journal of the Mechanics and Physics of Solids*, 36(1), 59–75.

Oliver, W.C. and Pharr, G.M. (1992). An improved technique for determining hardness and elastic modulus using load and displacement sensing indentation experiments. *Journal of Materials Research*, 7(6), 1564–1583.

Oliver, J., Hartmann, S., Cante, J., Weyler, R., Hernández, J. (2009). A contact domain method for large deformation frictional contact problems. Part 1: Theoretical basis. *Computer Methods in Applied Mechanics and Engineering*, 198, 2591–2606.

Olsson, E. and Larsson, P.-L. (2013). On force-displacement relations at contact between elastic-plastic adhesive bodies. *Journal of the Mechanics and Physics of Solids*, 61(5), 1185–1201.

Olsson, E. and Larsson, P.-L. (2016). A unified model for the contact behaviour between equal and dissimilar elastic-plastic spherical bodies. *International Journal of Solids and Structures*, 81, 23–32.

Özel, T. (2006). The influence of friction models on finite element simulations of machining. *International Journal of Machine Tools and Manufacture*, 46(5), 518–530.

Paggi, M. and Ciavarella, M. (2010). The coefficient of proportionality κ between real contact area and load, with new asperity models. *Wear*, 268(7/8), 1020–1029.

Paggi, M., Pohrt, R., Popov, V.L. (2014). Partial-slip frictional response of rough surfaces. *Scientific Reports*, 4, 5178.

Panagiotopoulos, P. (1975). A nonlinear programming approach to the unilateral contact-, and friction-boundary value problem in the theory of elasticity. *Archive of Applied Mechanics*, 44(6), 421–432.

Pei, L., Hyun, S., Molinari, J.F., Robbins, M.O. (2005). Finite element modeling of elasto-plastic contact between rough surfaces. *J. Mech. Phys. Solids*, 53, 2385–2409.

Peng, W. (2001). Contact mechanics of multilayered rough surfaces in tribology. PhD Thesis, Ohio State University, Columbus.

Pérez-Ràfols, F. and Almqvist, A. (2019). Generating randomly rough surfaces with given height probability distribution and power spectrum. *Tribology International*, 131, 591–604.

Pérez-Ràfols, F., Larsson, R., Almqvist, A. (2016). Modelling of leakage on metal-to-metal seals. *Tribology International*, 94, 421–427.

Pietrzak, G. (1997). Continuum mechanics modelling and augmented Lagrangian formulation of large deformation frictional contact problems. PhD Thesis, École Polytechnique Fédérale de Lausanne, Lausanne.

Pietrzak, G. and Curnier, A. (1999). Large deformation frictional contact mechanics: Continuum formulation and augmented Lagrangian treatment. *Computer Methods in Applied Mechanics and Engineering*, 177(3/4), 351–381.

Plouraboué, F., Geoffroy, S., Prat, M. (2004). Conductances between confined rough walls. *Physics of Fluids*, 16(3), 615–624.

Pohrt, R. and Li, Q. (2014). Complete boundary element formulation for normal and tangential contact problems. *Physical Mesomechanics*, 17(4), 334–340.

Pohrt, R. and Popov, V.L. (2012). Normal contact stiffness of elastic solids with fractal rough surfaces. *Physical Review Letters*, 108(10), 104301.

Pohrt, R. and Popov, V.L. (2015). Adhesive contact simulation of elastic solids using local mesh-dependent detachment criterion in boundary elements method. *Facta Universitatis, Series: Mechanical Engineering*, 13(1), 3–10.

Polonsky, I. and Keer, L.M. (1999). A numerical method for solving rough contact problems based on the multi-level multi-summation and conjugate gradient techniques. *Wear*, 231(2), 206–219.

Polonsky, I. and Keer, L.M. (2000). Fast methods for solving rough contact problems: A comparative study. *Journal of Tribology*, 122(1), 36–41.

Popov, V.L. and Pohrt, R., Li, Q. (2017). Strength of adhesive contacts: Influence of contact geometry and material gradients. *Friction*, 5(3), 308–325.

Popp, A. (2012). Mortar methods for computational contact mechanics and general interface problems. PhD Thesis, Technische Universität München, Munich.

Popp, A., Gitterle, M., Gee, M.W., Wall, W.A. (2010). A dual mortar approach for 3D finite deformation contact with consistent linearization. *International Journal for Numerical Methods in Engineering*, 83(11), 1428–1465.

Prakash, V. (1995). A pressure-shear plate impact experiment for investigating transient friction. *Experimental Mechanics*, 35(4), 329–336.

Prakash, V. and Clifton, R. (1993). Time resolved dynamic friction measurements in pressure-shear. In *Experimental Techniques in the Dynamics of Deformable Solids*, Ramesh, K.T. (ed.). ASME, New York.

Prodanov, N., Dapp, W.B., Müser, M.H. (2014). On the contact area and mean gap of rough, elastic contacts: Dimensional analysis, numerical corrections, and reference data. *Tribology Letters*, 53(2), 433–448.

Pullen, J. and Williamson, J. (1972). On the plastic contact of rough surfaces. *Proceedings of the Royal Society of London. A. Mathematical and Physical Sciences*, 327(1569), 159–173.

Puso, M.A. and Laursen, T.A. (2002). A 3D contact smoothing method using Gregory patches. *International Journal for Numerical Methods in Engineering*, 54(8), 1161–1194.

Puso, M.A. and Laursen, T.A. (2004). A mortar segment-to-segment contact method for large deformation solid mechanics. *Computer Methods in Applied Mechanics and Engineering*, 193(6/8), 601–629.

Puso, M.A., Laursen, T., Solberg, J. (2008). A segment-to-segment mortar contact method for quadratic elements and large deformations. *Computer Methods in Applied Mechanics and Engineering*, 197(6/8), 555–566.

Putignano, C., Afferrante, L., Carbone, G., Demelio, G. (2012). A new efficient numerical method for contact mechanics of rough surfaces. *Int. J. Solids Struct.*, 49(2), 338–343.

Putignano, C., Carbone, G., Dini, D. (2015). Mechanics of rough contacts in elastic and viscoelastic thin layers. *International Journal of Solids and Structures*, 69, 507–517.

Qiu, X., Huang, Y., Nix, W., Hwang, K., Gao, H. (2001). Effect of intrinsic lattice resistance in strain gradient plasticity. *Acta Materialia*, 49(19), 3949–3958.

Rabinowicz, E. (1965). *Friction and Wear of Materials*. John Wiley & Sons, New York.

Ranjith, K. and Rice, J. (2001). Slip dynamics at an interface between dissimilar materials. *Journal of the Mechanics and Physics of Solids*, 49(2), 341–361.

Ren, N. and Lee, S.C. (1994). The effects of surface roughness and topography on the contact behavior of elastic bodies. *Journal of Tribology*, 116(4), 804–810.

Renard, Y. (1998). Modélisation des instabilités liées au frottement sec des solides élastiques, aspects théoriques et numériques. PhD Thesis, LMC-IMAG Grenoble, Grenoble.

Renard, F., Candela, T., Bouchaud, E. (2013). Constant dimensionality of fault roughness from the scale of micro-fractures to the scale of continents. *Geophysical Research Letters*, 40(1), 83–87.

Renardy, M. (1992). Ill-posedness at the boundary for elastic solids sliding under Coulomb friction. *Journal of Elasticity*, 27(3), 281–287.

Rey, V., Anciaux, G., Molinari, J.-F. (2017). Normal adhesive contact on rough surfaces: Efficient algorithm for FFT-based BEM resolution. *Computational Mechanics*, 60(1), 69–81.

Rey, V., Krumscheid, S., Nobile, F. (2019). Quantifying uncertainties in contact mechanics of rough surfaces using the multilevel monte carlo method. *International Journal of Engineering Science*, 138, 50–64.

Rice, J.R. (2006). Heating and weakening of faults during earthquake slip. *Journal of Geophysical Research: Solid Earth*, 111, B05311.

Rice, J.R. and Tse, S.T. (1986). Dynamic motion of a single degree of freedom system following a rate and state dependent friction law. *Journal of Geophysical Research: Solid Earth*, 91(B1), 521–530.

Russ, J.C. (1994). *Fractal Surfaces*. Springer, New York.

Sabnis, P.A., Forest, S., Arakere, N.K., Yastrebov, V.A. (2013). Crystal plasticity analysis of cylindrical indentation on a Ni-base single crystal superalloy. *International Journal of Plasticity*, 51, 200–217.

Sahlin, F., Larsson, R., Almqvist, A., Lugt, P., Marklund, P. (2010). A mixed lubrication model incorporating measured surface topography. Part 1: Theory of flow factors. *Proceedings of the Institution of Mechanical Engineers, Part J: Journal of Engineering Tribology*, 224(4), 335–351.

Sainsot, P., Leroy, J.M., Villechaise, B. (1990). Paper VI (i) Effect of surface coatings in a rough normally loaded contact. *Tribology Series*, 17, 151–156.

Sang, Y., Dubé, M., Grant, M. (2001). Thermal effects on atomic friction. *Physical Review Letters*, 87(17), 174301.

Sayles, R. (1996). Basic principles of rough surface contact analysis using numerical methods. *Tribology International*, 29(8), 639–650.

Sayles, R. and Thomas, T. (1978). Computer simulation of the contact of rough surfaces. *Wear*, 49(2), 273–296.

Schatzman, M. (1978). A class of nonlinear differential equations of second order in time. *Nonlinear Analysis: Theory, Methods and Applications*, 2(3), 355–373.

Seabra, J. and Berthe, D. (1987). Influence of surface waviness and roughness on the normal pressure distribution in the hertzian contact. *Journal of Tribology*, 109(3), 462–469.

Sewerin, F. and Papadopoulos, P. (2017). An exact penalty approach for the finite element solution of frictionless contact problems. In *Proceedings of the 7th GACM Colloquium on Computational Mechanics*, October 11-13. University of Stuttgart, Stuttgart.

Shi, X. and Polycarpou, A.A. (2005). Measurement and modeling of normal contact stiffness and contact damping at the meso scale. *Journal of Vibration and Acoustics*, 127(1), 52–60.

Shvarts, A.G. (2019). Coupling mechanical frictional contact with interfacial fluid flow at small and large scales. PhD Thesis, PSL Research University, MINES ParisTech, Paris.

Shvarts, A.G. and Yastrebov, V.A. (2018a). Fluid flow across a wavy channel brought in contact. *Tribology International*, 126, 116–126.

Shvarts, A.G. and Yastrebov, V.A. (2018b). Trapped fluid in contact interface. *Journal of the Mechanics and Physics of Solids*, 119, 140–162.

Shyu, S., Chang, T., Saleeb, A. (1989). Friction-contact analysis using a mixed finite element method. *Computers & Structures*, 32(1), 223–242.

Simo, J. and Hughes, T. (1998). *Computational Inelasticity*. Springer, New York.

Simoes, F. and Martins, J. (1998). Instability and ill-posedness in some friction problems. *International Journal of Engineering Science*, 36(11), 1265–1293.

Song, Z. and Komvopoulos, K. (2013). Elastic-plastic spherical indentation: Deformation regimes, evolution of plasticity, and hardening effect. *Mechanics of Materials*, 61, 91–100.

Song, W., Li, L., Ovcharenko, A., Jia, D., Etsion, I., Talke, F.E. (2012). Plastic yield inception of an indented coated flat and comparison with a flattened coated sphere. *Tribology International*, 53, 61–67.

Song, H., Vakis, A., Liu, X., Van der Giessen, E. (2017). Statistical model of rough surface contact accounting for size-dependent plasticity and asperity interaction. *Journal of the Mechanics and Physics of Solids*, 106, 1–14.

Spence, D. (1968). Self similar solutions to adhesive contact problems with incremental loading. *Proceedings of the Royal Society of London. Series A. Mathematical and Physical Sciences*, 305(1480), 55–80.

Spence, D. (1975). The hertz contact problem with finite friction. *Journal of Elasticity*, 5, 297–319.

Spinu, S. and Frunza, G. (2015). The hysteretic behaviour of partial slip elastic contacts undergoing a fretting loop. *Journal of Physics: Conference Series*, 585, 012007.

Stanley, H.M. and Kato, T. (1997). An FFT-based method for rough surface contact. *J. Tribol-T ASME*, 119, 481–485.

Stingl, B., Ciavarella, M., Hoffmann, N. (2013). Frictional dissipation in elastically dissimilar oscillating hertzian contacts. *International Journal of Mechanical Sciences*, 72, 55–62.

Storåkers, B. and Elaguine, D. (2005). Hertz contact at finite friction and arbitrary profiles. *Journal of the Mechanics and Physics of Solids*, 53(6), 1422–1447.

Straffelini, G. (2001). A simplified approach to the adhesive theory of friction. *Wear*, 249(1/2), 78–84.

Stupkiewicz, S., Lengiewicz, J., Sadowski, P., Kucharski, S. (2016). Finite deformation effects in soft elastohydrodynamic lubrication problems. *Tribology International*, 93, 511–522.

Sun, L. (2012). Étude numérique de l'amorçage et de la propagation de fissures de fretting. PhD Thesis, MINES ParisTech, Paris.

Sun, Y., Bloyce, A., Bell, T. (1995). Finite element analysis of plastic deformation of various tin coating/substrate systems under normal contact with a rigid sphere. *Thin Solid Films*, 271(1/2), 122–131.

Svetlizky, I. and Fineberg, J. (2014). Classical shear cracks drive the onset of dry frictional motion. *Nature*, 509(7499), 205.

Svetlizky, I., Muñoz, D.P., Radiguet, M., Kammer, D.S., Molinari, J.-F., Fineberg, J. (2016). Properties of the shear stress peak radiated ahead of rapidly accelerating rupture fronts that mediate frictional slip. *Proceedings of the National Academy of Sciences*, 113(3), 542–547.

Swadener, J., George, E., Pharr, G. (2002). The correlation of the indentation size effect measured with indenters of various shapes. *Journal of the Mechanics and Physics of Solids*, 50(4), 681–694.

Tabor, D. (1951). *The Hardness of Metals*. Oxford University Press, Oxford.

Tan, D. (2003). Mesh matching and contact patch test. *Computational Mechanics*, 31(1/2), 135–152.

Tangena, A. and Wijnhoven, P. (1985). Finite element calculations on the influence of surface roughness on friction. *Wear*, 103(4), 345–354.

Taylor, R. and Papadopoulos, O. (1991). On a patch test for contact problems in two dimensions. In *Nonlinear Computational Mechanics*, Wriggers, P. and Wagner, W. (eds). Springer, Berlin.

Temizer, I. and Stupkiewicz, S. (2016). Formulation of the Reynolds equation on a time-dependent lubrication surface. *Proceedings of the Royal Society A: Mathematical, Physical and Engineering Sciences*, 472(2187), 20160032.

Temizer, I., Wriggers, P., Hughes, T. (2011). Contact treatment in isogeometric analysis with NURBS. *Computer Methods in Applied Mechanics and Engineering*, 200(9/12), 1100–1112.

Temizer, I., Wriggers, P., Hughes, T. (2012). Three-dimensional mortar-based frictional contact treatment in isogeometric analysis with NURBS. *Computer Methods in Applied Mechanics and Engineering*, 209, 115–128.

Thomas, T.R. (1999). *Rough Surfaces*. World Scientific Publishing Co., Singapore.

Thompson, M.K. and Thompson, J.M. (2010). Considerations for the incorporation of measured surfaces in finite element models. *Scanning*, 32(4), 183–198.

Tkalich, D., Yastrebov, V.A., Cailletaud, G., Kane, A. (2017). Multiscale modeling of cemented tungsten carbide in hard rock drilling. *International Journal of Solids and Structures*, 128, 282–295.

Torstenfelt, B.R. (1984). An automatic incrementation technique for contact problems with friction. *Computers & Structures*, 19(3), 393–400.

Vakis, A.I., Yastrebov, V.A., Scheibert, J., Nicola, L., Dini, D., Minfray, C., Almqvist, A., Paggi, M., Lee, S., Limbert, G. et al. (2018). Modeling and simulation in tribology across scales: An overview. *Tribology International*, 125, 169–199.

Venner, C.H. and Lubrecht, A. (1996). Numerical analysis of the influence of waviness on the film thickness of a circular EHL contact. *Journal of Tribology*, 118(1), 153–161.

Vlădescu, S.-C., Putignano, C., Marx, N., Keppens, T., Reddyhoff, T., Dini, D. (2019). The percolation of liquid through a compliant seal – An experimental and theoretical study. *Journal of Fluids Engineering*, 141(3), 031101.

Webster, M. and Sayles, R. (1986). A numerical model for the elastic frictionless contact of real rough surfaces. *Journal of Tribology*, 108(3), 314–320.

Wei, Z., Li, Z., Qian, Z., Chen, R., Dollevoet, R. (2016). 3D FE modelling and validation of frictional contact with partial slip in compression-shift-rolling evolution. *International Journal of Rail Transportation*, 4(1), 20–36.

West, M. and Sayles, R. (1987). A 3-dimensional method of studying 3-body contact geometry and stress on real rough surfaces. *Tribology Series*, 12, 195–200. doi: 10.1016/S0167-8922(08)71066-6.

Whitehouse, D.J. (2010). *Handbook of Surface and Nanometrology*. CRC Press, Boca Raton.

Willis, J. (1966). Hertzian contact of anisotropic bodies. *Journal of the Mechanics and Physics of Solids*, 14(3), 163–176.

Willis, J. (1967). Boussinesq problems for an anisotropic half-space. *Journal of the Mechanics and Physics of Solids*, 15(5), 331–339.

Wilson, W. and Sheu, S. (1988). Real area of contact and boundary friction in metal forming. *International Journal of Mechanical Sciences*, 30(7), 475–489.

Wriggers, P. (2006). *Computational Contact Mechanics*, 2nd edition. Springer, Berlin, Heidelberg.

Wriggers, P. (2008). *Nonlinear Finite Element Methods*. Springer, Berlin, Heidelberg.

Wriggers, P. and Krstulović-Opara, L. (2000). On smooth finite element discretizations for frictional contact problems. *ZAMM-Journal of Applied Mathematics and Mechanics/Zeitschrift für Angewandte Mathematik und Mechanik*, 80(S1), 77–80.

Wriggers, P. and Zavarise, G. (1993). Application of augmented Lagrangian techniques for non-linear constitutive laws in contact interfaces. *Communications in Numerical Methods in Engineering*, 9(10), 815–824.

Wriggers, P. and Zavarise, G. (2008). A formulation for frictionless contact problems using a weak form introduced by Nitsche. *Computational Mechanics*, 41(3), 407–420.

Wriggers, P., Krstulović-Opara, L., Korelc, J. (2001). Smooth C1-interpolations for two-dimensional frictional contact problems. *International Journal for Numerical Methods in Engineering*, 51, 1469–1495.

Yang, B., Laursen, T., Meng, X. (2005). Two dimensional mortar contact methods for large deformation frictional sliding. *International Journal for Numerical Methods in Engineering*, 62, 1183–1225.

Yastrebov, V.A. (2013). *Numerical Methods in Contact Mechanics*. ISTE Ltd, London, and John Wiley & Sons, New York.

Yastrebov, V.A. (2016). Sliding without slipping under coulomb friction: Opening waves and inversion of frictional force. *Tribology Letters*, 62(1), 1–8.

Yastrebov, V.A. (2019). The elastic contact of rough spheres investigated using a deterministic multi-asperity model. *Journal of Multiscale Modelling*, 10(1), 1841002.

Yastrebov, V.A., Durand, J., Proudhon, H., Cailletaud, G. (2011). Rough surface contact analysis by means of the finite element method and of a new reduced model. *Comptes Rendus Mécanique*, 339(7/8), 473–490.

Yastrebov, V.A., Anciaux, G., Molinari, J.-F. (2014). The contact of elastic regular wavy surfaces revisited. *Tribology Letters*, 56(1), 171–183.

Yastrebov, V.A., Anciaux, G., Molinari, J.-F. (2015). From infinitesimal to full contact between rough surfaces: Evolution of the contact area. *International Journal of Solids and Structures*, 52, 83–102.

Yastrebov, V.A., Anciaux, G., Molinari, J.-F. (2017a). On the accurate computation of the true contact-area in mechanical contact of random rough surfaces. *Tribology International*, 114, 161–171.

Yastrebov, V.A., Anciaux, G., Molinari, J.-F. (2017b). The role of the roughness spectral breadth in elastic contact of rough surfaces. *Journal of the Mechanics and Physics of Solids*, 107, 469–493.

Yoon, E.-S., Singh, R.A., Kong, H., Kim, B., Kim, D.-H., Jeong, H.E., Suh, K.Y. (2006). Tribological properties of bio-mimetic nano-patterned polymeric surfaces on silicon wafer. *Tribology Letters*, 21(1), 31–37.

Zahouani, H., Vargiolu, R., Loubet, J.-L. (1998). Fractal models of surface topography and contact mechanics. *Mathematical and Computer Modelling*, 28(4/8), 517–534.

Zavarise, G. and De Lorenzis, L. (2009a). A modified node-to-segment algorithm passing the contact patch test. *International Journal for Numerical Methods in Engineering*, 79(4), 379–416.

Zavarise, G. and De Lorenzis, L. (2009b). The node-to-segment algorithm for 2D frictionless contact: Classical formulation and special cases. *Computer Methods in Applied Mechanics and Engineering*, 198(41/44), 3428–3451.

Zavarise, G., Wriggers, P., Stein, E., Schrefler, B.A. (1992). Real contact mechanisms and finite element formulation – A coupled thermomechanical approach. *International Journal for Numerical Methods in Engineering*, 35(4), 767–785.

Zhang, H., Wang, H., Wriggers, P., Schrefler, B. (2005). A finite element model for contact analysis of multiple cosserat bodies. *Computational Mechanics*, 36(6), 444–458.

Zienkiewicz, O. and Taylor, R. (2000a). *The Finite Element Method, Volume 1: The Basis*, 5th edition. Butterworth-Heinemann, Oxford.

Zienkiewicz, O. and Taylor, R. (2000b). *The Finite Element Method, Volume 2: Solid Mechanics*, 5th edition. Butterworth-Heinemann, Oxford.

Zisis, T., Gourgiotis, P., Baxevanakis, K.P., Georgiadis, H. (2014). Some basic contact problems in couple stress elasticity. *International Journal of Solids and Structures*, 51(11/12), 2084–2095.

PART 2

Damage and Cracking

4

Numerical Methods for Ductile Fracture

Jacques BESSON
CNRS, Mines Paris PSL, France

4.1. Introduction

Numerous micromechanics-based damage models have been developed since the 1970s in order to simulate ductile fracture from the laboratory specimen to the industrial structure. Their implementation in structural analysis remains problematic, however, (size of calculations, damage localization, parameterization, etc.) and that they are not currently used directly to perform design calculations. They can be used in some relatively simple problems for prior design or analyzing observed fracture situations. The purpose of this chapter is to review the models and their utilization in structural analysis. We will only consider simulations using the finite element method. First, displacement-based formulations will be treated to address enhanced formulations for overcoming localization issues.

A brief reminder of the mechanisms of ductile fracture will first be introduced. The GTN model, which is the most widely used for simulating ductile tear, is then covered in detail and is later employed for illustrating the methods described. The localization phenomenon related to the softening nature of the constitutive equations with damage is then described. The displacement-based formulations will then be addressed, emphasizing their limitations. The chapter concludes with enhanced formulations for controlling localization and fracture energies.

Numerical Methods for Strong Nonlinearities in Mechanics,
coordinated by Jacques BESSON, Frédéric LEBON and Eric LORENTZ. © ISTE Ltd 2024.

4.2. Physical mechanisms of ductile fracture

Ductile fracture is often described as comprising three distinct phases (Garrison and Moody 1987). Cavities initiate first (nucleation) on defects (primary inclusions) before growing to eventually coalesce and thus form a macroscopic crack.

This situation prevails at high stress triaxiality[1] and is shown in Figure 4.1(a). Dimples can be seen on the fracture surfaces containing inclusions (indicated by arrows, CaS in this case), which are the site of damage nucleation. The cavities are coalesced through an internal necking mechanism. For lower triaxiality ratios, it is common to find much smaller sized dimples that have been initiated on hardening phases such as carbides Fe_3C for steels (see Figure 4.1(b)) or dispersoids for aluminum alloys (Bron et al. 2004; Tanguy et al. 2008). This mechanism corresponds to strain localization between cavities initiated on primary defects and leading to secondary nucleation (Becker and Smelser 1994). This mechanism is often called *void sheeting* fracture (Haynes and Gangloff 1998). At high triaxiality, the growth is fast so that the fracture occurs for strain levels lower than those required to initiate this second fracture mode.

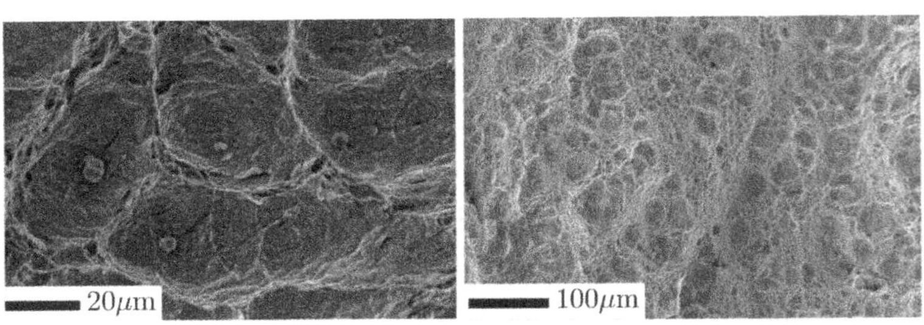

a) X100 steel b)

Figure 4.1. *Ductile fracture in pipeline steel (X100). (a) Fracture by nucleation on CaS-type inclusions, growth and coalescence (high stress triaxiality); (b) fracture by nucleation on CaS-type inclusions, growth and then nucleation on Fe_3C carbides (average stress triaxiality) (source: Luu 2006; Tanguy et al. 2008)*

1 Stress triaxiality is defined as the ratio σ_m/σ_{eq} where σ_m is the mean stress and σ_{eq} is the von Mises stress. It is equal to 0 in shear, $1/3$ in traction, from 0.8 to 1.5 for circumferentially notched specimens (including tensile specimens after pronounced necking) and from 1.5 to 4 for cracked specimens.

4.3. Some ductile fracture models

Ductile fracture is accompanied by large strains that must be taken into account regardless of the chosen formalism. Therefore, neglecting changes in the shape of the parts can, for example, lead to interpreting the ultimate stress in a tensile test by the occurrence of damage while this is generally not the case (with rare exceptions such as spheroidal graphite cast iron or certain aluminum alloys containing a lot of silicon or iron impurities). Section reduction usually explains the fact that the force reaches a maximum (Considère 1885). For example, the Hencky tensor (logarithmic strain tensor) defined as (Miehe et al. 2002) can be used as strain measure:

$$\underline{E} = \frac{1}{2} \log \left(\underline{F}^T . \underline{F} \right) \qquad [4.1]$$

where $\underline{F}$ is the transformation gradient. An additive decomposition of $\underline{E}$ can then be used, which can then be expressed as the sum of an "elastic" part and a "plastic" part so that a notation similar to that of small transformations may be adopted. In order to avoid complicating the rest of the chapter, large strain formulations will not be described in detail but the reader will be able to find them in the different articles cited.

A more comprehensive presentation of ductile fracture models will be found in Tvergaard (1990); Berdin et al. (2004); Besson (2009); Benzerga and Leblond (2010); Pineau et al. (2016).

4.3.1. *Rice and Tracey model and fracture criteria*

Despite its age, it seems important to recall the Rice and Tracey model (Rice and Tracey 1969). Indeed, it contains most of the ingredients needed to describe the ductile fracture. In its simplest version, the model describes the growth of a spherical cavity isolated in an infinite, rigid, perfectly plastic massif (von Mises plasticity). In this case, the growth rate of a cavity of radius R is given by:

$$\frac{\dot{R}}{R} = \alpha \exp\left(\beta \frac{\sigma_m^\infty}{\sigma_{\text{eq}}^\infty} \right) \dot{p}^\infty \quad \sigma_m^\infty = \frac{1}{3} \sigma_{kk}^\infty \qquad [4.2]$$

where σ_{kk}^∞ and $\sigma_{\text{eq}}^\infty$ are the trace and the von Mises invariant of the stress tensor ($\underline{\sigma}$) at infinity. $\dot{p}^\infty$ is the equivalent strain rate at infinity. The ratio $\tau = \sigma_m / \sigma_{\text{eq}}$ corresponds to the stress triaxiality. α and β are, respectively, equal to 0.283 and 1.5 but can be possibly adjusted. The model can be employed for defining a fracture criterion after integration on the possibly not proportional load path. This thus gives:

$$\log\left(\frac{R}{R_0} \right) = \int_{\text{history}} \alpha \exp\left(\beta \frac{\sigma_m}{\sigma_{\text{eq}}} \right) dp = \int_{p_c}^{p} \alpha \exp\left(\beta \frac{\sigma_m}{\sigma_{\text{eq}}} \right) dp \qquad [4.3]$$

where p_c is the critical strain at which damage starts. It can then be proposed that fracture begins at a material point for a "critical" value ($R/R_0|_c$), assumed to be material-specific, of the parameter R/R_0. The fracture criterion then becomes: $R/R_0 = R/R_0|_c$ (Marini et al. 1985). Although very simple, the model includes all of the necessary elements to the description of ductile fracture: damage growth nonlinearly depending on stress triaxiality, need for increasing plastic deformation to break matter, nucleation (represented here by parameter p_c), coalescence and fracture (represented here by parameter $R/R_0|_c$). The model can then be easily utilized for post-processing calculations of elastoplastic structures in order to determine the areas and moments of initiation. The approach can be generalized by introducing a damage indicator, D, whose evolution can be generically formulated by Defaisse et al. (2018):

$$\dot{D} = \text{FUNCTION}\,(\text{stress state}, p)\ \dot{p} \qquad [4.4]$$

The fracture occurs for $D = 1$. This approach has seen a renewed interest with ductile fracture criteria adapted to low triaxiality and integrating the effect of the stress tensor Lode parameter[2] on ductile fracture (Bao and Wierzbicki 2004; Bai and Wierzbicki 2008; Luo et al. 2012; Faleskog and Barsoum 2013; Gu and Mohr 2015; Marcadet and Mohr 2016).

4.3.2. *The Gurson–Tvergaard–Needleman model*

The Gurson model (Gurson 1977) revisits the ideas of the Rice and Tracey model but now considers a finite-sized hollow sphere consisting of a matrix and a spherical cavity (pore). This model therefore very naturally introduces a damage variable corresponding to the void volume fraction or porosity f. The model is based on an "upper bound"-type approach for a perfectly plastic von Mises matrix (yield point σ_0) and leads to the definition of a plasticity surface:

$$\Phi = \frac{\sigma_{\text{eq}}^2}{\sigma_0^2} + 2f \cosh\left(\frac{1}{2}\frac{\sigma_{kk}}{\sigma_0}\right) - 1 - f^2 \equiv 0 \qquad [4.5]$$

2 The Lode parameter of an arbitrary tensor $\underline{a}$ is defined as:

$$\mathcal{L}(\underline{a}) = \frac{27}{2}\frac{\det \underline{a}'}{a_{\text{eq}}}$$

where $\underline{a}'$ is the deviator of tensor $\underline{a}$ and a_{eq} is its von Mises invariant. $\mathcal{L}(\underline{a})$ is comprised between -1 and $+1$. The case where $\mathcal{L} = 0$ corresponds to a generalized plane strain state (von Mises plasticity) and often corresponds to a ductility minimum.

The approach used ensures that the normality rule applies to the homogenized pore/matrix set so that the plastic strain increment tensor is written as:

$$\dot{\underline{\varepsilon}}_p = \lambda \frac{\partial \Phi}{\partial \underline{\sigma}} \qquad [4.6]$$

where λ is the plastic multiplier. The evolution law of the damage variable is obtained by writing the conservation of mass; we then get:

$$\dot{f} = (1 - f)\text{trace}\left(\dot{\underline{\varepsilon}}_p\right) = (1 - f)\lambda\text{trace}\left(\frac{\partial \Phi}{\partial \underline{\sigma}}\right) \qquad [4.7]$$

It can therefore be seen that the definition of the threshold surface leads to the damage law without the need to add an additional evolution equation. This is directly related to the fact that damage corresponds to the growth of cavities. Matter fracture corresponds to zero stresses $\underline{\sigma} = \underline{0}$ verifying the plasticity condition $\Phi = 0$. This leads to the condition $f = 1$. This condition is of course unrealistic because the volume of matter becomes negligible compared to the cavity volume.

In order for the initial Gurson model to be used in structure analysis, Tvergaard and Needleman (1984) proposed a phenomenological extension now widely employed and frequently implemented in structural analysis codes. The initial success of the model is probably due to the fact that it was used to reproduce the *cup-cone* fracture phenomenon. This is currently the most widely used model. The model (called GTN model) preserves the micromechanical nature of the original approach by giving a physical meaning (i.e. measurable through observations) to the different variables but moves away from the rigorous framework of the "upper bound" approach. In the first place, the material is considered elastoplastic and incorporates isotropic hardening. The plasticity surface is then expressed as:

$$\Phi = \frac{\sigma_{\text{eq}}^2}{\sigma_F^2} + 2q_1 f_\star \cosh\left(\frac{q_2}{2}\frac{\sigma_{kk}}{\sigma_F}\right) - 1 - q_1^2 f_\star^2 \equiv 0 \qquad [4.8]$$

σ_F is the flow stress of the undamaged matrix depending on the hardening variable p verifying: $(1 - f)\dot{p}\sigma_F = \dot{\underline{\varepsilon}}_p : \underline{\sigma}$. q_1 and q_2 are two parameters that enable better reproducing cavity growth rates. They can be adjusted to computations of matrix/pore elementary cells (Koplik and Needleman 1988; Faleskog et al. 1998) or experiments. These cell computations lead to values of q_1 and q_2 close to 1.5 and 1. It should be noted in particular that q_1 is then significantly different from the value inferred from the Gurson analysis (i.e. $q_1 = 1$). An adjustment to experiments leading to very different values would indicate that the prevailing damage mechanism is not cavity growth and that the GTN model (with no nucleation) is probably not adapted.

Nucleation can be introduced into the model by introducing a "nucleation porosity" (f_n) whose law of evolution is very often described as (this is then called plastic deformation controlled nucleation):

$$\dot{f}_n = A_n(\dots)\dot{p} \qquad [4.9]$$

where the function A_n can depend on many parameters such as plastic deformation (Chu and Needleman 1980; Besson et al. 2000; Grange et al. 2000; Zhang et al. 2000) but also on the stress or strain state (Dalloz et al. 2009; Morgeneyer and Besson 2011; Petit et al. 2019). Porosity evolution then results from a contribution of growth (equation [4.7]) and nucleation: $\dot{f} = (1-f)\text{trace}\,(\dot{\underset{\sim}{\varepsilon}}_p) + A_n\dot{p}$. Coalescence, the last stage of damage before fracture, is taken into account through effective porosity $f_\star$ which is usually taken as:

$$f_\star = \begin{cases} f & \text{if } f < f_c \\ f_c + \dfrac{1/q_1 - f_c}{f_R - f_c}(f - f_c) & \text{otherwise} \end{cases} \qquad [4.10]$$

where f_c corresponds to porosity at the onset of coalescence and f_R is porosity at fracture. When $f = f_R$, then $f_\star = 1/q_1$, such that that the stress state $\underset{\sim}{\sigma} = \underset{\sim}{0}$ verifies the plasticity condition $\Phi = 0$. This stress state is then the only one admissible. The material can therefore no longer tolerate any load, which corresponds to fracture.

The "standard" GTN model has lead to many extensions aimed at integrating the physical deformation and damage mechanisms for a better description of fracture. Within the limited scope of this chapter, it is impossible to describe them. Here, below are given the main extensions with a nonexhaustive list of publications:

– introducing viscosity (Leblond et al. 1994);

– introducing a matrix that does not meet the von Mises criterion (anisotropy or other isotropic criterion) (Benzerga and Besson 2001; Tanguy et al. 2008; Cazacu et al. 2014);

– shape of cavities considered as ellipsoids (Gologanu et al. 1993, 1994; Danas and Aravas 2012; Madou and Leblond 2013; Madou et al. 2013; Cao et al. 2015);

– coalescence mechanism by internal necking (Thomason 1985a, 1985b; Gologanu et al. 2001a, 2001b; Benzerga 2002; Keralavarma et al. 2011; Hure and Barrioz 2016; Torki et al. 2017);

– introducing the effect of the third stress tensor invariant (Lode parameter) (Nahshon and Hutchinson 2008; Morgeneyer and Besson 2011);

– cavity growth in a single crystal (Han et al. 2013; Mbiakop et al. 2015; Paux et al. 2015; Ling et al. 2016).

However, the various extensions concern the growth and coalescence that can be considered by the mechanics of continuous media and homogenization models. It should be noted that nucleation models remains, for the moment, largely phenomenological. A micromechanical framework for nucleation would be useful. 3D tomography data are the components for the development of this framework (see, for example, Hannard et al. 2018).

4.3.3. *Other models*

There are of course many other models than the GTN model and its extensions for describing ductile fracture. These models are in general of a more phenomenological nature than models derived from the GTN approach that seek to foster a micromechanics framework. In this family, the Lemaitre models (Lemaitre 1985, 1996) can be found that are derived from the earlier work by Kachanov (1958). These models are often described as "continuous damage mechanics" models (Krajcinovic 1996; Lemaitre 1996). This terminology seem to be inappropriate however, because the models based on micromechanics also rely on the description of continuous damage. It therefore seems preferable to call them "phenomenological". These models introduce a scalar damage or tensor variable. In the latter case, it is then possible to describe an anisotropy of induced damage (Lemaitre et al. 2000; Lemaitre and Desmorat 2005). It is also possible to describe crack closure effects (Chaboche 1993; Halm and Dragon 1998). These models can be built in as part of strict thermodynamics framework (Germain et al. 1983) which, in particular, ensures positive dissipation. A detailed description of the approach will be found in Lemaitre and Desmorat (2005). In this context, the kinetics of damage is related to the volume energy restitution rate which corresponds to the derivative of the free energy with respect to the damage variable. The process is often reused in a simpler way, that is, without thermodynamic formalism and without taking the effect of damage on elastic behavior into account (Bonora 1997; Xue and Wierzbicki 2008). The Rousselier (1987) model recovers the thermodynamic approach while remaining very close to the GTN formalism. A comparison of the different approaches is proposed in Besson (2009).

4.4. Performing ductile fracture simulations with a finite elements code

4.4.1. *Calculation parameters*

In the rest of the chapter, two simple geometries will be used to illustrate the subject. The first is a notched axisymmetric specimen (NT, see Figure 4.2(a)) representing crack onset conditions in an uncracked structure (Hancock and Mackenzie 1976). The second is a *Compact Tension* specimen (CT, see Figure 4.2(b)) (ASTM 2008).

Figure 4.2 gives both the sizes of the specimens and an example of a finite element mesh. In all cases, simulations are performed with quadrangular linear elements (axisymmetric or plane deformation) with full integration (four Gaussian points). To avoid pressure fluctuations in the elements, a $\overline{\mathrm{B}}$ method is employed (Elguedj et al. 2008).

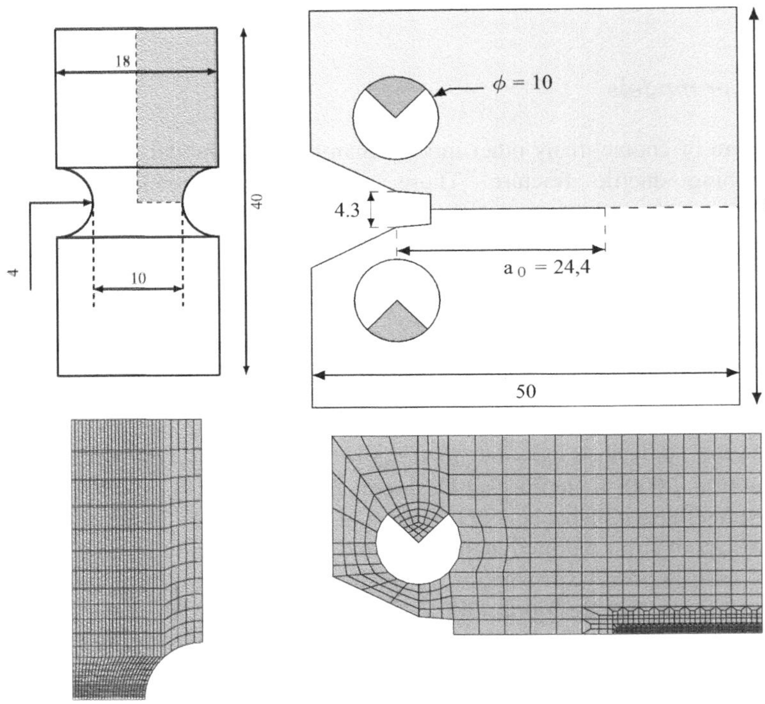

a) NT b) CT

Figure 4.2. *Mesh geometry and examples (minimum size 200 μm): (a) circumferentially notched specimen; (b) Compact Tension specimen*

In order to perform the simulations, the GTN model was employed. The material parameters used are collected in Table 4.1. It should be noted that the use of another model would lead to identical conclusions in terms of localization, mesh dependency, etc. Many computational codes include an implementation of the GTN model and there are many existing implementations in the form of "user law"[3].

3 As for example, the UMAT procedures of Abaqus. An example thereof can be found at the following link: https//hzg.de/institutes_platforms/materials_research/materials_mechanics/simulation/UMAT/index.php.de.

Strain hardening	$\sigma_F(p) = 510 + 295(1 - \exp(-9.6p))$ (MPa)
Model GTN	$f_0 = 0.001, q_1 = 1.47, q_2 = 1.05, f_c = 0.05, f_R = 0.25$
Regularization	$l_c = 0.2$ mm
Rice and Tracey model	$\alpha = 0.283, \beta = 1.5, p_c = 0$

Table 4.1. *GTN model parameters used in calculations for structural steel*

4.4.2. *Pressure control*

Ductile fracture simulations lead to large strains in the quasi-incompressible domain ($\det \underline{F} \approx 1$). Kinematics related to the finite element method make it difficult to verify this condition. This can lead to strong pressure oscillations between or inside elements. An illustration of this problem is given in Figure 4.3 for bilinear elements with four nodes/four Gaussian points. A standard formulation leads to strong stress oscillations near the crack tip, which is a heavily deformed area. The use of a formulation $\overline{B}$ (Hughes 1980) makes it possible to solve the problem. This formulation is very often used within the framework of damage mechanics. It is also possible to use square (2D) or brick (3D) elements at a Gaussian point with *hour glass* effect control. This solution is very often put into practice in explicit codes. The example in Figure 4.3 is of course an extreme case but the problem can also arise with displacement quadratic interpolation elements. In this case, mixed displacement–pressure–volume variation elements (Jinka et al. 1997; Bellet 1999; Feld-Payet et al. 2011; Zhang et al. 2018) or "mini" elements should be used (Gay et al. 1994). These latest solutions are seldom implemented in the context of damage mechanics and remain problematic (Auricchio et al. 2013). It is therefore advisable to verify the stress fields at the end of the calculations.

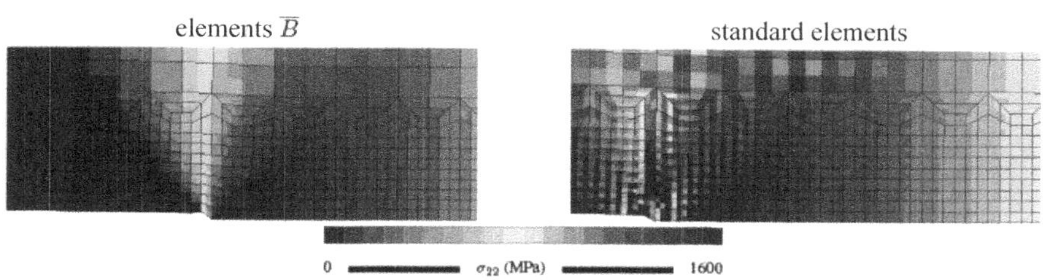

Figure 4.3. *Comparison of stress opening fields at crack tip for a $\overline{B}$ formulation and a standard formulation for bilinear elements with four nodes/four Gaussian points. Damageless elastoplastic calculation is shown. For a color version of this figure, see www.iste.co.uk/lebon/mechanics.zip*

4.4.3. *Application of the Rice and Tracey criterion*

The first application concerns the use of the Rice and Tracey criterion. It is first necessary to carry out an elastoplastic computation which is then post-processed. This procedure makes the application of a criterion relatively easy. In addition, parameter optimization of the rupture criterion does not require simulating parts many times; all that needs to be done is restart the post-processing, which is in general much faster.

The Rice and Tracey criterion (see coefficients in Table 4.1) was employed to post-process the computations on NT and CT specimens. Meshes using mesh sizes ranging from 25 to 200 μm were used. The maximum value of the Rice and Tracey criterion that is reached in the structure is plotted according to the load.

It can be seen that in the case of NT specimens mesh size convergence is quickly obtained. The damage field does not present strong gradients. The opposite situation occurs for CT specimens. Due to the presence of a crack, there exists a strong strain gradient (see, for example, the HRR fields describing the stress and strain fields at the tip of a crack (Hutchinson 1968; Rice and Rosengren 1968)). This explains the observed non-convergence. In practice, a mesh size can then be set to evaluate the damage criterion over a given distance. A special averaging of the fields can also be performed: by averaging either R/R_0 or plastic strain (and stresses), then integrating equation [4.3]. In practice, however, these approaches are not utilized.

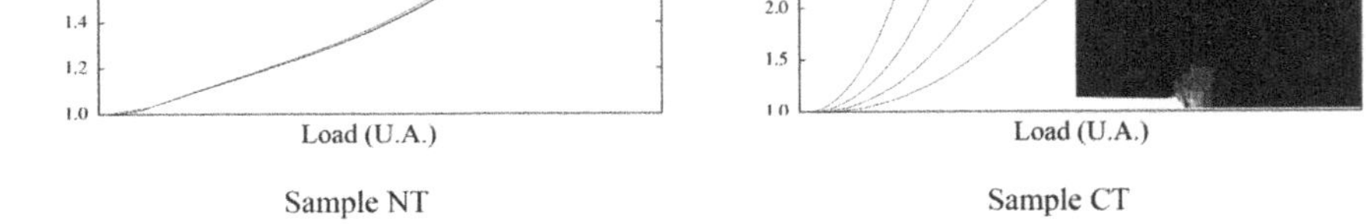

Figure 4.4. *Application of the Rice and Tracey criterion for NT and CT specimens with mesh sizes ranging from 25 to 200 μm. The maps represent for a mesh size of 50 μm the distribution of the damage (R/R_0) for the last calculated loading point. For a color version of this figure, see www.iste.co.uk/lebon/mechanics.zip*

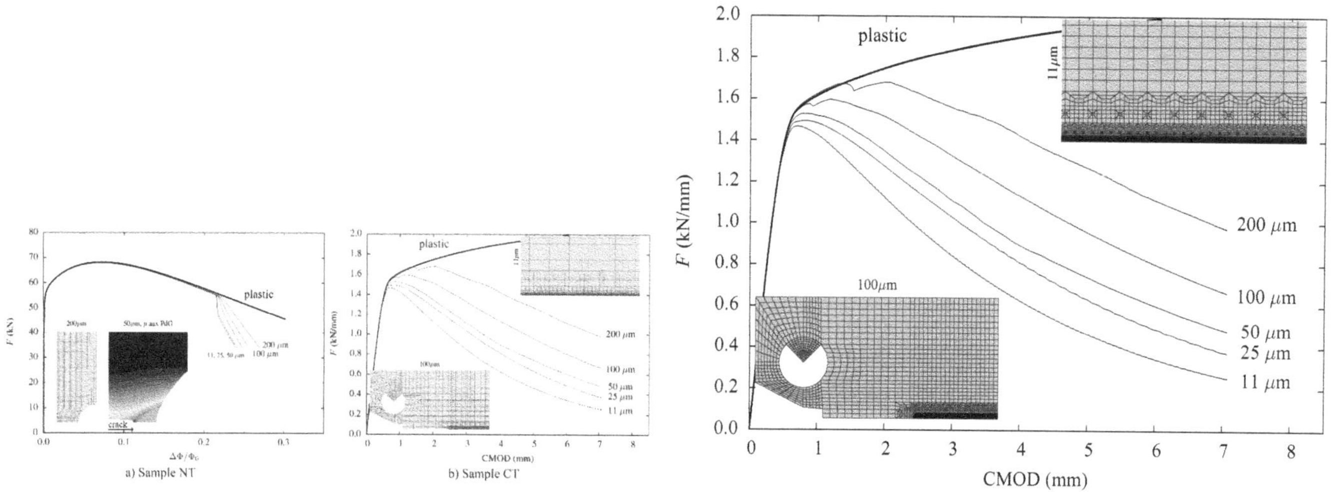

a) NT specimen b) CT specimen

Figure 4.5. *Application of the GTN model for NT and CT specimens for mesh sizes ranging from 11 to 200 μm. For a color version of this figure, see www.iste.co.uk/lebon/mechanics.zip*

4.4.4. *GTN model application*

We reuse here the same simulations but instead utilize the GTN model with the parameters of Table 4.1. A first series of computations shows the simulation of an axisymmetric specimen (Figure 4.5(a)) for which the ratio of the radius of the notch and the minimum diameter is 0.4. The latter is equal to 10 mm. Different mesh sizes are employed in the initiation and propagation area of the crack. Sizes range from 200 to 11 μm. It is first important to observe that independently of the mesh size, the elastoplastic response is unique: the calculations thus provide a "converged" result. When damage is taken into account, a "converged" response is again obtained up to the point where a sharp drop of the force F occurs. This point corresponds to the initiation of a crack at the core of the specimen. From this point on, the answers obtained depend on the mesh size utilized. This result is not surprising, since localization analysis (section 4.5) effectively predicts the formation of localized bands for softening materials. The figure also shows the cumulative plastic strain field (p) at the Gaussian points. Elements whose Gaussian points have reached the fracture condition (i.e. $f = f_F$) have been removed to make visualization easier.

It can be seen that an inclined band is formed in front of the crack tip that may correspond to the occurrence of a *cup-cone* fracture (Tvergaard and Needleman 1984), which in practice is not being formed in this case. The conditions for the occurrence of this mode of rupture are discussed in Besson et al. (2001b, 2003).

A second example of the application of the GTN model is presented in Figure 4.5(b) with a CT specimen. Here again, different mesh sizes are used. The calculation assuming a damageless material was also performed ("plastic" in Figure 4.5(b)). Here too, a very strong dependency on the mesh size utilized can be seen. For the same crack opening (*Crack Mouth Opening Displacement* (CMOD)), the force is lower when the mesh size is small. It also corresponds to a greater progression of the crack. An example of the opening stress field (σ_{22}) is shown in Figure 4.6. In this figure, the "broken" elements have been removed to facilitate the visualization of the crack. A maximum stress can be observed in front of the tip of the crack. The fracture process zone only extends over three to four elements as highlighted in Figure 4.7, which indicates the crack tip damage map.

The effect of mesh orientation is finally illustrated in Figure 4.8. The same NT-type specimen is meshed with a complete mesh, a half-mesh and finally with a deliberately inclined mesh. For the complete mesh with a horizontal mesh, it can be seen that the crack propagates in a row of elements. For the half-mesh, the crack therefore propagates in two rows of elements due to the imposed symmetry. With the inclined mesh, the crack follows a row of meshes and its path is therefore also inclined. It can therefore be seen that for a given model, the mesh design can influence the path of the crack. Diametral strength-contraction curves are also affected.

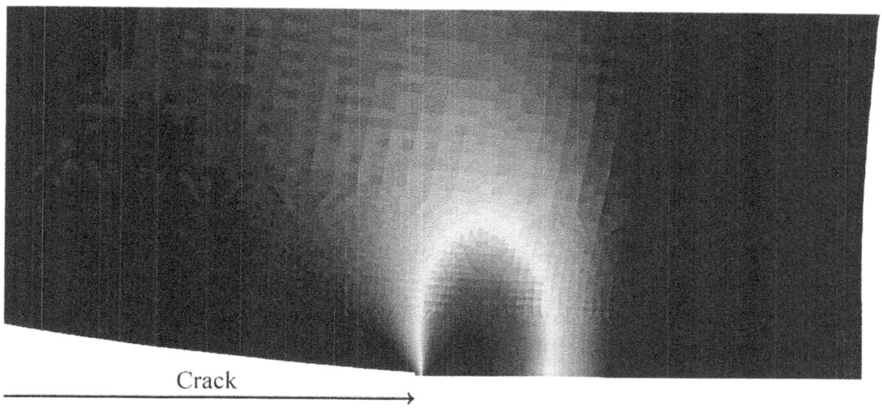

Figure 4.6. *Simulation of a Compact Tension (CT) specimen: opening stress at the tip of the propagating crack. For a color version of this figure, see www.iste.co.uk/lebon/mechanics.zip*

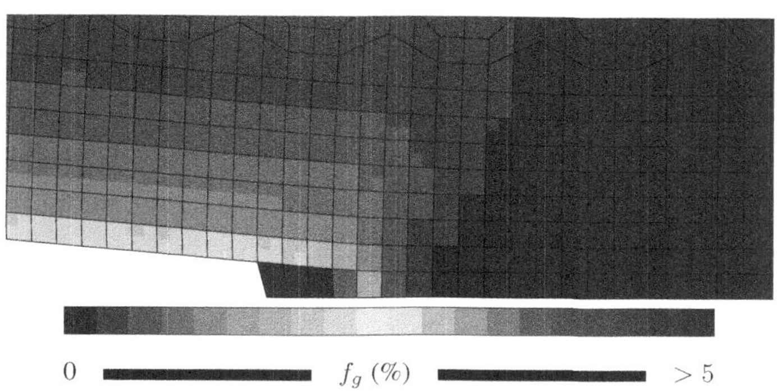

Figure 4.7. *Simulation of a Compact Tension (CT) specimen: crack tip damage during propagation. For a color version of this figure, see www.iste.co.uk/lebon/mechanics.zip*

4.4.5. *Pragmatic solution*

A common solution to avoid the problem of discretization dependency is to use only one formulation (type, size, orientation) for an element in areas where cracks propagate (Rousselier 1987; Xia and Shih 1995). The size of the element (for a given type) then becomes a model parameter since it directly influences the simulated responses. Finite element discretization then acts in two distinct different manners: (i) geometry representation and (ii) model parameter. Despite these limitations, damage models are routinely used to analyze small (Kofiani et al. 2013; Cortese et al. 2014; Shinohara et al. 2016) or intermediate size specimens (Besson et al. 2013; Novokshanov et al. 2015). Applications to real structures are much less frequent and mainly employ either the Rice and Tracey model, which then provides a criterion for eroding the elements (Oh et al. 2007) or the GTN model (Jayadevan et al. 2006; Nourpanah and Taheri 2011; Nicak et al. 2012; Tang et al. 2014; Medjo et al. 2015; Soret et al. 2017). Modeling can then be used to interpret testing but, currently, the lack of reliability of these simulations makes them unsuitable for performing safety assessment calculations. Because of the difficulty in controlling the crack path, simulations remain most often limited to cracks in a mode-I plane. However, a more complex cracking path can be prescribed in order to analyze changes of direction of the crack (Besson et al. 2013). With decoupled criteria models, it is possible to consider that fracture only occurs if the criterion is exceeded over a characteristic length/volume. This solution (analogous to the RKR criterion (Ritchie et al. 1973) in brittle fracture) is not used, however.

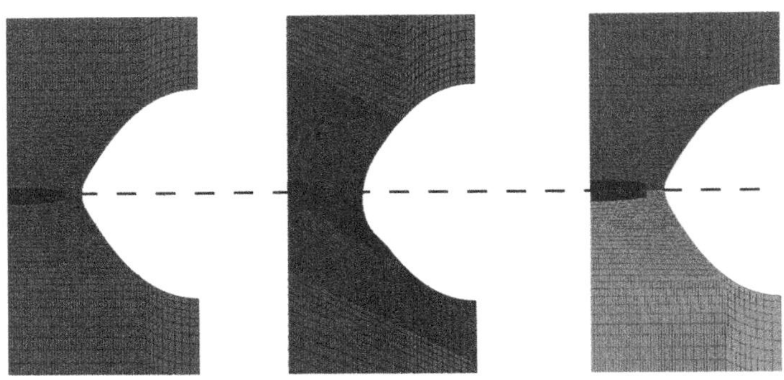

Full mesh Inclined mesh Half-mesh

Figure 4.8. *Effect of mesh orientation on the crack path. For a color version of this figure, see www.iste.co.uk/lebon/mechanics.zip*

4.5. Localization origin

Independently of their nature, damage models are constructed to represent matter degradation until fracture. The behavior therefore necessarily presents a softening character that leads to the emergence of bands in which plastic strain and damage are highly localized. We briefly present here the approach proposed by Rice (1976, 1980) to describe the occurrence of these bands.

For elastoplastic materials, the incremental constitutive equations of behavior can be expressed as:

$$\overset{\Delta}{\underline{\sigma}} = \underline{\underline{\mathcal{L}}}_t : \underline{D} \qquad [4.11]$$

$\underline{\underline{\mathcal{L}}}_t$ is the tangent elastoplastic tensor, $\underline{D}$ the strain rate and $\overset{\Delta}{\underline{\sigma}}$ the Jaumann derivative of the stress tensor. The tensor $\underline{\underline{\mathcal{L}}}_t$ naturally depends on the set of constitutive equations and is not explained here. Nonetheless, it depends only on the internal variables representing the state of the material but not on their derivatives.

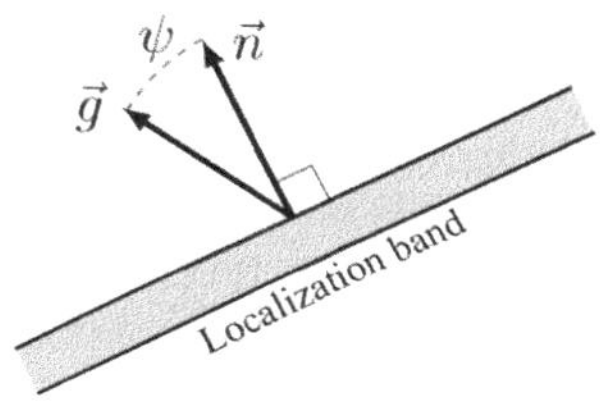

Figure 4.9. *Geometry of a localization band*

The localization is supposed to occur in a flat and infinite band characterized by its normal $\vec{n}$. Displacements are assumed to be discontinuous on both sides of the band. The direction of the displacement jump is denoted by $\vec{g}$ (see Figure 4.9). By expressing the continuity of displacement and equilibrium, it is possible to show (Rice 1976, 1980) that the jump of the strain rate tensor is proportional to:

$$\frac{1}{2}(\vec{g} \otimes \vec{n} + \vec{n} \otimes \vec{g}) \qquad [4.12]$$

and that the bifurcation condition is written as (Rice 1980; Mear and Hutchinson 1985):

$$\exists \vec{n}, \;\; \det(\underline{A}_t(\vec{n})) = 0 \qquad [4.13]$$

with:

$$\underline{A}_t(\vec{n}) = \vec{n}.\underline{\underline{\mathcal{L}}}_t.\vec{n} + \frac{1}{2}\left((\vec{n}.\underline{\sigma}) \otimes \vec{n} - \vec{n} \otimes (\vec{n}.\underline{\sigma}) + (\vec{n}.\underline{\sigma}.\vec{n})\underline{1} - \underline{\sigma}\right) \qquad [4.14]$$

where the second term of the right-hand member corresponds to the large strain formalism using the Jaumann derivative. $\vec{g}$ is then the eigenvector of $\underline{A}_t(\vec{n})$, corresponding to the zero eigenvalue. It is shown that localization becomes possible with softening behaviors. The critical strain hardening modulus allowing localization is, for a given triaxiality level, reached for lower cumulative strain values for a plane strain state, while an axisymmetric load leads to better resistance to localization. The analysis does not allow evaluating the thickness of the band once it is formed. In practice, in the usual finite element calculations, it will be as fine as possible; that is, localized in a row of elements as observed above.

The localization band is schematically shown in Figure 4.9. The normal to the band, $\vec{n}$, is in the plane defined by the eigenvectors corresponding to the maximal ($\vec{d}_I$) and minimal ($\vec{d}_{III}$) eigenvalues of the strain rate tensor, provided that the dilation remains low (Rudnicki and Rice 1975; Yamamoto 1978). The angle ψ characterizes the fracture mode that varies from single opening fracture ($\psi = 0$) to purely tangential fracture or shear ($\psi = \pi/2$). The mode $\psi = 0$ is in practice observed only for highly dilating materials. The mode $\psi \approx \pi/2$ is the most common. A "localization indicator" can be proposed from the previous analysis. It can be used to detect in computational post-processing the areas in which localization is possible (Billardon and Doghri 1989; Besson et al. 2001b, 2003). The indicator is obtained by minimizing $\det(\underline{A}_t(\vec{n}))$ on all possible orientations:

$$I_b = \min_{\vec{n},\, ||\vec{n}||=1} \det(\underline{A}_t(\vec{n})) \qquad [4.15]$$

The formation of a band can happen as soon as $I_b = 0$. The normal to the band is given by the vector $\vec{n}$ for which $I_b = 0$.

4.6. Regularization methods

In this section, some methods currently in development for avoiding discretization dependencies by introducing an internal length relative to the material are presented. What these methods have in common is that they allow the constitutive equations presented above to be preserved (admittedly with a few adaptations). This appears to be particularly interesting when the behaviors propose a physical approach based on micromechanics.

4.6.1. *Integral methods*

To regularize the calculations and avoid localization in thin bands, a first solution consisted of achieving a weighted average of one or more variables around each material point (Pijaudier-Cabot and Bazant 1987; Bazant and Pijaudier-Cabot 1988).

By considering a variable D, the associated nonlocal variable, $\overline{D}$, at point $\mathbf{x}$, is thus calculated as:

$$\overline{D}(\mathbf{x}) = \frac{\int_\Omega D(\mathbf{y})w(\mathbf{y}-\mathbf{x})\,d\Omega}{\int_\Omega w(\mathbf{y}-\mathbf{x})\,d\Omega} \quad [4.16]$$

where Ω is the solid under consideration and w is a weighting function such as:

$$\int_{\mathbb{R}^3} w(\mathbf{x})\,d\Omega = 1 \quad [4.17]$$

For example, consider the following function:

$$w(\mathbf{x}) = \frac{1}{(2\pi)^{3/2}\ell^3} \exp\left(-\frac{||\mathbf{x}||^2}{2\ell^2}\right) \quad [4.18]$$

or:

$$w(\mathbf{x}) = \frac{105}{32\pi^2 l^3}(1 - (||\mathbf{x}||/\ell)^2)^2 \quad [4.19]$$

where ℓ is a characteristic length. The second form makes it possible to have a bounded integration domain ($0 \leq ||\mathbf{x}||/\ell \leq 1$). The nonlocal variable $\overline{D}$ is then used to describe the degradation of the material. However, this approach remains difficult to implement within the framework of the finite element method. In particular, integrals require us to consider a set of elements (patch) around each material point so that the structure of the code must be modified and element-wise processing is no longer possible. Furthermore, following this approach, any pair of points separated by the same Euclidean distance interact in the same way. The integral method was, for example, used in Enakoutsa et al. (2007) for regularizing the GTN model. However, similar points but separated by a crack, a hole, or even a very damaged area should see their interaction reduced (Rastiello et al. 2018). A solution to this problem has recently been proposed. The calculation of the effective interaction distance is carried out based on an analogy with a wave propagation problem and travel time between two material points. The speed of the waves depends on damage and the signal is not transmitted when it meets a crack or a hole (Desmorat et al. 2015; Rastiello et al. 2018).

4.6.2. *Explicit or implicit gradient methods*

To deal with the problems of practically implementing integral methods, gradient-based methods have been developed (Peerlings et al. 1996; Geers et al.

1998; Engelen et al. 2003). They use a Taylor expansion of D. We then have (using the convention for the summation on repeated indexes):

$$\begin{aligned} D(\mathbf{y}) &= D(\mathbf{x}) + \frac{\partial D}{\partial x_i}(y_i - x_i) + \frac{1}{2!}\frac{\partial^2 D}{\partial x_i \partial x_j}(y_i - x_i)(y_j - x_j) \\ &+ \frac{1}{3!}\frac{\partial^3 D}{\partial x_i \partial x_j \partial x_k}(y_i - x_i)(y_j - x_j)(y_k - x_k) \\ &+ \frac{1}{4!}\frac{\partial^4 D}{\partial x_i \partial x_j \partial x_k \partial x_l}(y_i - x_i)(y_j - x_j)(y_k - x_k)(y_l - x_l) + \dots \end{aligned} \quad [4.20]$$

This expression is then used to evaluate the integral equation [4.16]. Nonetheless, the integral is considered to be computed for an infinite domain ($\mathbb{R}^3$):

$$\begin{aligned} \overline{D}(\mathbf{x}) &= D(\mathbf{x}) \int_{\mathbb{R}^3} w(\mathbf{y} - \mathbf{x})\, d\Omega + \frac{\partial D}{\partial x_i} \int_{\mathbb{R}^3} (y_i - x_i) w(\mathbf{y} - \mathbf{x})\, d\Omega \\ &+ \frac{1}{2!}\frac{\partial^2 D}{\partial x_i \partial x_j} \int_{\mathbb{R}^3} (y_i - x_i)(y_j - x_j) w(\mathbf{y} - \mathbf{x})\, d\Omega \\ &+ \frac{1}{3!}\frac{\partial^3 D}{\partial x_i \partial x_j \partial x_k} \int_{\mathbb{R}^3} (y_i - x_i)(y_j - x_j)(y_k - x_k) w(\mathbf{y} - \mathbf{x})\, d\Omega \\ &+ \dots \end{aligned} \quad [4.21]$$

Since the integrals are taken on $\mathbb{R}^3$, they are constant in expression [4.21]. In addition, odd terms are then zero. The following form is thus obtained for $\overline{D}$:

$$\overline{D}(\vec{x}) = D(\vec{x}) + c\nabla^2 D(\vec{x}) + d\nabla^4 D(\vec{x}) + \dots \quad [4.22]$$

where c and d are constants. By only considering the second order, the so-called "explicit gradient" model is obtained:

$$\boxed{\overline{D}(\vec{x}) = D(\vec{x}) + c\Delta D(\vec{x})} \quad [4.23]$$

The model is said to be explicit because $\overline{D}$ can be directly computed from the field D. However, this model remains difficult to use within the context of the finite element method because it requires the calculation of $\nabla^2 D(\vec{x})$, which can only be done by extrapolating the values of D at the Gaussian points to the nodes and then calculating the Laplacian. A solution to this problem was proposed by Peerlings et al.

(1996); Engelen et al. (2003). Consider the equation $\overline{D}(\vec{x}) = D(\vec{x}) + c\nabla^2 D(\vec{x})$ and by taking the Laplacian, we get:

$$\nabla^2\overline{D}(\vec{x}) = \nabla^2 D(\vec{x}) + c\nabla^4 D(\vec{x}) \Rightarrow \nabla^2 D(\vec{x}) = \nabla^2\overline{D}(\vec{x}) - c\nabla^4 D(\vec{x}) \qquad [4.24]$$

By injecting this last expression into [4.22], it then follows that:

$$\overline{D}(\vec{x}) = D(\vec{x}) + c\nabla^2\overline{D}(\vec{x}) - (d - c^2)\nabla^4 D(\vec{x}) + \dots \qquad [4.25]$$

that is if we neglect the fourth-order terms:

$$\boxed{\overline{D}(\vec{x}) = D(\vec{x}) + c\Delta\overline{D}(\vec{x})} \qquad [4.26]$$

This equation corresponds to an implicit gradient model.

NOTE. A 1D damage profile showing a localized area is plotted in Figure 4.10. The gradient changes sign when the maximum is reached. The maximum corresponds to a negative Laplacian. In the equation $\overline{D}(\vec{x}) = D(\vec{x}) + c\nabla^2 D(\vec{x})$, we thus have $\overline{D} < D$ $(c > 0)$, which properly reflects the regularizing nature of the model. The same is true when considering the implicit equation in [4.26].

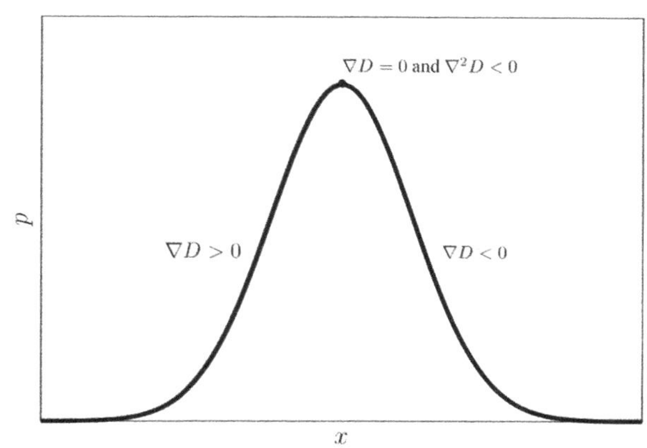

Figure 4.10. *Damage profile, gradient and Laplacian*

The implicit model version can be more easily used in the context of the finite element method. It is then necessary to introduce new degrees of freedom representing

$\overline{D}$. These degrees of freedom are introduced as nodal variables. The weak form of equation [4.26] can be obtained as follows (where D^* is a test field):

$$\int_\Omega \text{div}\left(c\nabla\overline{D}D^*\right)\,dV = \int_\Omega c\Delta\overline{D}D^*\,d\Omega + \int_\Omega c\nabla\overline{D}.\nabla D^*\,dV$$
$$= \int_{\partial\Omega} cD^*\nabla\overline{D}.\mathbf{n}\,dS \qquad \forall D^* \qquad [4.27]$$

A "natural" boundary condition is deduced:

$$\nabla\overline{D}.\mathbf{n} = 0 \text{ on } \partial\Omega \qquad [4.28]$$

we then have:

$$\int_\Omega c\Delta\overline{D}D^*\,dV = -\int_\Omega c\nabla\overline{D}.\nabla D^*\,dV \quad \forall D^* \qquad [4.29]$$

The weak form of equation [4.26] is then obtained which allows implantation in the framework of the finite element method[4]:

$$\int_\Omega \left((\overline{D} - D)D^* + c\nabla\overline{D}.\nabla D^*\right)\,dV = 0 \quad \forall D^* \qquad [4.30]$$

This formulation will be reused in the context of micromorphic models.

The constitutive laws must then be adapted. Considering a behavior in large transformations, the constitutive law must evaluate a measure of the stress tensor with respect to an increment of the transformation gradient $\underline{F}$ while updating the internal variables describing the condition of the material ($\mathbf{V}^i$). That is, for a time step $t \rightarrow t + \Delta t$:

$$(\underline{F}_{t+\Delta t}, \underline{F}_t, \mathbf{V}^i_t) \longrightarrow (\underline{S}_{t+\Delta t}, \mathbf{V}^i_{t+\Delta t}) \qquad [4.31]$$

where $\underline{S}$ represents a measure of the stress tensor (Cauchy, Kirchhoff, Piola-Kirchhoff-2, Piola-Kirchhoff-1, etc.). For implicit codes, the constitutive law must also provide the tangent operator:

$$\underline{\underline{T}} = \frac{\partial \underline{S}_{t+\Delta t}}{\partial \underline{F}_{t+\Delta t}} \qquad [4.32]$$

4 This implantation is not detailed in this chapter.

These notions can be extended for behaviors adapted to "internal length" formulations.

The generic interface (equation [4.31]) then becomes:

$$(\underline{F}_{t+\Delta t}, \underline{F}_t, \overline{D}_{t+\Delta t}, \overline{D}_t, \mathbf{V}_t^i) \longrightarrow (\underline{S}_{t+\Delta t}, D_{t+\Delta t}, \mathbf{V}_{t+\Delta t}^i) \quad [4.33]$$

The tangent operator can then formally be written as (Mediavilla et al. 2006c):

$$\mathbf{T} = \begin{pmatrix} \dfrac{\partial \underline{S}_{t+\Delta t}}{\partial \underline{F}_{t+\Delta t}} & \dfrac{\partial \underline{S}_{t+\Delta t}}{\partial \overline{D}_{t+\Delta t}} \\ \dfrac{\partial D_{t+\Delta t}}{\partial \underline{F}_{t+\Delta t}} & \dfrac{\partial D_{t+\Delta t}}{\partial \overline{D}_{t+\Delta t}} \end{pmatrix} \quad [4.34]$$

Implicit gradient models do not provide any rules to reformulate the constitutive law. Several choices have been proposed in the literature for ductile fracture. A first choice consists in using cumulative plastic deformation p. We notice that in the context of ductile fracture models, the evolution of all variables leading to softening (f_g, f_n and even the temperature if adiabatic heating is taken into account) is controlled by plastic deformation. The choice of this variable thus makes it possible to regularize the "low cost" behavior. This choice was made, for example, in Mediavilla et al. (2006b, 2006c) and Feld-Payet et al. (2011).

4.6.3. *Micromorphic models*

Another approach, very close in its implementation to the second implicit gradient method, can be considered. In the context of the thermodynamics of irreversible processes, the theory of the micromorphic continuum proposes that the solution be regularized by enhancing the kinematics and free energy with a micromorphic variable and its first gradient. This framework was reviewed by Forest (2009). The regularized internal variables (e.g. p the cumulative plastic strain that will be used thereafter) have a micromorphic counterpart (referred to here as p_χ), which is calculated by taking into account the interaction at the microscopic scale through its Laplacian. For an elastoplastic material (Mazière and Forest 2015), the expression of free energy becomes:

$$\psi = \frac{1}{2}(\underline{\varepsilon} - \underline{\varepsilon}_p) : \underline{\underline{\mathcal{E}}} : (\underline{\varepsilon} - \underline{\varepsilon}_p) + \psi_p(p) + \frac{1}{2}H_\chi(p - p_\chi)^2 + \frac{1}{2}A_\chi \nabla p_\chi . \nabla p_\chi \quad [4.35]$$

where A_χ and H_χ are two model parameters. $\underline{\underline{\mathcal{E}}}$ is the elasticity tensor and $\underline{\varepsilon}_e$ is the elastic strain tensor. ψ_p is the part of free energy describing strain hardening.

Therefrom, the associated thermodynamic forces can be deduced:

$$\underline{\sigma} = \frac{\partial \psi}{\partial \underline{\varepsilon}} = \underline{\underline{\mathcal{E}}} : (\underline{\varepsilon} - \underline{\varepsilon}_p) \tag{4.36}$$

$$R_\chi = \frac{\partial \psi}{\partial p} = R + H_\chi (p - p_\chi) \tag{4.37}$$

$$a_\chi = \frac{\partial \psi}{\partial p_\chi} = -H_\chi (p - p_\chi) \tag{4.38}$$

$$\vec{b}_\chi = \frac{\partial \psi}{\partial \nabla p_\chi} = A_\chi \nabla p_\chi \tag{4.39}$$

Strain hardening is now represented by R_χ instead of $R = \partial \psi_p / \partial p$ in the base model. In this way, the formulation of the constitutive law and in particular its coupling with the micromorphic variable are clearly defined. Dissipation over the entire volume can then be written as:

$$\mathcal{D} = \int_\Omega \left(\underline{\sigma} : \dot{\underline{\varepsilon}}_p - R_\chi \dot{p} - a_\chi \dot{p}_\chi - \vec{b}_\chi . \nabla \dot{p}_\chi \right) \, \mathrm{d}\Omega \tag{4.40}$$

Using equations [4.36]–[4.39] and the procedure described in [4.27], dissipation is reformulated as:

$$\mathcal{D} = \int_\Omega \left(\underline{\sigma} : \dot{\underline{\varepsilon}}_p - R_\chi \dot{p} - (A_\chi \Delta p_\chi + H_\chi (p - p_\chi)) \dot{p}_\chi \right) \, \mathrm{d}\Omega - \int_{\partial\Omega} A_\chi \nabla p_\chi . \vec{n} \, \dot{p}_\chi \, \mathrm{d}S \tag{4.41}$$

It is then assumed that there is no dissipation related to p_χ which therefore leads to:

$$A_\chi \Delta p_\chi + H_\chi (p - p_\chi) = 0 \tag{4.42}$$

Or even:

$$p_\chi = p + \frac{A_\chi}{H_\chi} \Delta p_\chi \tag{4.43}$$

which is equivalent to equation [4.26], as an implicit gradient model with $c = A_\chi / H_\chi$. In addition:

$$\nabla p_\chi . \vec{n} = 0 \text{ on } \partial\Omega \tag{4.44}$$

This general framework described by Forest (2009) was used in Brepols et al. (2017) and Diamantopoulou et al. (2017) in the field of ductile fracture. In Brepols et al. (2017), it is assumed that damage has gradient effects. In particular, damage and plasticity are considered two independent physical mechanisms. Each mechanism is characterized by its own threshold surface and a loading/unloading condition. Several 1D and 2D simulations are performed. The results presented are independent of the mesh size. In addition, this model can properly capture crack onset and propagation. In Diamantopoulou et al. (2017), damage is regularized. The comparison of the results obtained by the proposed model and by the local conventional model validates the implementation of the model. Dependency is effectively reduced.

In addition to the micromorphic approach proposed in Forest (2009), there are other models that are also micromorphic but derived from specific homogenization procedures (Gologanu et al. 2007; Enakoutsa and Leblond 2009; Bergheau et al. 2014; Huetter 2017), in order to account for nonlocal effects. These models are called the "second gradient model". In Gologanu et al. (2007), the Gurson homogenization approach is extended to constrained gradient theory in which an additional gradient term appears in the threshold function. This model (GLPD) can be classified as a constrained micromorphic theory because micro-deformation and macro-deformation are constrained to combine. The numerical results of Enakoutsa and Leblond (2009); Bergheau et al. (2014) show that the GLPD model can correctly reduce the parasitic effects induced by the localization of strain and damage. The problem with the GLPD model is a complex finite element implementation. In addition, the numerical implementation of the model requires the second derivatives of the shape function, therefore, the elements of the C^1 class.

Although this problem can be solved using penalized formulations, numerical implementation is still difficult because the strain tensor is considered as an additional nodal variable. More recently, a gradient formulation of the GLPD model was proposed in Huetter (2017) in order to combine the efficiency of implicit variable models and solid micromechanics-based models. However, no simulation is presented.

So-called micromorphic models use a "local" variable and its micromorphic counterpart. When it is desirable that the two variables be very close to use in practice the gradient of the "local" variable, a large penalty term should be used (H_χ in the example presented above). This can affect the convergence rate of the numerical solution in terms of the overall balance as well as the integration of laws of behavior. This problem is partly solved by the enhanced energy methods presented in section 4.6.4.

4.6.4. *Enhanced energy models*

By following Lorentz and Andrieux (1999); Lorentz and Benallal (2005); Lorentz and Godard (2011); Zhang et al. (2018) and Chen et al. (2020), the total free energy of the system is now written as:

$$\mathcal{F} = \int_{\Omega} \left(\frac{1}{2}(\underline{\varepsilon} - \underline{\varepsilon}_p) : \underline{\underline{\mathcal{E}}} : (\underline{\varepsilon} - \underline{\varepsilon}_p) + \psi_p(p) + \frac{1}{2} c \nabla p . \nabla p \right) \mathrm{d}\Omega \qquad [4.45]$$

where, by following Zhang et al. (2018), the nonlocal variable is considered as the cumulative plastic strain p. The use of such a formulation is difficult within the framework of the finite element method because p is generally evaluated at the Gaussian points so that we are facing the same difficulties as for the explicit gradient method. The solution then consists of introducing an additional variable (a in the remainder of the text). This variable is then defined at the nodes and its interpolation at Gaussian points should be as close as possible to p. The total free energy of the system is then expressed as:

$$\mathcal{F} = \int_{\Omega} \underbrace{\left(\frac{1}{2}(\underline{\varepsilon} - \underline{\varepsilon}_p) : \underline{\underline{\mathcal{E}}} : (\underline{\varepsilon} - \underline{\varepsilon}_p) + \psi_p(p) + \frac{1}{2} c \nabla a . \nabla a + l(a - p) \right)}_{\psi} \mathrm{d}\Omega \qquad [4.46]$$

thus with $\mathcal{F} = \int_{\Omega} \psi \, \mathrm{d}\Omega$. l is here a Lagrange multiplier used to ensure that $a = p$ in the weak sense. We note that in addition it is possible to use a penalty in $1/2\, r(a-p)^2$ to ensure the convergence of the numerical scheme as detailed in Zhang et al. (2018). We then obtain, as for the problem with the micromorphic model, the quantities associated with state variables:

$$\underline{\sigma} = \frac{\partial \psi}{\partial \underline{\varepsilon}} = \underline{\underline{\mathcal{E}}} : (\underline{\varepsilon} - \underline{\varepsilon}_p) \qquad [4.47]$$

$$R_{\mathrm{nl}} = \frac{\partial \psi}{\partial p} = R - l \qquad [4.48]$$

$$R_a = \frac{\partial \psi}{\partial a} = l \qquad [4.49]$$

$$R_l = \frac{\partial \psi}{\partial l} = a - p \qquad [4.50]$$

$$R_{\nabla} = \frac{\partial \psi}{\partial \nabla a} = c \nabla a \qquad [4.51]$$

The dissipation is then written as:

$$\begin{aligned}\mathcal{D} &= \int_\Omega \left(\underline{\sigma} : \underline{\varepsilon}_p - (R - l)\dot{p} - l\dot{a} - c\nabla a.\nabla\dot{a} - (a - p)\dot{l}\right)\,\mathrm{d}\Omega \\ &= \int_\Omega \left(\underline{\sigma} : \underline{\varepsilon}_p - (R - l)\dot{p} - (a - p)\dot{l} - (l - c\Delta a)\dot{a}\right)\,\mathrm{d}\Omega - \int_{\partial\Omega} c\nabla a.\vec{n}\,\dot{a}\,\mathrm{d}S\end{aligned} \quad [4.52]$$

As for the micromorphic model, it is deduced for the strong formulation that:

$$a = p \quad [4.53]$$

$$l - c\Delta a = 0 \quad [4.54]$$

$$\nabla a.\vec{n} = 0 \quad \text{on } \partial\Omega \quad [4.55]$$

Finally, note that the model proposed by Zhang et al. (2018) and Chen et al. (2020) also allows us to address pressure fluctuation problems related to the quasi-incompressibility of metals.

4.6.5. *Example*

This section presents the implementation of the enhanced energy model based on the reuse, to describe behavior, of the values of the material parameters given in Table 4.1. The nonlocal parameter c is rewritten as $c = l_c^2\sigma_Y$ so as to reveal an internal length l_c. σ_Y corresponds to the Cauchy stress at the onset of necking ($\sigma_Y = 730$ MPa in the example processed). We will take here $l_c = 0.2$ mm.

Reconsider the example of Figure 4.8 in which different meshes had been used for an NT specimen (full mesh, inclined full mesh size and half mesh taking into account the symmetries of the problem). The results are shown in Figure 4.11. It can be seen that the same cracking path is obtained in all cases.

The force-reduction diameter curves obtained for different mesh sizes (200, 100, 50 and 50 μm) are plotted in Figure 4.12. The results can be seen converging for 25 and 50 μm. Moreover, the result for a mesh size of 100 μm is already very close to these results. It is also observed that the simulation with the "standard" model (dashed curve) for a mesh size of 100 μm provides for a significantly earlier fracture onset. This clearly indicates that it is necessary to specifically calibrate the damaged parameters of the "nonlocal" model, and therefore it is impossible to directly reuse the parameters from the "standard" model calibration. Simulations with a damage-free model are also plotted in Figure 4.12. A difference between the "standard" model and the "nonlocal" model can be observed, which predicts slightly higher forces. This difference can be explained by hardening induced by plastic strain gradients. Actually, equation [4.48] gives the yield point modified by the presence of the gradient (using equation [4.54]):

$$R_{\text{nl}} = R - l = R - c\Delta a \approx R - c\Delta p \qquad [4.56]$$

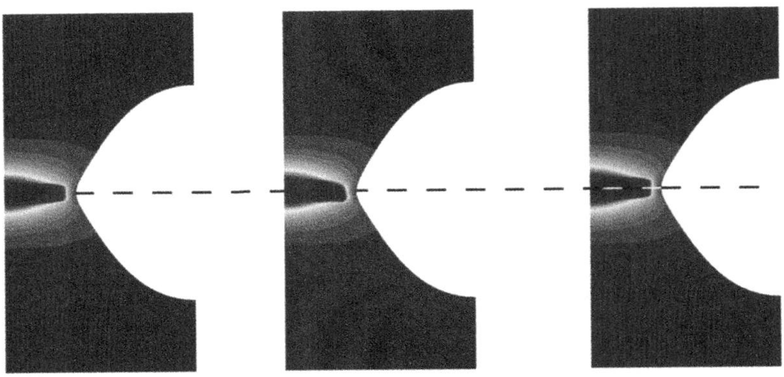

Figure 4.11. *Effect of mesh orientation on the crack path, and nonlocal enhanced energy model is shown. For a color version of this figure, see www.iste.co.uk/lebon/mechanics.zip*

As $\Delta p < 0$ in heavily deformed areas, the flow stress in the minimum cross-section of the specimen is then larger than in the "local" case. This leads to higher forces than in the "standard" case. This results is also achieved with micromorphic models but can be avoided with implicit gradient models. The "nonlocal" calculation carried out with $l_c = 0$ leads to the result obtained with the "standard" elastoplastic model.

Reconsider here the case of the CT specimen (Figure 4.2) with the "nonlocal" model already used above. The results are plotted in Figure 4.13 and compared with the damageless problem. Calculations were carried out for meshes equal to 200, 100 and 66 μm. For 200 μm, the progress of the crack is done sporadically because the mesh size is very coarse and does not properly discretize the internal length. The results for 100 and 66 μm are, on the other hand, very close, indicating the convergence of the model. By recalling the results for the local model for 100 μm (Figure 4.5), it can be seen that it leads to much lower forces.

The damage stress fields (66 μm mesh size) are shown in Figure 4.14. The correct stress continuity can be noticed, which was obtained through controlling pressure fields (see Zhang et al. 2018; Chen et al. 2020) as well as the development of significant damage beyond a row of elements unlike what is observed for a "local" model (Figure 4.7).

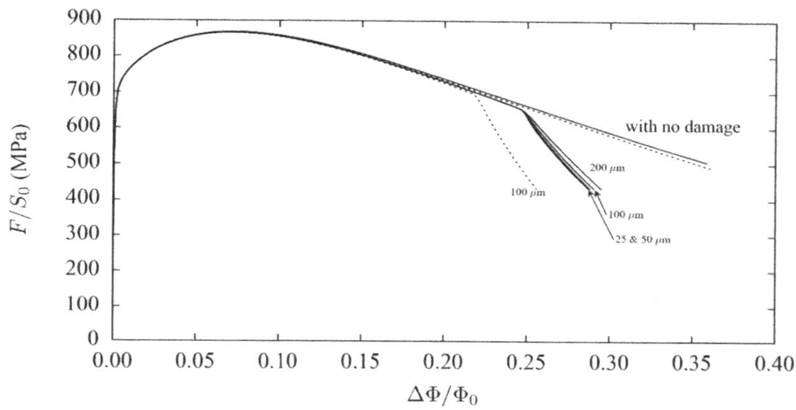

Figure 4.12. *Normalized force-diameter reduction response (S_0 initial section). Continuous curves: nonlocal model (l_c = 0.2 mm). Dashed curves: standard model*

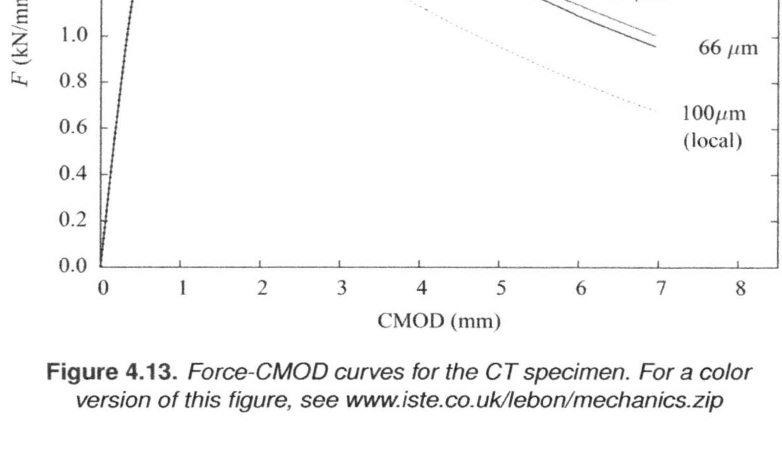

Figure 4.13. *Force-CMOD curves for the CT specimen. For a color version of this figure, see www.iste.co.uk/lebon/mechanics.zip*

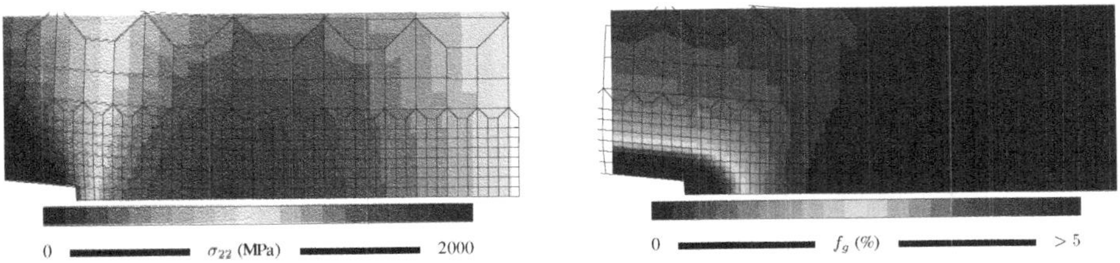

Figure 4.14. *Gaussian point values of opening stresses (σ_{22}) and damage (f_g). The elements for which porosity at fracture is reached have been removed. For a color version of this figure, see www.iste.co.uk/lebon/mechanics.zip*

4.7. Conclusion

To conclude this analysis, an example is proposed summarizing the main issues met in the field of ductile fracture modeling and simulation.

The example concerns tests on notched or pre-cracked thin sheets. The onset of fracture at the base of the notch or pre-crack results from the formation of a triangular region (i.e. the normal to the crack corresponds to the loading direction). In this area, large dimples are observed (Bron et al. 2004). Once the initial triangle is formed, the crack tilts and becomes slanted forming an angle of about 45° with the loading direction (Besson et al. 2013). In this area, the primary dimples are observed with smaller secondary dimples.

Figure 4.15 shows a cross-section by synchrotron radiation tomography of a tear test on an interrupted Kahn (ASTM 2001) specimen carried out on an aluminum plate 2139 T8 in the transition area between flat fracture and slanted fracture. The situation is shown schematically in Figure 4.16.

The problem was qualitatively analyzed in Bron et al. (2004); Bron and Besson (2006); Besson et al. (2013) using a finite element simulation. In inclined regions, the mean stress, $\sigma_{kk}/3$, is low and the strain state corresponds to conditions of flat strain in the propagation direction. The fracture of the samples is essentially controlled by a phenomenon of plastic instability, namely the localization of the plastic strain into a shear band. Near the base of the notch of the heavily notched samples, the average stress is higher and promotes cavity growth. In addition, the strain state differs greatly from the plane strain conditions thus delaying localization (Needleman and Rice 1978; Besson et al. 2001b). These two factors lead to internal necking between the primary cavities. Shear bands are avoided and secondary dimples are not observed.

Simulating the transition from flat to inclined fracture remains a difficult task. It was first carried out in Mathur et al. (1996) for studying dynamic crack growth, but no attempt was made to compare the results with actual tests. The transition was also modeled in Besson et al. (2001a) but it was not possible to simultaneously match the trajectories of the cracks and the load-displacement curves.

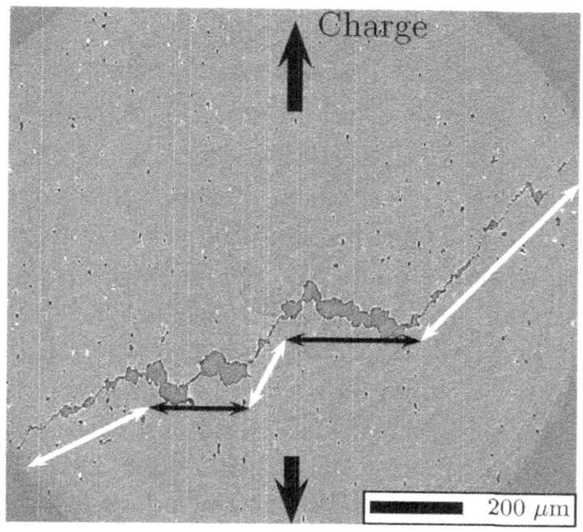

Figure 4.15. *Cross-section by synchrotron radiation tomography of an interrupted Kahn test showing the transition between a flat fracture (horizontal black arrows) and the slanted fracture (white arrows) (with the permission of T.F. Morgeneyer/aluminum sheet 2139 T8)*

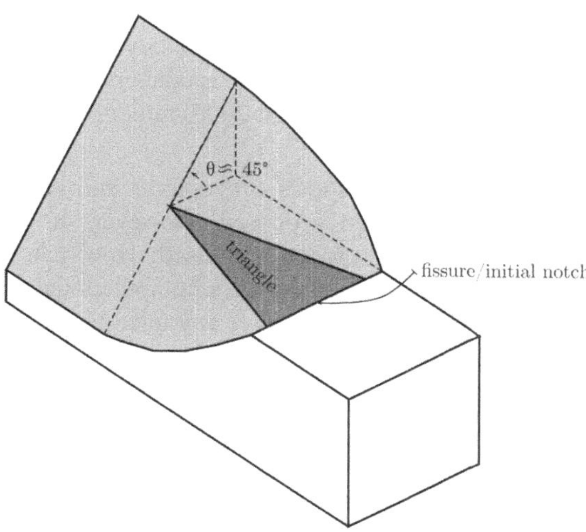

Figure 4.16. *Schematic view of the transition flat fracture/slanted fracture in sheet metal*

More recently, the transition between flat and inclined fracture was obtained by Xue (2007) and Xue and Wierzbicki (2008) using an explicit finite element simulation but also lacking a comparison with experiments.

The transition between flat and inclined fracture in sheet metal therefore raises several questions and highlights the need for research in the field of ductile fracture:

1) Slant fracture occurs at a relatively low stress triaxiality. Models, such as the GTN model, are representative of damage growth for a significant triaxiality for which cavity growth is the mechanism controlling damage kinetics. Low-triaxiality rupture has now been widely studied (Bao and Wierzbicki 2004; Barsoum and Faleskog 2007a, 2007b; Xue 2007; Marcadet and Mohr 2016; Roth and Mohr 2016). In particular, it was shown that, depending on the material, ductility can decrease when triaxiality shifts from $\frac{1}{3}$ (pure tension) to 0 (pure shear) (Bao and Wierzbicki 2004; Barsoum and Faleskog 2007a). These studies also highlighted the role of the Lode parameter on ductility (Zhang et al. 2001; Xue 2007; Defaisse et al. 2018). Consequently, the GTN model was modified to introduce a softening term depending on the Lode parameter (Nahshon and Hutchinson 2008) and there are many other models to replicate this phenomenon.

2) The problem of the origin of localization and fracture by *void sheeting* should be raised. Does localization occur first so that a second population of cavities appears inside the band or does nucleation of a second population trigger localization? Since the second cavity population associated with the *void sheeting* mechanism is in most cases not visible outside the localization band (Defaisse et al. 2018), the first explanation seems correct. However, secondary nucleation is not easy to detect, the question therefore remains open. The models proposed by Fabrègue and Pardoen (2008) and (Nahshon and Hutchinson 2008) rather correspond to the second hypothesis.

3) It can be seen that numerical models for reproducing the transition between flat and inclined fracture have been proposed. However, it is clear that their implementation is limited if local models are used. How can we be sure, for example, that the slant fracture obtained is not a computational artifact? Only the use of regularized models can help us to remove any uncertainty. These calculations have to be performed in 3D, and this results in very large resource requirements (CPU, storage, etc.). A solution could be found in using remeshing and parallel computing. Remeshing allowing for refining the mesh in the areas where the crack propagates, but also a de-refinement (Mediavilla et al. 2006a, 2006b; Feld-Payet et al. 2015) when the crack has propagated, should make it possible to reproduce large crack extensions by retaining the fine elements to represent the fields at the tip of the crack while maintaining a problem size enabling in practice computations to be performed within a reasonable time. An example of the implementation of this strategy is shown in Figure 4.17.

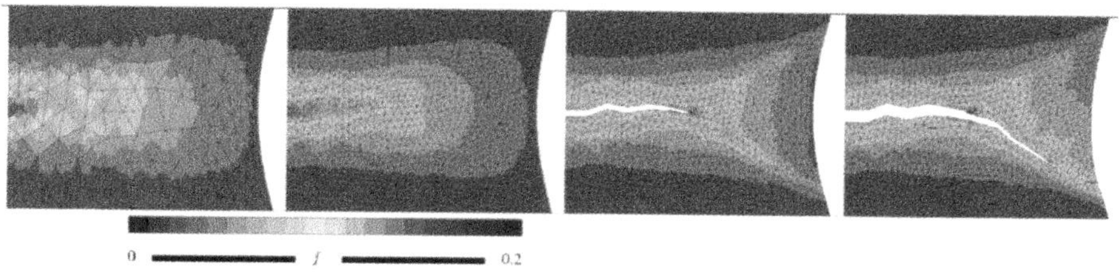

Figure 4.17. *Refinement example of the mesh being computed. The aim here is to model the so-called cup-cone fracture (source: Tvergaard and Needleman 1984; Besson et al. 2001b). For a color version of this figure, see www.iste.co.uk/lebon/mechanics.zip*

4.8. References

ASTM (2001). Standard B871-01.2001: Standard test method for tear testing of aluminium alloy products. Technical Report, ASTM-B871, ASTM-International, West Conshohocken.

ASTM (2008). Standard test method for measurement of fracture toughness. Technical Report, ASTM-1820, ASTM-International, West Conshohocken.

Auricchio, F., Beirao Da Veiga, L., Lovadina, C., Reali, A., Taylor, R., Wriggers, P. (2013). Approximation of incompressible large deformation elastic problems: Some unresolved issues. *Comput. Mech.*, 52, 1153–1167.

Bai, Y. and Wierzbicki, T. (2008). A new model of metal plasticity and fracture with pressure and Lode dependence. *Int. J. Plasticity*, 24(6), 1071–1096.

Bao, Y. and Wierzbicki, T. (2004). On fracture locus in the equivalent strain and stress triaxiality space. *Int. J. Mech. Sci.*, 46(1), 81–98.

Barsoum, I. and Faleskog, J. (2007a). Rupture mechanisms in combined tension and shear – Experiments. *Int. J. Solids Structures*, 44(6), 1768–1786.

Barsoum, I. and Faleskog, J. (2007b). Rupture mechanisms in combined tension and shear – Micromechanics. *Int. J. Solids Structures*, 44(17), 5481–5498.

Bazant, Z. and Pijaudier-Cabot, G. (1988). Non local continuum damage. Localization, instability and convergence. *J. Applied Mech.*, 55, 287–294.

Becker, R. and Smelser, R. (1994). Simulation of strain localization and fracture between holes in an aluminum sheet. *J. Mech. Phys. Solids*, 42(5), 777–796.

Bellet, M. (1999). Finite element analysis of compressible viscoplasticity using a three-field formulation. Application to metal powder hot compaction. *Comp. Meth. Appl. Mech. Engng.*, 175(1/2), 19–40.

Benzerga, A. (2002). Micromechanics of coalescence in ductile fracture. *J. Mech. Phys. Solids*, 50, 1331–1362.

Benzerga, A. and Besson, J. (2001). Plastic potentials for anisotropic porous solids. *Eur. J. Mech./A*, 20A(3), 397–434.

Benzerga, A. and Leblond, J.-B. (2010). Ductile fracture by void growth to coalescence. *Advances in Applied Mechanics*, 44, 169–305.

Berdin, C., Besson, J., Bugat, S., Desmorat, R., Feyel, F., Forest, S., Lorentz, E., Maire, E., Pardoen, T., Pineau, A. et al. (2004). *Local Approach to Fracture*. Presses de l'École des Mines, Paris.

Bergheau, J.-M., Leblond, J.-B., Perrin, G. (2014). A new numerical implementation of a second-gradient model for plastic porous solids, with an application to the simulation of ductile rupture tests. *Computer Methods in Applied Mechanics and Engineering*, 268, 105–125.

Besson, J. (2009). Damage of ductile materials deforming under multiple plastic or viscoplastic mechanisms. *Int. J. Plasticity*, 25, 2204–2221.

Besson, J., Devillers-Guerville, L., Pineau, A. (2000). Modeling of scatter and size effect in ductile fracture: Application to thermal embrittlement of duplex stainless steels. *Eng. Fract. Mech.*, 67(2), 169–190.

Besson, J., Brocks, W., Chabanet, O., Steglich, D. (2001a). Ductile rupture of aluminum sheet materials. *European Journal of Finite Elements*, 10, 401–415.

Besson, J., Steglich, D., Brocks, W. (2001b). Modeling of crack growth in round bars and plane strain specimens. *Int. J. Solids Structures*, 38(46/47), 8259–8284.

Besson, J., Steglich, D., Brocks, W. (2003). Modeling of plane strain ductile rupture. *Int. J. Plasticity*, 19(10), 1517–1541.

Besson, J., McCowan, C., Drexler, E. (2013). Modeling flat to slant fracture transition using the computational cell methodology. *Eng. Fract. Mech.*, 104, 80–95.

Billardon, R. and Doghri, I. (1989). Prediction of macro-crack initiation by damage localization. *C. R. Acad. Sci. Paris*, 308(4), 347–352.

Bonora, N. (1997). A nonlinear CDM model for ductile failure. *Eng. Fract. Mech.*, 58(1/2), 11–28.

Brepols, T., Wulfinghoff, S., Reese, S. (2017). Gradient-extended two-surface damage-plasticity: Micromorphic formulation and numerical aspects. *Int. J. Plasticity*, 97, 64–106.

Bron, F. and Besson, J. (2006). Simulation of the ductile tearing for two grades of 2024 aluminum alloy thin sheets. *Eng. Fract. Mech.*, 73, 1531–1552.

Bron, F., Besson, J., Pineau, A. (2004). Ductile rupture in thin sheets of two grades of 2024 aluminum alloy. *Materials Science and Engineering: A*, 380, 356–364.

Cao, T., Maziere, M., Danas, K., Besson, J. (2015). A model for ductile damage prediction at low stress triaxialities incorporating void shape change and void rotation. *Int. J. Solids Structures*, 63, 240–263.

Cazacu, O., Revil-Baudard, B., Chandola, N., Kondo, D. (2014). New analytical criterion for porous solids with tresca matrix under axisymmetric loadings. *Int. J. Solids Structures*, 51, 861–874.

Chaboche, J. (1993). Development of continuum damage mechanics for elastic solids sustaining anisotropic and unilateral damage. *Int. J. Damage Mech.*, 2, 311–329.

Chen, Y., Lorentz, E., Besson, J. (2020). Properties of a nonlocal GTN model within the context of small – Scale yielding. *Int. J. Plasticity*, 130, 102701.

Chu, C. and Needleman, A. (1980). Void nucleation effects in biaxially stretched sheets. *J. Eng. Mater. Technol.*, 102, 249–256.

Considère, A. (1885). Mémoire sur l'emploi du fer et de l'acier dans les constructions. *Annales des ponts et chaussées*, 574–775.

Cortese, L., Coppola, T., Campanelli, F., Campana, F., Sasso, M. (2014), Prediction of ductile failure in materials for onshore and offshore pipeline applications. *Int. J. Damage Mech.*, 23(1), 104–123.

Dalloz, A., Besson, J., Gourgues-Lorenzon, A.-F., Sturel, T., Pineau, A. (2009). Effect of shear cutting on ductility of a dual phase steel. *Eng. Fract. Mech.*, 76(10), 1411–1424.

Danas, K. and Aravas, N. (2012). Numerical modeling of elasto-plastic porous materials with void shape effects at finite deformations. *Composites Part B: Engineering*, 43(6), 2544–2559.

Defaisse, C., Mazière, M., Marcin, L., Besson, J. (2018). Ductile fracture of an ultra-high strength steel under low to moderate stress triaxiality. *Eng. Fract. Mech.*, 194, 301–318.

Desmorat, R., Gatuingt, F., Jirásek, M. (2015). Nonlocal models with damage-dependent interactions motivated by internal time. *Eng. Fract. Mech.*, 142, 255–275.

Diamantopoulou, E., Liu, W., Labergere, C., Badreddine, H., Saanouni, K., Hu, P. (2017). Micromorphic constitutive equations with damage applied to metal forming. *Int. J. Damage Mech.*, 26(2), 314–339.

Elguedj, Y., Bazilevs, Y., Calo, V., Hughes, T. (2008). $\overline{\text{B}}$ and $\overline{\text{F}}$ projection methods for nearly incompressible linear and non-linear elasticity and plasticity using higher-order NURBS. *Computer Methods in Applied Mechanics and Engineering*, 197, 2732–2762.

Enakoutsa, K. and Leblond, J.-B. (2009). Numerical implementation and assessment of the glpd micromorphic model of ductile rupture. *Eur. J. Mech./A*, 28, 445–460.

Enakoutsa, K., Leblond, J., Perrin, G. (2007). Numerical implementation and assessment of a phenomenological nonlocal model of ductile rupture. *Computer Methods in Applied Mechanics and Engineering*, 196(13/16), 1946–1957.

Engelen, R., Geers, M., Baaijens, F. (2003). Nonlocal implicit gradient-enhanced elasto-plasticity for the modelling of softening behaviour. *Int. J. Plasticity*, 19, 403–433.

Fabrègue, D. and Pardoen, T. (2008). A constitutive model for elastoplastic solids containing primary and secondary voids. *J. Mech. Phys. Solids*, 56, 719–741.

Faleskog, J. and Barsoum, I. (2013). Tension – Torsion fracture experiments – Part I: Experiments and a procedure to evaluate the equivalent plastic strain. *Int. J. Solids Structures*, 50(25), 4241–4257.

Faleskog, J., Gao, X., Shih, C. (1998). Cell model for nonlinear fracture analysis – I. Micromechanics calibration. *Int. J. Frac.*, 89, 355–373.

Feld-Payet, S., Feyel, F., Besson, J. (2011). Finite element analysis of damage in ductile structures using a nonlocal model combined with a three-field formulation. *Int. J. Damage Mech.*, 20, 655–680.

Feld-Payet, S., Chiaruttini, V., Besson, J., Feyel, F. (2015). A new marching ridges algorithm for crack path tracking in regularized media. *Int. J. Solids Structures*, 71, 57–69.

Forest, S. (2009). Micromorphic approach for gradient elasticity, viscoplasticity, and damage. *J. Eng. Mech.*, 135, 117–131.

Garrison, W.M. and Moody, N.R. (1987). Ductile fracture. *J. Phys. Chem. Solids*, 48(11), 1035–1074.

Gay, C., Montmitonnet, P., Coupez, T., Chenot, J. (1994). Test of an element suitable for fully automatic remeshing in 3D elastoplastic simulation of cold forging. *J. Mater. Processing Technol.*, 45, 683–688.

Geers, M., de Borst, R., Brekelmans, W., Peerlings, R. (1998). Strain-based transient-gradient damage model for failure analyses. *Comp. Meth. Appl. Mech. Engng*, 160, 133–153.

Germain, P., Nguyen, Q., Suquet, P. (1983). Continuum thermodynamics. *J. Applied Mech.*, 5, 1010–1020.

Gologanu, M., Leblond, J., Devaux, J. (1993). Approximate models for ductile metals containing non-spherical voids – Case of axisymmetric prolate ellipsoidal cavities. *J. Mech. Phys. Solids*, 41(11), 1723–1754.

Gologanu, M., Leblond, J., Devaux, J. (1994). Approximate models for ductile metals containing non-spherical voids – Case of axisymmetric oblate ellipsoidal cavities. *J. Eng. Mater. Technol.*, 116, 290–297.

Gologanu, M., Leblond, J., Devaux, J. (2001a). Theoretical models for void coalescence in porous ductile solids. II. Coalescence "in columns". *Int. J. Solids Structures*, 38, 5595–5604.

Gologanu, M., Leblond, J., Perrin, G., Devaux, J. (2001b). Theoretical models for void coalescence in porous ductile solids. I. Coalescence "in layers". *Int. J. Solids Structures*, 38, 5581–5594.

Gologanu, M., Leblond, J.-B., Perrin, G., Devaux, J. (2007). *Continuum Micromechanics*. Springer, Cham.

Grange, M., Besson, J., Andrieu, E. (2000). An anisotropic Gurson model to represent the ductile rupture of hydrided Zircaloy-4 sheets. *Int. J. Frac.*, 105(3), 273–293.

Gu, G. and Mohr, D. (2015). Anisotropic Hosford-Coulomb fracture initiation model: Theory and application. *Eng. Fract. Mech.*, 147, 480–497.

Gurson, A.L. (1977). Continuum theory of ductile rupture by void nucleation and growth: Part I – Yield criteria and flow rules for porous ductile media. *J. Eng. Mater. Technol.*, 99, 2–15.

Halm, D. and Dragon, A. (1998). An anisotropic model of damage and frictional sliding for brittle materials. *Eur. J. Mech./A*, 17A(3), 439–460.

Han, X., Besson, J., Forest, S., Tanguy, B., Bugat, S. (2013). A yield function for single crystals containing voids. *Int. J. Solids Structures*, 50, 2115–2131.

Hancock, J. and Mackenzie, A. (1976). Mechanisms of ductile failure in high-strength steels subjected to multi-axial stress states. *J. Mech. Phys. Solids*, 24, 147–160.

Hannard, F., Simar, A., Maire, E., Pardoen, T. (2018). Quantitative assessment of the impact of second phase particle arrangement on damage and fracture anisotropy. *Acta Mater.*, 148(15), 456–466.

Haynes, M. and Gangloff, R. (1998). Temperature-dependent void-sheet fracture in Al-Cu-Mg-Ag-Zr. *Metall. Mater. Trans. A*, 29A, 1599–1613.

Huetter, G. (2017). A micromechanical gradient extension of gurson's model of ductile damage within the theory of microdilatational media. *Int. J. Solids Structures*, 110/111, 15–23.

Hughes, T. (1980). Generalization of selective integration procedures to anisotropic and nonlinear media. *Int. J. Numer. Meth. Engng*, 15, 1413–1418.

Hure, J. and Barrioz, P.-O. (2016). Theoretical estimates for flat voids coalescence by internal necking. *Eur. J. Mech./A*, 60, 217–226.

Hutchinson, J.W. (1968). Plastic stress and strain fields at a crack tip. *J. Mech. Phys. Solids*, 16, 337–342.

Jayadevan, K., Berg, E., Thaulow, C., Østby, E., Skallerud, B. (2006). Numerical investigation of ductile tearing in surface cracked pipes using line-springs. *Int. J. Solids Structures*, 43(7/8), 2378–2397.

Jinka, A., Bellet, M., Fourment, L. (1997). A new three-dimensional finite element model for the simulation of powder forging processes: Application to hot forming of P/M connecting rod. *Numerical Methods in Engineering*, 40(21), 3955–3978.

Kachanov, L. (1958). Time of the rupture process under creep conditions. *Isv. Akad. Nauk. SSR. Otd Tekh. Nauk.*, 8, 26–31.

Keralavarma, S., Hoelscher, S., Benzerga, A. (2011). Void growth and coalescence in anisotropic plastic solids. *Int. J. Solids Structures*, 48(11/12), 1696–1710.

Kofiani, K., Nonn, A., Wierzbicki, T. (2013). New calibration method for high and low triaxiality and validation on SENT specimens of API X70. *Int. J. of Pressure Vessels and Piping*, 111, 187–201.

Koplik, J. and Needleman, A. (1988). Void growth and coalescence in porous plastic solids. *Int. J. Solids Structures*, 24(8), 835–853.

Krajcinovic, D. (1996). *Damage Mechanics*. Elsevier, Amsterdam.

Leblond, J., Perrin, G., Suquet, P. (1994). Exact results and approximate models for porous viscoplastic solids. *Int. J. Plasticity*, 10(3), 213–235.

Lemaitre, J. (1985). A continuous damage mechanics model for ductile fracture. *J. Eng. Mater. Technol.*, 107, 83–89.

Lemaitre, J. (1996). *A Course on Damage Mechanics*. Springer, Berlin.

Lemaitre, J. and Desmorat, R. (2005). *Engineering Damage Mechanics*. Springer, Berlin.

Lemaitre, J., Desmorat, R., Sausay, M. (2000). Anisotropic damage laws of evolution. *Eur. J. Mech./A*, 19A, 187–208.

Ling, C., Besson, J., Forest, S., Tanguy, B., Latourte, F., Bosso, E. (2016). An elastoviscoplastic model for porous single crystals at finite strains and its assessment based on unit cell simulations. *Int. J. Plasticity*, 84, 58–87.

Lorentz, E. and Andrieux, S. (1999). A variational formulation for nonlocal damage models. *Int. J. Plasticity*, 15(2), 119–138.

Lorentz, E. and Benallal, A. (2005). Gradient constitutive relations: Numerical aspects and application to gradient damage. *Computer Methods in Applied Mechanics and Engineering*, 194(50–52), 5191–5220.

Lorentz, E. and Godard, V. (2011). Gradient damage models: Toward full-scale computations. *Computer Methods in Applied Mechanics and Engineering*, 200, 1927–1944.

Luo, M., Dunand, M., Mohr, D. (2012). Experiments and modeling of anisotropic aluminum extrusions under multi-axial loading – Part II: Ductile fracture. *Int. J. Plasticity*, 32/33, 36–58.

Luu, T. (2006). Déchirure ductile des aciers à haute résistance pour gazoducs (X100). PhD Thesis, École des Mines de Paris, Paris.

Madou, K. and Leblond, J.-B. (2013). Numerical studies of porous ductile materials containing arbitrary ellipsoidal voids – I: Yield surfaces of representative cells. *Eur. J. Mech./A*, 42, 480–489.

Madou, K., Leblond, J.-B., Morin, L. (2013). Numerical studies of porous ductile materials containing arbitrary ellipsoidal voids II: Evolution of the length and orientation of the void axes. *Eur. J. Mech./A*, 42, 490–507.

Marcadet, S. and Mohr, D. (2016). Critical hardening rate model for predicting path-dependent ductile fracture. *Int. J. Frac.*, 200, 77–98.

Marini, B., Mudry, F., Pineau, A. (1985). Experimental study of cavity growth in ductile rupture. *Eng. Fract. Mech.*, 22(6), 989–996.

Mathur, K., Needleman, A., Tvergaard, V. (1996). Three dimensional analysis of dynamic ductile crack growth in a thin plate. *J. Mech. Phys. Solids*, 44(3), 439–464.

Mazière, M. and Forest, S. (2015). Strain gradient plasticity modeling and finite element simulation of lüders band formation and propagation. *Continuum Mech. Thermodyn.*, 27, 83–104.

Mbiakop, A., Constantinescu, A., Danas, K. (2015). An analytical model for porous single crystals with ellipsoidal voids. *J. Mech. Phys. Solids*, 84, 436–467.

Mear, M. and Hutchinson, J. (1985). Influence of yield surface curvature on flow localization in dilatant plasticity. *Mech. Mater.*, 4, 395–407.

Mediavilla, J., Peerlings, R., Geers, M. (2006a). An integrated continuous-discontinuous approach towards damage engineering in sheet metal forming processes. *Eng. Fract. Mech.*, 73(7), 895–916.

Mediavilla, J., Peerlings, R., Geers, M. (2006b). Discrete crack modelling of ductile fracture driven by non-local softening plasticity. *International Journal for Numerical Methods in Engineering*, 66(4), 661–688.

Mediavilla, J., Peerlings, R., Geers, M. (2006c). A nonlocal triaxiality-dependent ductile damage model for finite strain plasticity. *Computer Methods in Applied Mechanics and Engineering*, 195(33–36), 4617–4634.

Medjo, B., Rakin, M., Gubeljak, N., Matvienko, Y., Arsić, M., Šarkoćević, Ž., Sedmak, A. (2015). Failure resistance of drilling rig casing pipes with an axial crack. *Eng. Fail. Anal.*, 58(2), 429–440.

Miehe, C., Apel, N., Lambrecht, M. (2002). Anisotropic additive plasticity in the logarithmic strain space: Modular kinematic formulation and implementation based on incremental minimization principles for standard materials. *Computer Methods in Applied Mechanics and Engineering*, 191, 5383–5425.

Morgeneyer, T.F. and Besson, J. (2011). Flat to slant ductile fracture transition: Tomography examination and simulations using shear-controlled void nucleation. *Scripta Metall. Mater.*, 65, 1002–1005.

Nahshon, K. and Hutchinson, J. (2008). Modification of the Gurson model for shear failure. *Eur. J. Mech./A*, 27A, 1–17.

Needleman, A. and Rice, J. (1978). Limits to ductility set by plastic flow localization. In *Mechanics of Sheet Metal Forming*, Koistinen, E.D. (ed.). Plenum Publishing Corporation, New York.

Nicak, T., Schendzielorz, H., Keim, E., Meier, G. (2012). STYLE: Study on transferability of fracture material properties from small scale specimens to a real component. In *Proceedings of the ASME Pressure Vessels and Piping Conference*, Toronto.

Nourpanah, N. and Taheri, F. (2011). Ductile crack growth and constraint in pipelines subject to combined loadings. *Eng. Fract. Mech.*, 78(9), 2010–2028.

Novokshanov, D., Döbereiner, B., Sharaf, M., Münstermann, S., Lian, J. (2015). A new model for upper shelf impact toughness assessment with a computationally efficient parameter identification algorithm. *Eng. Fract. Mech.*, 148, 281–303.

Oh, C., Kim, Y., Baek, J., Kim, W. (2007). Development of stress-modified fracture strain for ductile failure of API X65 steel. *Int. J. Frac.*, 143, 119–133.

Paux, J., Morin, L., Brenner, R., Kondo, D. (2015). An approximate yield criterion for porous single crystals. *Eur. J. Mech./A*, 51, 1–10.

Peerlings, R., De Borst, R., Brekelmans, W., De Vree, J., Spee, I. (1996). Some observations on localisation in non-local and gradient damage models. *Eur. J. Mech./A*, 15A(6), 3391–3403.

Petit, T., Besson, J., Ritter, C., Colas, K., Helfen, L., Morgeneyer, T. (2019). Effect of hardening on toughness captured by stress-based damage nucleation in 6061 aluminum alloy. *Acta Mater.*, 180, 349–365.

Pijaudier-Cabot, G. and Bazant, Z.P. (1987). Nonlocal damage theory. *J. Engrg. Mech.*, 113, 1512–1533.

Pineau, A., Benzerga, A.A., Pardoen, T. (2016). Failure of metals I: Brittle and ductile fracture. *Acta Mater.*, 107, 424–483.

Rastiello, G., Giry, C., Gatuingt, F., Desmorat, R. (2018). From diffuse damage to strain localization from an Eikonal Non-Local (ENL) continuum damage model with evolving internal length. *Computer Methods in Applied Mechanics and Engineering*, 331, 650–674.

Rice, J.R. (1976). The localisation of plastic deformation. In *Proc. 14th Int. Conf. Theoretical and Applied Mechanics*. IUTAM, Delft/Amsterdam.

Rice, J.R. (1980). The mechanics of earthquake rupture. In *Proceedings of the International School of Physics "Enrico Fermi"*. IOS Press, Amsterdam.

Rice, J.R. and Rosengren, G.F. (1968). Plane strain deformation near a crack tip in a power-law hardening material. *J. Mech. Phys. Solids*, 16, 1–12.

Rice, J.R. and Tracey, D.M. (1969). On the ductile enlargement of voids in triaxial stress fields. *J. Mech. Phys. Solids*, 17, 201–217.

Ritchie, R., Knott, J., Rice, J. (1973). On the relationship between critical tensile stress and fracture toughness in mild steel. *J. Mech. Phys. Solids*, 21, 395–410.

Roth, C. and Mohr, D. (2016). Ductile fracture experiments with locally proportional loading histories. *Int. J. Plasticity*, 79, 328–354.

Rousselier, G. (1987). Ductile fracture models and their potential in local approach of fracture. *Nucl. Eng. Des.*, 105, 97–111.

Rudnicki, J. and Rice, J. (1975). Conditions for the localization of deformation in pressure-sensitive dilatant materials. *J. Mech. Phys. Solids*, 23, 371–394.

Shinohara, Y., Madi, Y., Besson, J. (2016). Anisotropic ductile failure of a high-strength line pipe steel. *Int. J. Frac.*, 197, 127–145.

Soret, C., Madi, Y., Gaffard, V., Besson, J. (2017). Local approach to fracture applied to the analysis of a full size test on a pipe containing a girth weld defect. *Eng. Fail. Anal.*, 82, 404–419.

Tang, H., Fairchild, D., Panico, M., Crapps, J., Cheng, W. (2014). Strain capacity prediction of strain-based pipelines. In *Proceedings of the 10th International Pipeline Conference – 2014*. ASME, Calgary.

Tanguy, B., Luu, T., Perrin, G., Pineau, A., Besson, J. (2008). Plastic and damage behavior of a high strength X100 pipeline steel: Experiments and modelling. *Int. J. of Pressure Vessels and Piping*, 85(5), 322–335.

Thomason, F. (1985a). A three-dimensional model for ductile fracture by the growth and coalescence of microvoids. *Acta Metall.*, 33(6), 1087–1095.

Thomason, F. (1985b). Three-dimensional models for the plastic limit-loads at incipient failure of the intervoid matrix in ductile porous solids. *Acta Metall.*, 33(6), 1079–1085.

Torki, M., Tekoğlu, C., Leblond, J.-B., Benzerga, A. (2017). Theoretical and numerical analysis of void coalescence in porous ductile solids under arbitrary loadings. *Int. J. Plasticity*, 81, 160–181.

Tvergaard, V. (1990). Material failure by void growth to coalescence. *Advances in Applied Mechanics*, 27, 83–151.

Tvergaard, V. and Needleman, A. (1984). Analysis of the cup-cone fracture in a round tensile bar. *Acta Metall.*, 32, 157–169.

Xia, L. and Shih, C.F. (1995). Ductile crack growth – I. A numerical study using computational cells with microstructurally-based length scales. *J. Mech. Phys. Solids*, 43, 233–259.

Xue, L. (2007). Damage accumulation and fracture initiation in uncracked ductile solids subject to triaxial loading. *Int. J. Solids Structures*, 44, 5163–5181.

Xue, L. and Wierzbicki, T. (2008). Ductile fracture initiation and propagation modeling using damage plasticity theory. *Eng. Fract. Mech.*, 75(11), 3276–3293.

Yamamoto, H. (1978). Conditions for shear band localization in the ductile fracture of void-containing materials. *Int. J. Frac.*, 14(4), 347–365.

Zhang, Z., Thaulow, C., Ødegård, J. (2000). A complete Gurson model approach for ductile fracture. *Eng. Fract. Mech.*, 67(2), 155–168.

Zhang, K., Bai, J., Francois, D. (2001). Numerical analysis of the influence of the lode parameter on void growth. *Int. J. Solids Structures*, 38(32/33), 5847–5856.

Zhang, Y., Lorentz, E., Besson, J. (2018). Ductile damage modelling with locking-free regularised gtn model. *International Journal for Numerical Methods in Engineering*, 113(13), 1871–1903.

5

Quasi-brittle Fracture Modeling

Éric LORENTZ
EDF Lab Paris-Saclay, EDF R&D/ERMES, Palaiseau, France

5.1. What are the approaches for predicting quasi-brittle fracture?

In the industrial field, the risk of brittle or quasi-brittle fracture is thoroughly examined as it is often associated with an abrupt event. More precisely, the term *quasi-brittle fracture* refers to a mechanism of material degradation under monotonic loading in which the energy consumed by fracture is moderate, as opposed to a *ductile fracture*[1]. The dissipative phenomena prior to the occurrence of crack develop in a region known as *fracture process zone* (FPZ), whose extension remains *confined*. Materials subject to quasi-brittle fracture include geomaterials (rocks and concretes), ceramics, glass, graphite or metals at low temperatures.

A first modeling approach consists of defining a failure threshold in stress or strain space and examining, for each loading level, whether this threshold is reached at a point in the structure under study (Cowan 1952; Rinehart 1966). This approach focuses on the conditions under which damage initiates and does not address the subsequent redistributions of mechanical fields. This is relevant if the initiation leads to global fracture of the structure or if any damage is prohibited for regulatory issues, too pessimistic otherwise. In addition, this approach faces a limitation when there is a prior crack in the structure: the latter actually results in stress singularity, so that the fracture threshold is reached at the onset of loading. Pragmatic solutions are then usually implemented by excluding an area around the crack tip (whose size

1 This definition is deliberately simplistic at this stage. Strictly speaking, the brittle nature also depends on other properties of the material, in particular stiffness and fracture stress.

Numerical Methods for Strong Nonlinearities in Mechanics,
coordinated by Jacques BESSON, Frédéric LEBON and Eric LORENTZ. © ISTE Ltd 2024.

becomes a new model parameter). It should be noted that extensions were proposed to account for the statistical nature of brittle fracture, associated with the probability of finding (or not) a microscopic critical defect, in size or orientation (Beremin et al. 1983). An ergodicity assumption makes it possible to reduce the problem to the calculation of weighted spatial averages which can also, in some cases, provide solutions to the problem of stress singularities at crack tips. However, again, only the onset phase of fracture is modeled.

An alternative modeling approach for configurations with cracks is provided by linear elastic fracture mechanics (LEFM) in which the FPZ is assimilated to a point, the behavior of the rest of the structure being considered linear elastic (Griffith 1921; Irwin 1957; Leblond 2003). It is therefore a priori appropriate for structures whose size is large compared with the actual FPZ. Widely used in industrial practice for metallic materials, it is reduced to the calculation of an energy restitution rate, that is, the energy returned by the structure during infinitesimal crack propagation relatively to the crack area created, to be compared with a toughness or fracture (surface) energy, characteristic of the material; when the energy restitution rate, which increases with loading, reaches the material fracture energy, the crack is likely to initiate. It should be noted that in elasticity, the energy restitution rate is equal to the path integral J introduced in Cherepanov (1967) and Rice (1968), which provides a convenient way to calculate it. In the absence of history effects other than crack development, this theory makes it possible to treat not only the onset phase but also the subsequent propagation. However, because of its simplicity, it also faces several limitations, including (i) the necessary presence of an initial crack; (ii) a simplistic consideration of the stress state in the crack zone (typically, stress triaxiality is not taken into account); and (iii) a perfectible transferability from laboratory specimen measurements to industrial structures, which leads to, among other things, a misrepresentation of scale effects.

The first of these limitations was solved using the variational fracture approach (Francfort and Marigo 1998) in which not only infinitesimal but also finite crack propagations are considered. The problem is then expressed as the minimization of an energy with respect to displacements and potential crack configurations. In such form, crack creation becomes possible, contrary to the energy restitution rate formulation. Since minimization with respect to crack configurations can be numerically challenging, regularized formulations have been introduced (Bourdin et al. 2000) in which cracks are "thickened". These are the phase-field fracture models, which have grown in popularity in recent years. Nevertheless, issues (ii) and (iii) still persist; they are inherent to the fact that the FPZ has been restricted to a single point.

Other models seek to combine the energy restitution rate and a stress threshold. Concerning the onset, the works by Leguillon (2002) and Charlotte et al. (2006) can be mentioned, for example, which propose a criterion covering both situations with and without initial cracks. Finer models of the FPZ have been used for the same purpose. These are the cohesive zone models (Barenblatt 1959; Dugdale 1960; Hillerborg et al. 1976): the FPZ is represented as a surface through which displacement discontinuities can develop, which are related to the corresponding stress vector via a dedicated constitutive relation. The latter encompasses a fracture energy and a critical stress, thus bridging the gap between stress and energy formulations. These models will not be addressed in this study. Nevertheless, it can be noted here that they cover all the stages of failure (initiation, onset, propagation and final failure), that they are comparable with other material non-linearities, but that they generally require knowledge of the geometric crack path, which constitutes a major issue (theoretical and practical) when it cannot be a priori established (Oliver and Huespe 2004; Fraternali 2007; Jäger et al. 2008; Oliver et al. 2014; Ferté et al. 2016). In particular, a search for the path along all finite element faces leads to strong mesh dependency (de Borst 2003; Negri 2005), which compromise the predictive nature of the approach.

Finally, damage models represent the last family of approaches for predicting quasi-brittle fracture (see, Marigo 1981; Mazars 1986; Lemaitre et al. 2009). Probably the most complex in numerical terms, they consist of describing the local degradation state of the material and its evolution. These models are also potentially the richest since they cover a priori without restriction any potential distribution of damage, and therefore any fracture scenario. In addition, these models are expressed as constitutive relations, which allows couplings with other physical, mechanical or not, phenomena to be taken into account. However, for quasi-brittle materials, all these constitutive models share a common physically unavoidable characteristic: they exhibit softening, that is to say that the sustained level of stress – the capacity of resistance – decreases with the evolution of damage. Other material properties can also gradually degrade, such as stiffness, and irreversible strains are likely to appear. Softening is at the origin of several issues:

– Since the stresses are bounded through the damage threshold, limit load may appear, which refers to the existence or not of solutions.

– When a point is damaged, the stresses are reduced. Since it is a priori similar in its vicinity because of structure equilibrium, the surrounding points – which had not reached the threshold – are not damaged. A competition between neighboring material points can take place; damage is spatially concentrated – this is referred to as *damage localization* – and bifurcations or losses of uniqueness may occur.

– According to the above-mentioned mechanism, the areas that are damaged dissipate some energy but the surrounding areas experience a stress decrease, hence releasing elastic energy. If the latter prevails over the dissipated energy, the structural response becomes unstable.

– The phenomenon of macroscopic damage localization develops at a scale interacting with that of the underlying microstructures at the macroscopic material point. The behaviors of neighboring material points become interdependent, with the consequence that constitutive laws are no longer expressed only on the basis of local quantities.

– Softening can lead to total degradation of the material point. How can the constitutive model, which is a volume-based model in essence, be consistent with the reality of a crack (surface)?

Modeling choices and numerical resolution methods must adapt to these characteristics, knowing that so far, these physical, mathematical and numerical problems have not yet found fully satisfactory answers. As a practical matter, therefore, it is recommended that one of the other approaches discussed in this section be preferred when the problem is suitable. However, some situations nonetheless require the use of damage models; this will be the subject of the rest of this chapter. First, the emergence of internal lengths in the constitutive laws is reconsidered. Then an overview of the formulations capable of expressing such non-local behavior is presented. Other issues common to quasi-brittle materials are then examined: anisotropy, the unilateral nature of microcracks and the ultimate fracture behavior. Finally, some of the numerical issues related to the simulation of quasi-fragile structures are exposed.

5.2. Materials with internal lengths

5.2.1. *Localization and non-locality*

Damage to a material usually leads to a decrease in its strength capacity. The range of admissible stresses, that is those that the material can withstand, is led to shrink as damage increases. This is called softening or negative strain hardening, as opposed to hardening observed in the presence of plasticity. This physical characteristic common to a large number of damage mechanisms is at the origin of the localization phenomenon of damage in bands of moderate thickness; this is evidenced by the identification of damage sites through acoustic measurements or by digital image correlation (DIC) (see, for example, Alam et al. (2014) for an experimental study on concrete). This localization is explained as follows (see Figure 5.1 for a one-dimensional illustration)[2]:

1) As a result of excessive loading, damage begins to appear in a small area of the structure.

2) Stress is brought to decrease in this area because of softening.

3) Stress equilibrium in the structure also leads to a decrease in stress around the damaged area.

4) An elastic unloading occurs on the periphery of the damaged area, thus confining the damage to a band.

5) In the localization band (in red in Figure 5.1), damage and strain continue to increase (red curve) while strain decreases in neighboring areas by elastic unloading (blue curve).

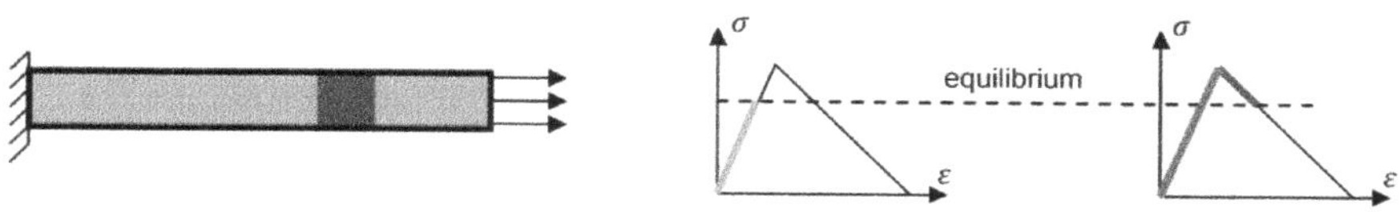

Figure 5.1. *Damage localization (in red) in a uniaxial tensile bar. For a color version of this figure, see www.iste.co.uk/lebon/mechanics.zip*

This damage and strain concentration therefore appears as a consequence not only of softening at the level of the material point but also of the equilibrium of stresses at the structural level. It has a major consequence on the expression of constitutive relations at the macroscopic level. The latter systematically results from a homogenization – explicit or not – of the response of the underlying microstructures at the macroscopic material point under consideration. However, the usual approaches to homogenization are based on a fundamental assumption of *scale separation* between mean macroscopic fields and microscopic fields. Unfortunately, in the presence of localization, gradients of the damage field develop at a scale comparable to that of microstructures. Typically, coming back to the example of concrete, the width of the damage band observed in Alam et al. (2014) is 7 cm, which corresponds to about three times the size of the largest aggregates (i.e. the

2 The example of uniaxial tension in a bar, although very simple, is well suited to identify the characteristics and pathologies of a localized distribution of damage. It illustrates the damage profile transverse to the direction of cracking (but not propagation because it is perpendicular to the bar). It will be used several times in this study, in one form or another.

size of microstructures). The fundamental character of the assumption of scale separation appears in the definition of an elementary representative volume (ERV), as illustrated in Figure 5.2. Classical homogenization assumes that a volume three times larger than the chosen ERV remains representative or, in other words, that the mean fields vary little from one ERV to its neighbor, which corresponds to the situation in Figure 5.2(a): a dashed cell indeed corresponds to an ERV. Unfortunately, localization leads more naturally to a situation of the type shown in Figure 5.2(b), in which an ERV cannot be isolated.

Contrary to what classical homogenization would predict, a length related to the size of the microstructures – or the distance between the damage sites – emerges at the macroscopic scale. The resulting constitutive relation can be characterized as *non-local*, that is, the stresses and evolution of the internal variables at a point depend on strains and values of the internal variables in a whole neighborhood of that point. This influence distance (or influence distances, in an anisotropic case) is a parameter of the non-local constitutive model.

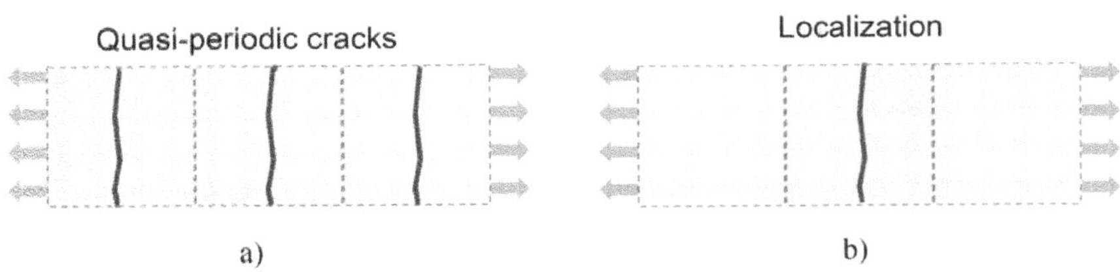

Figure 5.2. *Problems related to the definition of an elementary representative volume in the presence of localization. For a color version of this figure, see www.iste.co.uk/lebon/mechanics.zip*

Although this study exclusively focuses on localized damage, it should be noted that there are nevertheless situations in which the localization mechanism is deferred or even inactive. As a matter of fact, localization is fundamentally based on a repercussion of stress reduction resulting from softening in the entire vicinity of damage sites. In other words, damage must result in a relaxation of the surrounding stresses. However, when structural arrangements such as steel reinforcements in reinforced concrete are intended to take some of the (tensile strength) forces and transfer them to the surrounding areas, evenly spaced cracks can be seen to emerge, as shown in Figure 5.3. At a higher scale, this repetitive arrangement of cracks can be interpreted as diffuse or spread damage, as opposed to localized damage. Another case concerns crazing. Shrinkage strains (for example, of thermal or hydric origin) located near a wall cause tension on this area of the structure; shallow cracks appear but they only relax the stresses very locally, the extension of the neighborhood involved being related to the depth of the cracks. This results in a network of evenly spaced cracks which, again, can be interpreted on a higher scale as diffuse damage. When these cracks propagate in thickness, some of them end up shielding their neighbors which then stop progressing; at the macroscopic scale, a transition from a diffuse state to a localized state can be observed (Sicsic et al. 2014).

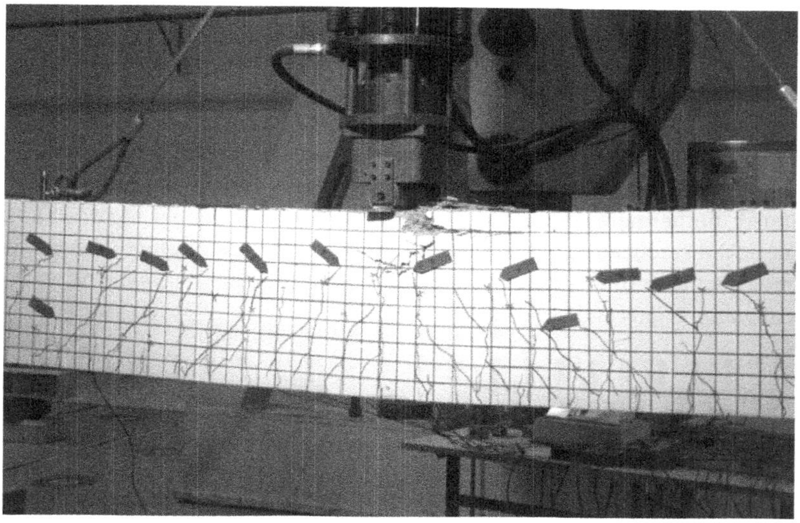

Figure 5.3. *Reinforced concrete: distribution of cracks (in red and blue) due to reinforcement (source: S. Ghavamian). For a color version of this figure, see www.iste.co.uk/lebon/mechanics.zip*

5.2.2. *Risks of ignoring the non-local nature inherent to damage*

Although the physical basis of homogenization advocates non-local constitutive models, it may be tempting, for practical reasons, to maintain the usual framework of local constitutive relations. Unfortunately, a mathematical analysis of the linear tangent problem – the one expressed in terms of velocity and not of displacement – at the structural scale indicates a possible loss of ellipticity, unlike the situation in which positive strain hardening constitutive relations are involved (see, for example, Benallal et al. (1993) or Chapter 4 of this book). As a reminder, the ellipticity criterion in the volume (which is supplemented by boundary and interface criteria) requires that at any point of the structure and for any direction n (in small strains):

$$\forall\, x \in \Omega \quad \forall\, n \quad \det(n \cdot \mathbb{H}(x) \cdot n) > 0 \qquad [5.1]$$

where $\mathbb{H}$ denotes the tangent operator of loading behavior ($\dot{\sigma} = \mathbb{H}:\dot{\varepsilon}$) and $n \cdot \mathbb{H} \cdot n$ is the acoustic tensor. With no ellipticity, the problem can become mathematically ill-posed and the existence of solutions (compatible with physical intuition) is questioned.

If the continuous problem in space is ill-posed, any approximation of its solutions by spatial discretization is also certain to fail. Typically, the use of finite elements results in solutions that do not converge with mesh refinement: the energy dissipated by damage tends toward zero, the thickness and orientation of the localization bands are dictated by those of the elements. An illustration of these pathologies is shown in Figure 5.4. The practice that consists of freezing the size of the elements (to introduce a scale parameter) is certainly appealing for its simplicity yet objectionable because the idea of continuous mechanical model is then abandoned; furthermore, the notion in which a mesh is no longer regarded as a tool of numerical discretization but as a description of physics is implicitly accepted. The predictive character of the model is then lost because the mesh sets the orientation of the localization bands and as such the crack path.

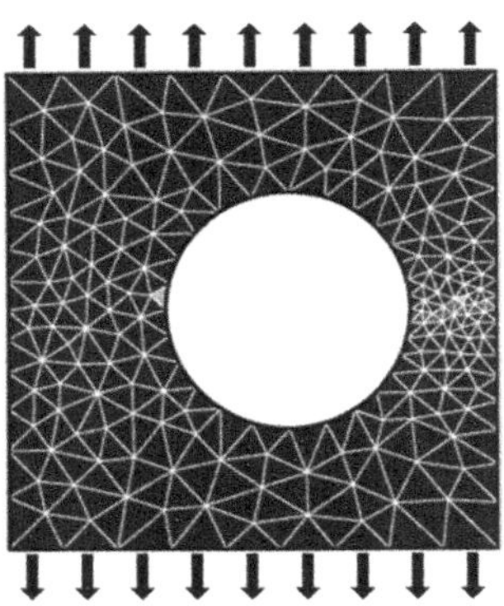
a)

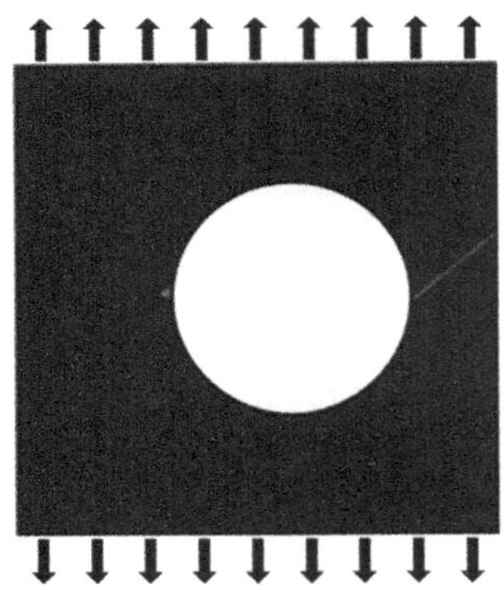
b)

Figure 5.4. *Location band dependency: (a) on mesh size and (b) mesh orientation. For a color version of this figure, see www.iste.co.uk/lebon/mechanics.zip*

Similarly, approaches link some of the parameters of the constitutive model (typically, the volume dissipation) to the mesh size in order to preserve a non-zero dissipated energy after mesh refinement. This method is attributed to Hillerborg, who emphasized the fundamental importance of preserving surface energy (Hillerborg et al. 1976). This is also referred to as *crack band* models (Bazant and Oh 1983). Through mesh refinement, it is expected to converge toward the solution of a well-posed problem (which is therefore not the initial continuous problem). Nonetheless, the dependency of the mesh on the orientation persists since the latter does not depend on the mesh size but on its orientation, which means that discrete problems do not in fact converge to any well-posed continuous problem.

The only possible recourse would be to postulate the mesh according to expected crack directions (or localization bands). However, if this information is available a priori, it makes sense to opt for cohesive zone models, better suited when the crack path is given.

5.2.3. *Limitations of a localization: regularization by viscosity*

Attempts seeking to avoid the non-local nature of damage have highlighted the challenge of properly controlling the energy dissipated. Zero fracture energy would cause extreme structural instabilities, as shown in section 5.2.3. Within the framework of quasi-static modeling, this would result in strain jumps in time, thus infinite strain rates $\dot{\varepsilon}$ in localization: bands of zero thickness. The introduction of viscosity into the constitutive law, whether real or for regularization purposes, opposes these excessive strain rates and offers a form of controlling the dissipated energy (Simo 1989). Also, the tangent problem (formulated in terms of rate) generally becomes well-posed due to viscosity, which makes it possible to express the evolution of the problem variables according to the current state (Cauchy's problem) and to circumvent the mathematical issues mentioned in section 5.2.2. This explains why the introduction of viscosity appears to be a practical means of smoothing damage problems (see, Needleman 1988; Allix and Deü 1997).

Is the solution physically admissible? By still remaining in a quasi-static framework, a first observation of a dimensional nature can be made: viscosity introduces a characteristic time but no length[3]. The width of the damage localization bands is therefore necessarily controlled by the characteristic dimensions of the structure and the loading rate. Linear perturbation analyses (Fressengeas and Molinari 1985; Anand et al. 1987) consolidated this observation since they indicated that beyond a certain mechanical state, any perturbation of the current solution is amplified (instability), regardless of the spatial scale at which it develops. These results were consistent with the one-dimensional numerical experiments carried out in Needleman (1988), where localization is triggered by imperfections whose size determines that of the width of the localization bands.

3 In dynamics, the velocity of the waves combined with the characteristic time of viscosity leads to the emergence of a length, provided that the loads are fast enough.

In the same spirit, a simple example makes it possible to understand how viscosity can contribute to the regularization of the problem and what are the limits according to the loading rate. To this end, we consider a bar, under tension and whose geometric defect locally concentrates the stress. The material is visco-plastic and softening, where τ denotes the characteristic viscosity time. The problem data set is gathered in Figure 5.5, where the Macauley brackets denote the positive part of a number: $\langle x \rangle \equiv (x + |x|)/2$. We want to know if the damage (expressed here in terms of plastic deformation p) remains confined to the level of the geometric defect or if it overflows over the entire bar. Three dimensionless parameters play a key role in this problem: the loading rate relative to the characteristic time of viscosity $\dot{U}^0\, \tau / U^0$ (U^0 being the displacement at the onset of damage), the amplitude of the perturbation ΔA and its scale relatively to the size of the structure γ.

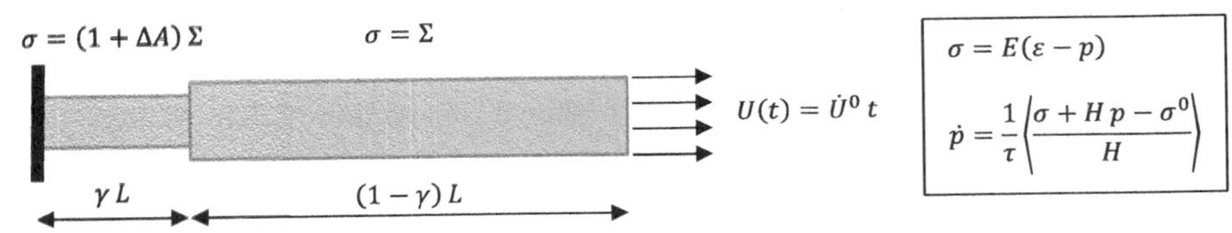

Figure 5.5. *Model problem for evaluating the effectiveness or the lack thereof of viscosity regularization*

This problem admits an analytical solution that is not exposed here. On the other hand, the following elements can be retained:

1) Due to the concentration (or amplification) of stresses in the defect, this is where the onset occurs. The rest of the bar has not yet reached the damage threshold; its behavior is elastic.

2) At the time of initiation, the evolution of strains and stresses remains governed by elasticity (this is the tangent operator) and the current mechanical state, therefore zero plastic strain: the stress in the defect therefore temporarily still increases.

3) This is a major difference with the problem without viscosity for which, as soon as damage occurs, softening takes place and leads to a decrease in stresses throughout the structure; the elastic area can then never be damaged. Here, as the stress continues to increase in the defect, the same occurs in the elastic area of the bar because of the equilibrium equation. This is the driving force behind viscosity regularization.

4) Softening eventually takes place in the defect and the stress decreases. The viscosity therefore delayed the decrease in the stress. The peak stress is all the higher, the more pronounced the viscous effects, that is, for large stress rates or

characteristic viscosity times. In the post-peak phase, the damage increases exponentially in time (until failure).

5) Damage remains confined to the defect if the peak stress is not high enough to cause a stress, in the nominal area of the bar, above the damage threshold. It depends on the peak stress (too low) and the perturbation amplitude (too strong). For a perturbation size γ that tends toward zero, damage remains confined in the defect at the following condition (from the analytical solution):

$$\frac{\dot{U}^0\,\tau}{U^0} \leq \frac{\Delta A}{\ln(1/\gamma)} \qquad [5.2]$$

If the material is not naturally viscous with respect to the loading rates, the introduction of a low artificial viscosity does not make it possible to regularize the problem on its own, since the scale of the localization bands is set by that of imperfections. This has an impact on finite element calculations: the width and path of the location bands remain set by the mesh, as evidenced by an example in Forest and Lorentz (2004). In any case, the strong (experimentally observed) gradients of macroscopic fields necessarily refer to a non-local constitutive model, whether viscous or not, because of the lack of scale separation.

5.2.4. *Characterization of the internal length: toward a fracture model*

The use of a non-local model requires knowledge of one (or more) internal length that should be a priori identified. This is all the more restrictive since the quantities usually accessible through standardized tests do not have a direct link with this internal length. On the other hand, it determines the size of the FPZ, that is, the area where the damage is active, shown in blue in Figure 5.6.

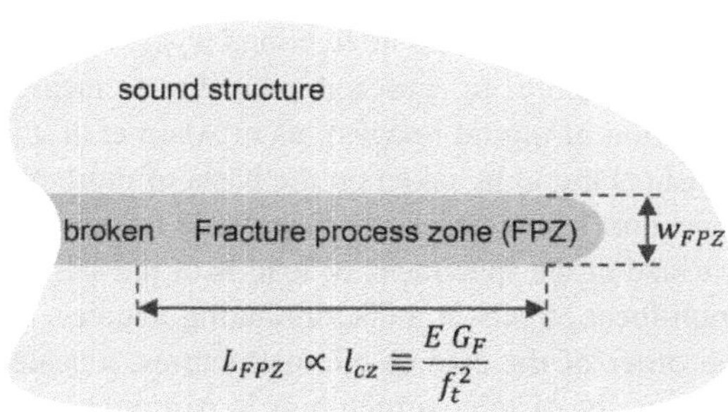

Figure 5.6. *Fracture process zone. For a color version of this figure, see www.iste.co.uk/lebon/mechanics.zip*

Indeed, the FPZ can be geometrically characterized by its width w_{FPZ}, measured in the transverse direction to the localization band, and by its length L_{FPZ}, measured in the direction of propagation.

The length L_{FPZ} results from the distribution of the stress field in front of the broken area (shown in orange in Figure 5.6). Results from cohesive model theory indicate that its size is of the order of magnitude of the cohesive length l_{cz} defined as follows in Hillerborg et al. (1976):

$$l_{cz} \equiv \frac{E\, G_F}{f_t^2} \qquad [5.3]$$

where E denotes Young modulus of the material, G_F is the fracture energy (the energy consumed to create a crack of 1 m^2) and f_t is the tensile strength (the maximal tensile stress that the material can withstand). For concrete, for example, empirical formulae originating from the code-model (FIB 2013) allow the length l_{cz} to be evaluated between 75 and 40 cm for concretes with compressive strength ranging from 30 to 50 MPa.

The width w_{FPZ} results from the phenomenon of macroscopic localization competing with the interaction between the microstructures causing the non-local character. It therefore depends on the size of the microstructures. An experimental study on concrete (Hadjab et al. 2007) leads, for example, to $w_{FPZ} \approx 2.75\, d_a$ with d_a the size of the largest aggregates (12 mm in their study). The width w_{FPZ} is thus a direct manifestation of the non-local character of which it provides the representative scale. It can be noted that w_{FPZ} also partly determines the fracture *surface* energy G_F since, by definition, the latter can be expressed as the product $\omega\, w_{FPZ}$, where ω is the average *volume* fracture energy (unfortunately not measurable by usual tests).

To determine the scale of non-local interactions w_{FPZ}, that is, the internal length of the model, it is possible to proceed via direct measurements (acoustic measurements, correlation of digital images), as in Alam et al. (2014). On the other hand, if indirect measures are to be taken on the basis of more global quantities (via an identification procedure), it is necessary to provide for an experimental program whose characteristic size of the specimens or load is of the same order of magnitude as w_{FPZ} to excite non-local effects in a discriminating manner. Unfortunately, w_{FPZ} is essentially of the order of the size of microstructures, a scale at which material heterogeneity (the disparity of microstructures) is particularly sensitive. The scale effects that we want to bring forward for identification can a priori be taken for experimental scattering (see the interlocking observed in Alam et al. (2014)). Moreover, regardless of these identification problems, the width w_{FPZ} may be too small compared to that of the structures studied to be usable in a simulation because it would require an excessively fine spatial discretization.

This observation raises the following question: is it possible to avoid taking w_{FPZ} into account? It is a priori possible under two conditions: (i) the scale w_{FPZ} is not excited by the loading nor the geometry of the structure and (ii) we are not looking for the details of damage distribution in the localized area, but only its effect on the rest of the structure, such as peak force, extension of the damaged area, residual stiffness, etc. It is then proposed to neglect w_{FPZ}, which refers to a cohesive zone model where the FPZ is concentrated on a surface and the fracture energy becomes a parameter in itself and not the product of an underlying energy density and width. This choice is justified by the fact that the predictions at the scale of the structure of a cohesive formulation prove quantitatively satisfactory, *at least* for quasi-brittle materials such as concrete (Petersson 1981; Elices et al. 2002; Hoover and Bazant 2014). The major disadvantage remains in the difficulty of predicting the crack path, which should be hopefully solved with a non-local damage model that allows the choice of an internal length (a scale w_{FPZ}) arbitrarily small without altering the predictions at the structural scale. It is therefore necessary that this non-local constitutive model leads to results close to those of a cohesive model when its internal length tends toward zero: the intrinsic ability of damage models to determine the crack path (the location of the points that are damaged) can be thus exploited while maintaining the relevant predictive capacities of cohesive models.

As for the internal length, it should be then simply taken small enough to have results that are no longer sensitive to it (i.e. close to convergence), but also not too small for computational cost reasons (spatial discretization must be all the finer as the internal length is small). The internal length becomes a numerical parameter, unrelated to the actual physical scale and therefore no longer needs to be identified. The feasibility of such an approach has been demonstrated in Lorentz et al. (2011, 2012), Freddi and Iurlano (2017) and Wu (2017) for concrete structures; an illustration thereof is provided in section 5.3.5. *Thick level set* (TLS) type models follows the same logic since an arbitrary form of damage distribution in the localization band is postulated, under the constraint of leading to the material effective fracture energy (Moës et al. 2011). Chapter 6 is devoted to them.

For large structures, the introduction of an internal length, also quite large, may seem to be tempting. Unfortunately, the internal length is constrained by an upper bound whose value depends on the model used but which remains of the order of the cohesive length l_{cz}. It results from a stability issue in the damage area which stipulates that the strain increases in amplitude for the points that are damaged. This implies in particular that the fracture energy density ω be greater than the elastic volume energy stored during the load increase phase. The first is about G_F/w_{FPZ}, the second about $f_t^2/2E$, which given equation [5.3] refers to:

$$w_{FPZ} \leq \frac{1}{2}\beta\, l_{cz} \qquad [5.4]$$

A model-dependent correction factor β applies to this relation because the stability constraint must be checked at any point on the localization band and at any time during the damage phase. It should be noted that the existence of such a maximal permissible length has long been identified for the Hillerborg type of approaches discussed in section 5.2.2, where it determines the size of the elements that can be used, independently of the other limitations which have been identified.

This idea of dealing with the internal length of the model as a small numerical parameter also inspired phase field formulations (see Bourdin et al. (2008) and associated references for further study). Most of the corresponding work is established with a problem where the length of the FPZ L_{FPZ} is also small compared to the characteristic structural and loading lengths. Neglecting it is tantamount to assimilating the FPZ to a single point (Sicsic and Marigo 2013): it is the framework of Griffith's linear fracture mechanics whose variational formulation was proposed in Francfort and Marigo (1998) and which corresponds to an infinite critical stress f_t (only the notion of fracture energy subsides). Phase field models regularize for numerical reasons the treatment of cracks by "thickening" them via an auxiliary field: if we picture the healthy material in white and the crack in black, this auxiliary field reflects the intermediate shades of gray[4]. The thickness of the cracks corresponds to the small-scale parameter, intended to tend toward zero. In section 5.3.4, it will be shown to what extent these formulations are analogous to a class of non-local damage models.

Finally, the approach outlined in this section underlines a question: are we talking about a non-local damage model or a regularized crack model? As long as there are no attempts to describe the fine distribution of the damage field (namely inside the localization band), the question remains philosophical in nature. At the scale of the targeted quantities of interest, the value of the internal length does not *ultimately* matter, as long as it remains small compared to the other geometric

4 The analogy is not trivial because these formulations originally emanate from a class of image processing problems (Mumford and Shah 1989).

quantities of the structure and load. Also, it does not matter which non-local model class is chosen, as long as the convergence toward a cohesive fracture model is assured.

5.3. Non-local formulations

Strictly speaking, knowledge of microstructures should make it possible to recover a non-local macroscopic constitutive model. Unfortunately, a homogenization theory adapted to this context still seems out of reach, despite a number of exciting works (Drugan and Willis 1996; Gologanu et al. 1997; Kouznetsova et al. 2002). Another approach could consist of conducting the entire simulation at the microscopic scale, for which the problem is a priori well-posed. If this may prove fruitful in an academic environment, it is probably premature from an industrial point of view, due, on the one hand, to the exorbitant computational cost and, on the other hand, to the lack of knowledge of the actual microstructures (which would lead to a multiplication of calculations for different possible realizations of microstructural distributions). The most pragmatic approach to date therefore remains phenomenological by proposing constitutive relations that incorporate non-local effects.

It is proposed here to illustrate some families of non-local formulations, adopting a restrictive framework for pedagogical reasons, even if the concepts presented are generic enough to be extended to a wide variety of material behaviors and loading. It is thus assumed that damage is responsible for the entire mechanical dissipation, thermal effects can be ignored, external loading is conservative, the operating regime is quasi-static and the assumption of small perturbations (small displacements, small rotations and small deformations) can be considered.

5.3.1. *Formulation of the mechanical problem at the structural scale*

5.3.1.1. *Equilibrium and state equations*

The non-local character of the material behavior makes it necessary to go beyond the scale of the material point to consider the structure as a whole at the cost of a certain formalism. In the framework of the previous assumptions, it is chosen to describe its state using displacement $u \in \mathcal{C}$ and damage fields a, where $\mathcal{C}$ designates all the kinematically admissible displacements.

We introduce $\mathcal{F}(u, a)$, the Helmholtz free energy of the structure, $\mathcal{P}_e$ the mechanical power transmitted to the system by the external environment, $\mathcal{W}_e(u)$ the potential of external forces and $\mathcal{C}^0$ the (vector) space of the admissible infinitesimal displacements (or rates up to the unit), which do not cause the support reactions related to the kinematic boundary conditions to work ($\mathcal{C}^0$ is the vector space tangent to $\mathcal{C}$ in u). The internal forces and associated equilibrium equations are characterized by the following variational equation, where the symbol $\star$ refers to the duality product:

$$\forall\ \delta u \in \mathcal{C}^0 \qquad \frac{\partial \mathcal{F}}{\partial u} \star \delta u = \frac{\partial \mathcal{W}_e}{\partial u} \star \delta u \qquad [5.5]$$

The differential $\mathcal{S} \equiv \partial \mathcal{F}/\partial u$ is a linear form that represents the internal forces in the structure (the stresses). According to some authors, the variational formulation of equilibrium [5.5] is called the principle of virtual work (PVW). Typically, for a local model, for which the Helmholtz free energy is the sum of point contributions Φ depending only on the values of the strain $\varepsilon = \hat{\varepsilon}(u)$ and damage fields a at these points, the PVW takes the following form, where we recognize the usual expression of the stress tensor $\sigma \equiv \partial \Phi/\partial \varepsilon$:

$$\begin{aligned} &\mathcal{F}_{loc}(u,a) = \int_\Omega \Phi\big(\varepsilon(u)(x), a(x)\big)\, dx \\ &\rightarrow \quad \forall\ \delta u \in \mathcal{C}^0 \quad \int_\Omega \frac{\partial \Phi}{\partial \varepsilon} : \hat{\varepsilon}(\delta u)\, d\Omega = \frac{\partial \mathcal{W}_e}{\partial u} \delta u \end{aligned} \qquad [5.6]$$

The mechanical power $\mathcal{P}_e$, to which the power of the support reactions can contribute when subjected to imposed rates, is also related to the variation of the free energy (equilibrium equation taking into account the support reactions):

$$\frac{\partial \mathcal{F}}{\partial u} \star \dot{u} = \mathcal{P}_e \qquad [5.7]$$

The mechanical dissipation $\mathcal{D}$ is then defined as follows in an isothermal situation:

$$\mathcal{D} \equiv \mathcal{P}_e - \dot{\mathcal{F}} = \mathcal{P}_e - \frac{\partial \mathcal{F}}{\partial u} \star \dot{u} - \frac{\partial \mathcal{F}}{\partial a} \star \dot{a} = -\frac{\partial \mathcal{F}}{\partial a} \star \dot{a} \qquad [5.8]$$

where [5.7] has been applied to reduce the expression. The differential $\mathcal{A} \equiv -\partial \mathcal{F}/\partial a$ is a linear form that represents the thermodynamic force associated with damage, also called the driving force of damage. For a local model as in [5.6], the dissipation takes the following form:

$$\mathcal{F}_{loc}(u,a) = \int_{\Omega} \Phi\big(\varepsilon(u)(x), a(x)\big)\, dx$$
$$\rightarrow \quad \mathcal{D} = \int_{\Omega} -\frac{\partial \Phi}{\partial a}\, \dot{a}\, d\Omega \qquad [5.9]$$

Here, the usual expression of the driving force of damage $A \equiv -\partial\Phi/\partial a$ can be recognized. In terms of vocabulary, the definitions of internal forces $\mathcal{S}$ and driving force $\mathcal{A}$ are called state equations:

$$\mathcal{S} \equiv \frac{\partial \mathcal{F}}{\partial u}(u,a)\ ;\ \mathcal{A} = -\frac{\partial \mathcal{F}}{\partial a}(u,a) \qquad [5.10]$$

5.3.1.2. *Evolution equations*

The last thing that now remains to be provided is the evolution law which sets $\dot{a}$ according to the state of the structure. In general, the field $\dot{a}$ depends on the fields u, a and $\dot{u}$. However, taking into account the expression of dissipation, it may be interesting to express the evolution of the field $\dot{a}$ according to the driving force $\mathcal{A}$, even if it means introducing a dependency with respect to u and a if the phenomenology so requires.

It is thus formally denoted by:

$$\dot{a} \leftarrow \mathcal{G}(\mathcal{A}\ ;\ u,a) \qquad [5.11]$$

The purpose of the notation $\leftarrow \mathcal{G}$ is to highlight that $\mathcal{G}$ is not a functional but a process, as for example the solution of a nonlinear system, which may possibly require a time differentiation of the driving force $\mathcal{A}$. A concrete example thereof will be seen below. The damage evolution will be local if $\dot{a}(x)$ depends only on the mechanical state in x, which is always formally denoted by:

$$\text{Local evolution: } \forall\, x \in \Omega \quad \dot{a}(x) \leftarrow \Gamma\big(A(x)\ ;\ \varepsilon(x), a(x)\big) \qquad [5.12]$$

Again, Γ does not designate a function but, more generally, an algebraic system for determining $\dot{a}$ and potentially strongly nonlinear (unilateral constraints, threshold effects, etc.).

For example, Γ can represent the following nonlinear system that combines a flow equation, a threshold, an irreversibility condition and a consistency condition, similar to classical practices in plasticity:

$$\dot{a} = \lambda\, N(A\,;\varepsilon,a) \qquad [5.13]$$

$$g(A;\ \varepsilon,a) - v(\lambda) \leq 0\,;\ \lambda \geq 0\,;\ \lambda\left[g(A;\ \varepsilon,a) - v(\lambda)\right] = 0 \qquad [5.14]$$

The positive scalar λ reflects the intensity of the damage rate and represents a system unknown of [5.13]–[5.14] in addition to $\dot{a}$. The function N sets the flow direction; it can be multivalued to represent a singular flow regime (e.g. edge regime). The threshold function g characterizes the reversibility (or elasticity) domain of the model: if $g < 0$, then $\dot{a} = 0$. Finally, v is a function that accounts for a possible viscosity. Its introduction is inspired by a proposal by Chaboche and Cailletaud (1996) to represent viscous and non-viscous behaviors in the same framework. If v is identically zero, we find a threshold evolution law (associated if $N = \partial g/\partial A$) for rate-independent materials[5]. The system solution then requires, when $\lambda \neq 0$, the time differentiation of the condition $g = 0$, that is, $\dot{g} = 0$. On the other hand, if v is invertible on $\mathbb{R}^+$, then equation [5.14] effectively leads to a threshold viscous behavior and does not require time differentiation of g:

$$\lambda = v^{-1}\big(\langle g(A;\ \varepsilon,a)\rangle\big) \qquad [5.15]$$

It is common to seek to construct laws of evolution such that dissipation $\mathcal{D}$ is positive, so as to satisfy the second law of thermodynamics. However, it can be noted that by choosing to ignore the temperature variations, this restriction is no longer fully motivated. As a matter of fact, thermodynamics means change of form of energy, without application in a purely mechanical framework, as here. Typically, the energy dissipated by damage contributes to heat production, unlike the free energy stored by damage. However, different choices of partitions between stored free energy and (possibly negative) dissipated energy can lead rigorously to the same mechanical constitutive law because, often, only their sum has a mechanical influence. The choices therefore differ only in the production of heat; they are indistinguishable without resorting to the examination of the temperature distribution. On the other hand, physical motivations can lead to imposing the positivity of the sum of the powers, stored and dissipated by damage (this is the power actually consumed by damage).

5 In this case, the solution of the system during the (lasting) damage phase ($\lambda \neq 0$) involves the introduction of the consistency equation expressed in rate $\dot{g} = 0$ to eliminate λ.

5.3.1.3. *Generalized standard material framework*

The framework of generalized standard materials (Halphen and Nguyen 1975; Germain et al. 1983) proposes a restricted form of evolution laws. The evolution equation [5.11] is governed by a dissipation potential $\widehat{\mathcal{D}}(\dot{a})$ according to the following relation:

$$\mathcal{A} = \frac{\partial \widehat{\mathcal{D}}}{\partial \dot{a}}(\dot{a}) \qquad [5.16]$$

In the presence of rate-independent material behavior, the dissipation potential $\widehat{\mathcal{D}}$ is a non-differentiable function and relation [5.16] is generalized using the notion of subdifferential (Ekeland and Temam 1974). The evolution law in [5.16] guarantees that dissipation $\mathcal{D}$ is positive, provided that $\widehat{\mathcal{D}}$ is convex and minimal in zero.

On a numerical level, the framework of generalized standard materials combined with a temporal discretization of the evolution equation of [5.16] via an implicit Euler scheme makes it possible to express the problem on a time step Δt as the minimization of the total energy functional $\mathcal{E}$ with respect to displacement and damage fields (Mialon 1986; Francfort and Mielke 2006):

$$\mathcal{E}(u, a) \equiv \mathcal{F}(u, a) + \Delta t\, \widehat{\mathcal{D}}\left(\frac{\Delta a}{\Delta t}\right) - \mathcal{W}_e(u) \qquad [5.17]$$

The optimality condition with respect to u actually leads to the equilibrium equation in [5.5], while the optimality condition with respect to a leads to the evolution equation of [5.16].

The interpretation of the time discretized problem as the minimization of an energy makes it possible to establish attractive numerical properties. Therefore, coercivity and strict convexity of $\mathcal{E}$ with respect to u ensure the existence and uniqueness of the displacement solution with a given damage field (well-posed elastic problem). Coercivity and strict convexity of $\mathcal{E}$ with respect to a ensure the existence and uniqueness of the damage solution to a given displacement field (well-posed constitutive integration problem).

On the other hand, in the case of damage models, the total energy is not convex with respect to the pair (u, a); otherwise, the solution would be unique, whereas experience generally indicates the presence of bifurcations (solution non-uniqueness).

Note that the energy framework of generalized standard materials also makes it possible to propose stability criteria for solutions, inspired by those of nonlinear elastic systems[6] (Nguyen 1993; Benallal and Marigo 2006).

5.3.2. *Some non-local model classes*

5.3.2.1. *Localization limiters*

To build a non-local constitutive law, a first phenomenological variant consists of involving a regularization or smoothing operator $\mathcal{R}$ that transforms a field defined on Ω, say f, into another, smoother field, which will be denoted by $\bar{f} \equiv \mathcal{R}(f)$; the operator $\mathcal{R}$ is sometimes referred to as the localization limiter. Smoothing can be achieved using a convolution product, as in Bazant and Pijaudier-Cabot (1988):

$$\bar{f}(x) = \int_{\Omega} f(y)\, w_d(x,y)\, dy \qquad [5.18]$$

where the kernel w_d can typically be a Gaussian and depends on a characteristic length d. A second smoothing option, considered more numerically efficient, relies on the solution of a partial differential equation (Peerlings et al. 1996):

$$\begin{cases} \bar{f} - d^2\, \nabla^2 \bar{f} = f \text{ in } \Omega \\ \partial \bar{f}/\partial n = 0 \text{ in } \partial\Omega \end{cases} \qquad [5.19]$$

This is a special case of [5.18] with w_d the Green operator.

The smoothed variables can be displacement, strain, damage or the driving force of the damage. They can be involved instead of local values in the expression of free energy, in one or other of the state laws [5.10] or in the evolution law [5.11]. All this leads to many variants that have been published in the literature, the most popular of which is what is described in Peerlings et al. (1996). The latter is based on a regularization of the driving force of damage in the evolution equation formally expressed as in [5.12] by the system Γ:

$$\sigma = \frac{\partial \Phi}{\partial \varepsilon}(\varepsilon, a)\,;\; A = -\frac{\partial \Phi}{\partial a}(\varepsilon, a)\,;\; \dot{a} \leftarrow \Gamma(\bar{A}\,;\, \varepsilon, a) \qquad [5.20]$$

6 A steady state is characterized as a potential energy minimum (Lejeune–Dirichlet theorem).

However, not all of these variants are necessarily relevant. The operational nature of some of them is thus called into question in Lorentz and Andrieux (2003): a case-specific analysis is necessary. To illustrate this recommendation, the formulation in [5.20] can be reused which actually leads to good properties for a certain type of damage models. However, here we choose to apply it to a linear strain hardening or softening (it does not matter) elastoplastic behavior whose state variables are strain ε, plastic strain ε^p and the strain hardening variable κ.

State, consistency and flow equations of the non-local model are written respectively as:

$$\sigma = E(\varepsilon - \varepsilon^p) \; ; \; A_{\varepsilon^p} = \sigma \; ; \; A_\kappa = -H\,\kappa \qquad [5.21]$$

$$F = \|A_{\varepsilon^p}\| + \bar{A}_\kappa - \sigma^0 \; ; \; F \leq 0 \; ; \; \lambda \geq 0 \; ; \; F\,\lambda = 0 \qquad [5.22]$$

$$\dot{\varepsilon}^p = \lambda \frac{\partial F}{\partial A_{\varepsilon^p}} \; ; \; \dot{\kappa} = \lambda \qquad [5.23]$$

Regularization is applied to the driving force associated with the strain hardening A_κ in the definition of the threshold function [5.22]. By merely considering the situation of a bar in uniaxial tension as in Figure 5.1, the stress is uniform along the bar because of the equilibrium ($d\sigma/dx = 0$) and the condition of coherence in the active plastic zone ($\lambda \neq 0$) is written as, with uniform H:

$$F = \sigma - H\,\bar{\kappa} - \sigma^0 = 0 \qquad [5.24]$$

By adopting a regularization operator $\mathcal{R}$ such as proposition [5.18] with a reasonably regular function w_d or [5.19] the one-dimensional field $\bar{\kappa} = \mathcal{R}(\kappa)$ is necessarily continuous. However, when the yield strength σ^0 is not continuous (for example, for two materials), then the equation $F = 0$ admits no solution, even in the presence of positive strain hardening.

The same problem arises for a non-local elastic model where the strain is smoothed:

$$\sigma = E\,\bar{\varepsilon} \qquad [5.25]$$

In a uniaxial situation and in the presence of a discontinuous rigidity E, the problem does not admit any solution for the same reason as previously. This point has also led some authors to combine local and smoothed contributions using weighted sums $\gamma\bar{f} + (1-\gamma)f$ instead of $\bar{f}$ (Jirasek and Rolshoven 2003) or multiplicatively using ad hoc strain hardening (Engelen et al. 2003).

5.3.2.2. *Second gradient models*

A second variant for constructing non-local laws consists of introducing the strain gradient into the Helmholtz free energy.

This is referred to as second-gradient (displacement) models (Fernandes et al. 2008; Lorentz et al. 2008; Enakoutsa and Leblond 2009):

$$\mathcal{F}(u,a) \equiv \int_{\Omega} \left[\Phi(\varepsilon,a) + \frac{1}{2} c \, \nabla\varepsilon : \nabla\varepsilon\right] d\Omega \qquad [5.26]$$

where the parameter c has the dimension of a force and indirectly introduces an internal length. It is understood that the strain gradient term in [5.26] penalizes strong strain gradients and thus controls the localization.

It should already be noted that a variant of this model (Fernandes et al. 2008) merely introduces the volume strain gradient $\nabla(\operatorname{tr}\varepsilon)$ instead of the full strain tensor, the objective being in particular to simplify the numerical processing of the model.

Reconsidering to the general case [5.26], the linear form that represents the internal forces is written according to [5.10]:

$$\mathcal{S} \star \delta u = \int_{\Omega} [\sigma : \delta\varepsilon + \mathrm{S} : \nabla\delta\varepsilon] \, d\Omega \text{ where } \sigma \equiv \frac{\partial\Phi}{\partial\varepsilon} \text{ and } \mathrm{S} \equiv c \, \nabla\varepsilon \qquad [5.27]$$

An additional third-order stress tensor S can be seen to appear in the representation of the internal forces.

A more classical interpretation of the internal forces using an integration by parts of [5.27], where n denotes the outgoing normal to the domain:

$$\mathcal{S} \star \delta u = \int_{\Omega} \tilde{\sigma} : \delta\varepsilon \, d\Omega + \int_{\partial\Omega} (\mathrm{S} \cdot n) : \delta\varepsilon \, ds \text{ where } \tilde{\sigma} = \sigma - \operatorname{div} \mathrm{S} \qquad [5.28]$$

Therefore, the tensor $\tilde{\sigma}$ operates similarly to the usual stress tensor in a local model. On the other hand, the second term in [5.28] generates additional boundary conditions that will not be exposed here because their expression is more technical (emergence of surface differential operators).

Finally, in this strain gradient formulation, the driving force of damage retains the expression it would have in a local model:

$$\mathcal{A} \star \delta a = \int_{\Omega} A\, \delta a\, d\Omega \text{ where } A \equiv -\frac{\partial \Phi}{\partial a} \qquad [5.29]$$

A local evolution law is then chosen as in (12): $\dot{a} \leftarrow \Gamma(A;\ \varepsilon, a)$.

An example of the application of such a non-local model is given as illustration in Figure 5.7. It is the numerical simulation of the occurrence of damage in a rock following the excavation of a gallery. Damage is modeled by a strain softening plasticity model that accounts for confinement effects. The problem symmetry with respect to the vertical axis is exploited to reduce the problem size and thus the duration of the computation.

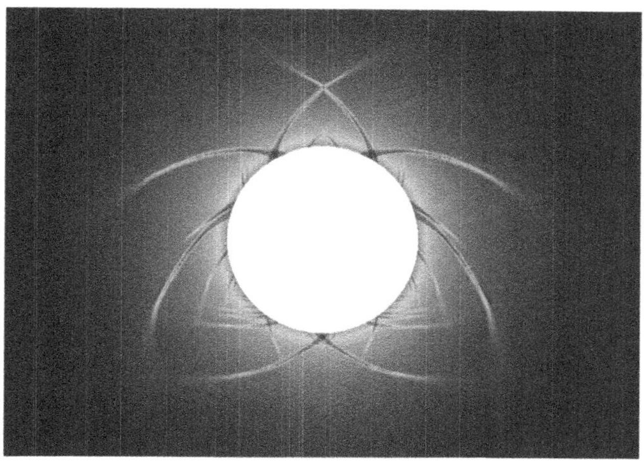

Figure 5.7. *Simulation of localization bands during the excavation of a tunnel by means of a volume strain gradient plasticity model (credit: EDF R&D). For a color version of this figure, see www.iste.co.uk/lebon/mechanics.zip*

5.3.2.3. *Damage gradient*

In the previous non-local formulation, the control of localization is ensured by a penalization of strong strain gradients. Since damage is the mechanism responsible for softening, it is also possible to choose to control localization by penalizing the damage gradient in the free energy (Muhlhaus and Aifantis 1991; Frémond and

Nedjar 1996; Svedberg and Runesson 1997; Comi 1999; Lorentz and Andrieux 1999; Pham et al. 2010):

$$\mathcal{F}(u,a) \equiv \int_{\Omega} \left[\Phi(\varepsilon,a) + \frac{1}{2} c\, \nabla a \cdot \nabla a\right] d\Omega \tag{5.30}$$

By applying definition [5.10], it can be seen that the internal forces remain modeled by the usual stress tensor, accompanied by the also usual boundary condition:

$$\begin{aligned} \mathcal{S} \star \delta u &= \int_{\Omega} \sigma : \delta\varepsilon \, d\Omega \\ &= \int_{\partial\Omega} (\sigma \cdot n) \cdot \delta u \, ds - \int_{\Omega} \operatorname{div} \sigma \cdot \delta u \, d\Omega \text{ where } \sigma \equiv \frac{\partial \Phi}{\partial \varepsilon} \end{aligned} \tag{5.31}$$

On the other hand, the driving force of damage has a more complicated expression than in the case of a local model:

$$\mathcal{A} \star \delta a = \int_{\Omega} [A\, \delta a - c\, \nabla a \cdot \nabla \delta a] \, d\Omega \text{ where } A \equiv -\frac{\partial \Phi}{\partial a} \tag{5.32}$$

With integration by parts, it is possible to find a pointwise interpretation of the driving force of the damage, as well as a boundary term that leads to an additional boundary condition:

$$\mathcal{A} \star \delta a = \int_{\Omega} \tilde{A}\, \delta a \, d\Omega - \int_{\partial\Omega} (\nabla a \cdot \mathrm{n})\, \delta a \, ds \text{ where } \tilde{A} \equiv A + \operatorname{div}(c\, \nabla a) \tag{5.33}$$

Starting from the assumption that no dissipation originates from the boundary of the structure, independently of the evolution of the damage, the limitations of which will be seen in section 5.3.2.5, then relation [5.33] leads to the following boundary condition:

$$\nabla a \cdot n = 0 \text{ on } \partial\Omega \tag{5.34}$$

Also, if we keep an evolution equation based on a local system Γ, then the "non-local" driving force $\tilde{A}$ simply replaces A:

$$\forall\, x \in \Omega \quad \dot{a}(x) \leftarrow \Gamma\left(\tilde{A}(x)\,;\, \varepsilon(x), a(x)\right) \qquad [5.35]$$

Finally, it is observed that the model is reduced to an additional boundary condition and the presence of a non-local term (the Laplacian of the damage field if c is constant in space) in the evolution law of damage.

In [5.30], the control of localization is ensured by the damage gradient term: the equilibrium equation of [5.5] with definition [5.31] of the stresses as well as the evolution equation of [5.35] govern the fine spatial distribution of the damage field in the localization bands.

By comparison, the Thick level set approach already mentioned above restricts from the outset all admissible damage fields to shapes with localization in which the fine spatial distribution is a priori imposed. Knowledge of the boundary of the damaged area then completely determines the damage field. The driving force of damage (see definition [5.10]) becomes a linear form of this geometric boundary so that the problem is reduced to moving boundaries equations, which require specific numerical techniques.

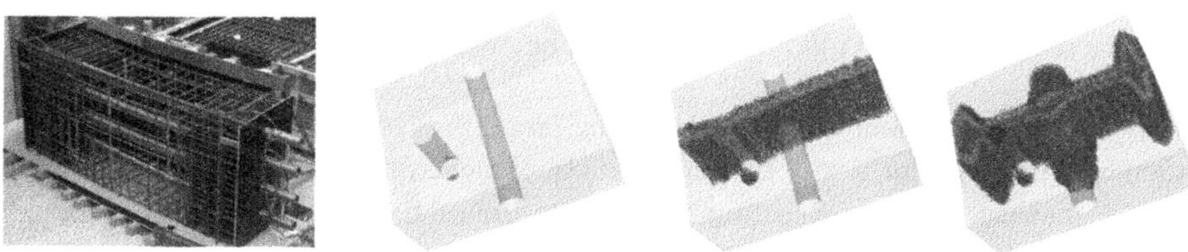

Figure 5.8. *Simulation of a fracture test on a pre-stressed concrete wall under internal pressure by a damage gradient model: initiation, through crack and then steel/concrete decohesion (source: EDF R&D). For a color version of this figure, see www.iste.co.uk/lebon/mechanics.zip*

As before, an example of the application of this non-local model is given as an illustration in Figure 5.8. It allows for the examination of crack paths in a reinforced and pre-stressed concrete structure subjected to hoop tensile loading. The initiation of damage at the foot of the vertical pre-stressing tendon is observed. It is followed by the development of a localization band which goes through the wall, bypassing the tendons. Finally, there is decohesion between the concrete and the horizontal tendon, on the one hand, and between the concrete and the reinforcements at the intrados and the extrados, on the other hand. Regarding this last step, a more detailed simulation is necessary when it is desirable to observe the "fish bones"-like decohesion cracks in an individualized way.

5.3.2.4. *Micromorphic media*

Unlike the formulations examined so far, micromorphal media rely on the introduction of additional kinematic degrees of freedom. The latter, which will be generically denoted by χ, can be, for example, dilatations (Pijaudier-Cabot and Burlion 1996), rotations (Cosserat media) (de Borst and Sluys 1991; Steinmann and Willam 1991) or strains (Forest 2009). For the latter, a typical case is provided by the following free energy:

$$\mathcal{F}(u,a,\chi) \equiv \int_{\Omega} \left[\Phi(\varepsilon(u),a) + \frac{1}{2} c \, \nabla\chi : \nabla\chi + \frac{1}{2} h \, (\varepsilon - \chi)^2\right] d\Omega \qquad [5.36]$$

Potential localization of χ is controlled by the term in $\nabla\chi$ while the last term ensures a relation between χ and strain, thus allowing indirect control of the localization of the latter. This is of course only true if the energy mobilized by this term in $\varepsilon - \chi$ "outweighs" the tendency to localize inherent to ε in a softening model: typically, for a given field χ, the model in (u, a) must be hardening and no longer softening, which requires a coefficient h high enough to "outweigh" the negative strain softening modulus that would be observed with the local free energy.

When the coefficient h is intended to be large, the last term of [5.36] ensures in practice the constraint $\varepsilon = \chi$ by penalization. We then recognize the family of the second gradient models of section 5.3.2.2. On the other hand, when h is bounded, χ can also be interpreted as a new internal variable, which yields the models of section 5.3.2.3 with internal variable gradients.

If, similarly to strain, it is assumed that micromorphic degrees of freedom do not contribute to dissipation, regardless of the evolution of the system, then it is deduced by reasoning analogous to the one led in [5.8] that:

$$\begin{aligned} &\forall\, \delta\chi \quad \frac{\partial \mathcal{F}}{\partial \chi} \star \delta\chi = 0 \\ \Rightarrow \quad &\int_{\Omega} [c\, \nabla\chi : \nabla\delta\chi - h\,(\varepsilon - \chi)\, \delta\chi] d\Omega = 0 \end{aligned} \qquad [5.37]$$

Implicitly, it has also been assumed that micromorphic degrees of freedom do not contribute to the power of external forces. However, this is a possibility offered by the theory of extending the PPV [5.5] to this new kinematic variable, which would lead us to introduce a source term in the potential of external forces, in the volume of the structure and/or at the boundaries; we choose here to ignore this option. By integration by parts in expression [5.37], a state equation and a boundary condition on χ is derived (where, for simplicity, it was considered that c and h are uniform in space):

$$\begin{cases} \chi - \frac{c}{h}\nabla^2\chi = \varepsilon \text{ in } \Omega \\ \nabla\chi \cdot n = 0 \text{ on } \partial\Omega \end{cases} \qquad [5.38]$$

This is the expression of one of the regularization operators introduced in section 5.3.2.1 and, by adopting the same notations, it can be written that $\chi = \bar{\varepsilon}$. To find out how this smoothed strain operates in the equilibrium and evolution equations, the linear forms of the stresses and the driving force of damage still have to be expressed:

$$\mathcal{S} \star \delta u = \int_{\Omega} \tilde{\sigma}{:}\, \delta\varepsilon \, \mathrm{d}\Omega \text{ where } \tilde{\sigma} = \frac{\partial\Phi}{\partial\varepsilon} + h(\varepsilon - \bar{\varepsilon}) \qquad [5.39]$$

$$A \star \delta a = \int_{\Omega} A\, \delta a \, \mathrm{d}\Omega \text{ where } A = -\frac{\partial\Phi}{\partial a} \qquad [5.40]$$

The driving force of damage and the evolution equation remain local, while the equilibrium equation involves an extended stress tensor $\tilde{\sigma}$ which depends not only on (ε, a) but also on $\bar{\varepsilon}$. It can thus be seen that the micromorphic formulation can also be interpreted as a localization limiter model, as in section 5.3.2.1, in which the strain would be smoothed and would intervene in the definition of stress, in addition to the local strain.

5.3.2.5. *Boundary conditions*

So far, we have not insisted on the choice of boundary conditions, systematically opting for the simplest form. Such a choice can however lead to spurious results. Therefore, in the context of localization limiter models in section 5.3.2.1, Borino et al. (2003) brought a boundary layer effect forward and proposed a correction of the weighting function w_d so as to take into account a local contribution of the fields at the boundary of the structure through the introduction of a Dirac function. This specificity of the boundaries is again explained by the homogenization process underlying the expression of a macroscopic constitutive model. It is indeed easy to understand that in the vicinity of the boundary, an assumption of quasi-periodicity no longer applies; microstructures play a special role therein (see, for example, Bazant et al. 2010). In non-local gradient models, a surface energy term would thus appear in the Helmholtz free energy, with a direct impact on boundary conditions. Nonetheless, despite that the origin of the problem is clearly identified, a rigorous homogenization method at the boundary and with no scale separation does not yet seem to be available, so that more phenomenological choices are necessary.

To illustrate the point, the damage gradient models of section 5.3.2.3 can be reused. For a localization band which leads to a free boundary, the zero normal gradient boundary condition [5.34] requires that its path be perpendicular to the boundary there. We might assume that this is a penalizing constraint for the orientation of damage, but in practice, this is simply reflected by a sudden change in orientation of the localization band, at a scale corresponding to that of the internal length.

Figure 5.9 can thus be referred to which presents the predicted damage fields for two concrete specimens: despite the orthogonality constrain at the boundary, the damage bands which initiate at the notches have an orientation compatible with the actual crack paths (shown in gray in Figure 5.9(a) and in white in Figure 5.9(b)).

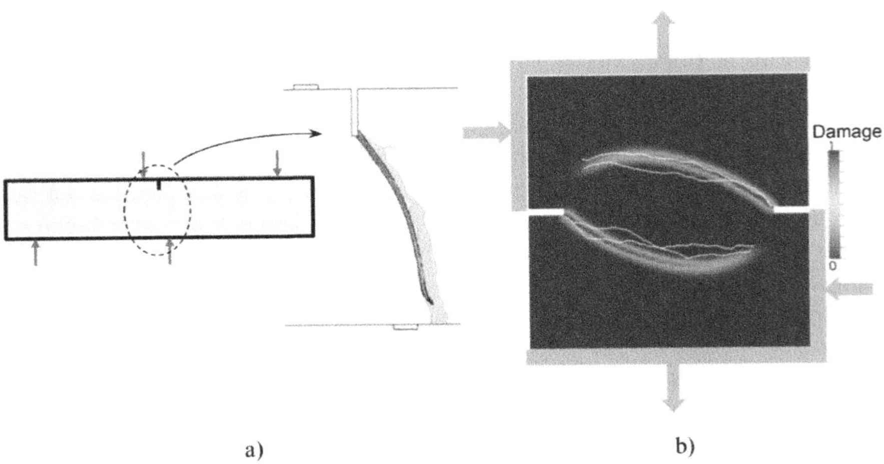

Figure 5.9. *Concrete notched specimens: orientation of the localization bands in the vicinity of the notches. For a color version of this figure, see www.iste.co.uk/lebon/mechanics.zip*

On the other hand, the problems are much more serious when the damage develops along a boundary loaded with force or prescribed displacement (interface crack). The profile of the damage band observed in the transverse direction to the boundary is actually shown schematically in Figure 5.10. It namely presents a "localization half-band" which effectively makes it possible to mechanically reduce the effect of loading (similarly to a crack) but at an energy cost a priori twice as low as that of a localization band inside the structure (see a typical profile in Figure 5.11). The damage along the boundary is then given preference to, wrongly, over that in the bulk. This has led some authors to propose a boundary condition of zero damage at the edge, $a = 0$ along $\partial\Omega$ (see, for example, Amor et al. 2009).

Although this approach effectively addresses the problem at hand, it may have an impact on damage areas that reach the boundary since they can no longer initiate or spread to the boundary. Another option, resulting from the mathematical smoothing of a phase field Griffith fracture model (Bourdin et al. 2000), introduces a fictional domain that covers the actual body domain and "slightly" goes beyond, so as to let the localization band settle completely and not just halfway. This, however, complicates the treatment of kinematic boundary conditions in the fictional area. Finally, in a more pragmatic manner, the model parameters can also be adjusted in the vicinity of the boundaries so that the energy consumed by damage is twice that in the bulk, in order for the energy cost of a half-band of localization along the boundary to be equal to that of a complete localization band inside the structure.

As can be seen, these are still open questions that may find a definitive answer using Robin-type boundary conditions based on surface energy that supplements the energy cost of damage along the boundary.

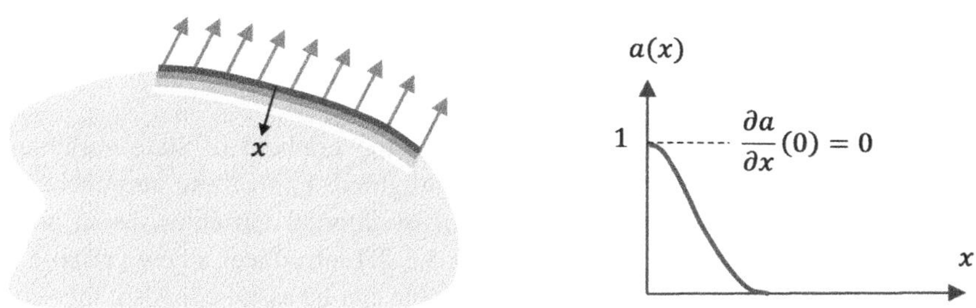

Figure 5.10. *Damage profile in a localization band which develops along a loaded boundary in the presence of a zero normal gradient-type boundary condition. For a color version of this figure, see www.iste.co.uk/lebon/mechanics.zip*

5.3.3. *Qualitative analysis of non-local formulations*

Many phenomenological variants allow non-local effects to be introduced into a constitutive model, as discussed in section 5.3.2. It is difficult to recommend one in particular, given that none is better than the others, and considering that their properties also depend on the constitutive behavior characterized by Φ and Γ. On the other hand, some recommendations can be made on the basis of the following pragmatic questions:

– Does the problem remain well-posed during the strain hardening phase?

– Is the problem well-posed during the strain softening phase?

– Does the model actually describe the propagation of localized damaged areas comparable to cracks and characterized by realistic dissipated energy?

– Are the numerical formulations for addressing the non-local problem robust and effective enough?

Essentially, the proposed approaches can be grouped into two distinct categories: those based on a regularization operator (section 5.3.2.1) and those that introduce gradient terms of state variables in the Helmholtz free energy, these terms being generally quadratic (sections 5.3.2.2–5.3.2.4). The former unfortunately require a case-by-case analysis, as any variant can lead to profound changes in their properties, as evidenced by the illustration of a non-local plasticity model in section 5.3.2.1 (equations [5.21]–[5.23]). An analysis methodology was proposed in Lorentz and Andrieux (2003) when the constitutive model can be included within the framework of generalized standard materials; it makes it possible to quickly eliminate a number of these formulations because it does not result in a well-posed problem during hardening and softening phases. However, in the absence of generic properties, this family of models will not be examined in detail here.

Concerning energy approaches enhanced by the gradient of state variables, a number of common characteristics can be highlighted. First, these approaches can generally be reduced to the case of gradient of internal variables, as in section 5.3.2.3. The micromorphic example in section 5.3.2.4 introduces a new variable χ to which the status of non-dissipative internal variable can be assigned. Also, the second gradient model of section 5.3.2.2 can also be written as a micromorphic case where the new variable χ will have to remain equal to the strain by way of an equality constraint $\varepsilon = \chi$ in the Helmholtz free energy. Hereafter, the notation χ is maintained to designate the state variable (the internal variable) whose gradient is introduced into the free energy.

The numerical impact of the corresponding non-local terms will be exposed in detail in section 5.5.1. At this stage, it can be retained that they result in the introduction of additional nodal degrees of freedom in a finite element formulation, possibly using a Lagrangian relaxation. In particular, the algorithmic separation between the processing of equilibrium at the structural level (linear algebra with large number of unknowns) and the treatment of the constitutive law at the level of the point of integration (small nonlinear systems) is preserved. We therefore remain within a well-established numerical framework, which is reassuring regarding robustness and performance.

We now examine the question of the well-posed nature of these models in the presence of localization. Since the non-local term is quadratic in $\nabla\chi$, the control of this variable is ensured: a pathological localization of χ would result in a non-integrable square gradient, thus an infinite free energy, which is excluded. It is still nonetheless unclear whether the control of χ is sufficient to ensure the well-posed nature of the problem. For this purpose, we simply have to examine whether the nature of the problem with a fixed value of χ is well posed or not. This is the criterion that makes it possible to decide whether the choice of the variable χ is relevant for controlling the localization. Therefore, in a second gradient model, setting χ amounts to setting ε and examining the problem of damage evolution with fixed strain, which is well posed; or still, in a damage gradient model where $a = \chi$ is the only internal variable, it must be shown that the fixed damage elasticity problem is well posed, which is a priori the case, however, with a reservation related to the loss of rigidity of the totally damaged points. Finally, a more nuanced example is that of the micromorphic formulation [5.36] with fixed χ, for which we have to study the properties of the model whose free energy density is written as:

$$\widetilde{\Phi}(\varepsilon, a) \equiv \Phi(\varepsilon, a) + \frac{1}{2} h\, \varepsilon^2 \qquad [5.41]$$

It is well-posed, provided that h is sufficiently large. This is easy to understand if we revisit the ellipticity criterion in [5.1] of the tangent operator of the problem $\widetilde{\mathbb{H}} = \mathbb{H} + h\,\mathbb{I}$, where $\mathbb{H}$ is the tangent operator corresponding to Φ only and $\mathbb{I}$ the fourth-order identity tensor (which explains why it was not necessary to retain the affine term in ε in the expression of $\widetilde{\Phi}$).

The reasoning used above also makes it possible to show that non-local terms do not raise problems in the hardening phase. As a matter of fact, the problem with a given value of χ is well-posed because of the hardening character. On the other hand, they lead to the possible development of boundary layers and limit the excessively high gradients resulting from loading or geometry: the results obtained with the non-local formulation are quantitatively different from those obtained without non-local terms, especially when the gradients are high.

With regard to the questioning proposed above, we still need to examine to what extent areas of localized damage reflect the behavior of cracks. The answer partly depends on the asymptotic character of the damage model when the latter tends toward its ultimate value, which corresponds to the failure of the material point. A detailed analysis is provided in sections 5.3.4, 5.3.5 and 5.4.3. It is nonetheless interesting to emphasize now an essential point related to the bounded nature (or not) of the variable χ. A "true" crack allows displacement discontinuities. If the damaged area is to reflect the response of a crack, this results in strains that are all the higher for larger displacement jumps and for a small internal length. It follows that their gradients are also very high. This may also be the case for certain internal model variables (for example, plastic strain). The variable χ might be concerned by this property (especially if speaking of strain), the quadratic control term in $\nabla\chi$ opposes this tendency. The only possible response of the model is then a transverse widening of the localization band to be able to increase χ (as in a crack) without increasing its gradient. A crack that opens or slides hence artificially widens the width of the localization band as well as the energy dissipated. This problem can be directly attributed to the unbounded character of the variable χ for a "true crack".

On the other hand, the problem does not occur with a bounded variable, as a damage variable comprised between 0 and 1. It should be noted that this difficulty is also observed in models with regularization operator: as a result of the averaging effect, very high strains in the localization band damage points further and further away (see Geers et al. 1998). One implemented strategy that will allow avoiding it consists of trying to bound the term in gradient using a dependency on the value of χ:

$$\frac{1}{2}c\,\nabla\chi^2 \rightarrow \frac{1}{2}c(\chi)\,\nabla\chi^2\,;\ \lim_{\chi\to\infty} c(\chi) = 0 \qquad [5.42]$$

Establishing a balance between localization control (c not too small), on the one hand, and limitation of band widening (c tends toward zero), on the other hand, is potentially challenging. This can be achieved by interpreting [5.42] via a change of variable $\xi(\chi)$ to find a constant coefficient:

$$c(\chi) \equiv c_0\left(\frac{d\xi}{d\chi}\right)^2 \Rightarrow \frac{1}{2}c(\chi)\,\nabla\chi^2 = \frac{1}{2}c_0\,\nabla\xi^2 \qquad [5.43]$$

Knowledge of $c(\chi)$ defines the change of variable to be adopted. If $c(\chi)$ is chosen so that ξ remains bounded, the parasitic widening of the localization band disappears, which explains the operational nature of this strategy. It then remains to study the properties of this new model in ξ with constant coefficient c_0 and in particular its asymptotic behavior when ξ tends toward its ultimate value.

Finally, even though these different formulations do not yet provide definitive answers to the non-local modeling of damage, they make it already possible to achieve operational results, as evidenced by the illustrations in Figures 5.7 and 5.8. Damaged areas are effectively similar to cracks whose "thickness" is controlled by the internal length of the model. The results are not influenced by the orientation and refinement of the mesh, provided that the latter is sufficient to properly discretize the fields within the localization bands.

5.3.4. *Phase field models and damage gradient models*

The concept of phase field model has been introduced in section 5.2.4. This is a spatial regularization of fracture models based on an extension of Griffith's theory (Francfort and Marigo 1998).

We propose to illustrate here the formal links between phase-field and damage gradient models, even if their underlying philosophy is different since the scale of non-local interaction is interpreted a priori as a consequence of microstructures and not as a small mathematical parameter.

To this end, we start by presenting the simplest of gradient damage models. An energetic formulation is then proposed. Finally, by reversing the philosophy and considering the width of the localization bands as a small adjustable parameter, in line with section 5.2.4, the link with the functionalities of phase field models is established.

5.3.4.1. *An elementary model with damage gradient*

The mechanical state is described by strain ε and scalar damage a comprised between 0 (sound material) and 1 (totally damaged material). By adapting the framework of gradient models proposed in section 5.3.2.3, the local contribution to strain energy Φ is introduced by choosing a form with separate variables:

$$\Phi(\varepsilon, a) \equiv \rho(a)\, w(\varepsilon) \; ; \; w(\varepsilon) \equiv \frac{1}{2}\left[\lambda \; \mathrm{tr}(\varepsilon)^2 + 2\mu \, \varepsilon : \varepsilon\right] \; ; \; \rho(a) \equiv (1-a)^2 \qquad [5.44]$$

The function $w(\varepsilon)$ corresponds to the elastic energy without damage effect, where we have denoted λ and μ the Lamé coefficients. Also, the function $\rho(a)$

accounts for the progressive loss of stiffness with damage; the interest of exponent 2 will later be seen in section 5.4.3.2, comparing with the more common affine form.

The equations of state are used to define the stress and the driving force of damage, which involves a term of non-local interactions:

$$\sigma \equiv \frac{\partial \Phi}{\partial \varepsilon} = (1-a)^2 \frac{\partial w}{\partial \varepsilon} \ ; \ \tilde{A} \equiv -\frac{\partial \Phi}{\partial a} + \mathrm{div}(c\, \nabla a) \\ = 2\,(1-a)\, w(\varepsilon) + \mathrm{div}(c\, \nabla a) \qquad [5.45]$$

The equations of state also stipulate the boundary condition relating to the damage field:

$$\nabla a \cdot n = 0 \text{ on } \partial\Omega \qquad [5.46]$$

It is finally considered that the evolution of damage is governed by the threshold function and the following Kuhn and Tucker conditions, where the first of the inequalities reflects the irreversible nature of damage:

$$g(A) \equiv A - \omega \qquad [5.47]$$

$$\dot{a} \geq 0 \ ; \ g(\tilde{A}) \leq 0 \ ; \ \dot{a}\, g(\tilde{A}) = 0 \qquad [5.48]$$

This completes the equations of behavior. They are based on four parameters: the two Lamé coefficients for elasticity, the threshold ω that represents here the local (volumetric) contribution to the energy consumed by damage (the area under the local response curve $\sigma - \varepsilon$) and the coefficient c that weights non-local interactions.

5.3.4.2. *Incremental energy formulation*

This constitutive model belongs to the framework of generalized standard materials. As discussed in section 5.3.1.3, it can be reformulated in energy terms. To this end, the approach proposed in Marigo (2014) can be followed. First, the variational formula of [5.48] is written as:

$$\dot{a} \geq 0 \ ; \ \forall\, \delta a \geq 0 \ \int_{\Omega} (\tilde{A} - \omega)\, \delta a \, d\Omega \\ \leq 0 \ ; \ \int_{\Omega} (\tilde{A} - \omega)\, \dot{a} \, d\Omega = 0 \qquad [5.49]$$

The total energy functional is then introduced as follows:

$$\mathcal{E}(u,a) \equiv \mathcal{F}(u,a) + \int_{\Omega} \omega\, a\, d\Omega - \mathcal{W}_e(u) \qquad [5.50]$$

Given this definition and that of $\tilde{A}$, the variational formulation [5.49] is written as:

$$\dot{a} \geq 0\,;\ \forall\, \delta a \geq 0\ \ \frac{\partial \mathcal{E}}{\partial a}(u,a) \star \delta a \geq 0\,;\ \frac{\partial \mathcal{E}}{\partial a}(u,a) \star \dot{a} = 0 \qquad [5.51]$$

Finally, an implicit temporal discretization is performed. We denote by a_t the damage field at the beginning of the time increment considered Δt and simply u and a the solution fields at the end of the time increment.

The variational formulation of the evolution problem is written as:

$$a - a_t \geq 0\,;\ \forall\, \delta a \geq 0\ \ \frac{\partial \mathcal{E}}{\partial a}(u,a) \star \delta a \geq 0\,;\ \frac{\partial \mathcal{E}}{\partial a}(u,a) \star (a - a_t) = 0 \qquad [5.52]$$

These are the Kuhn and Tucker conditions that characterize a solution of the minimization problem with respect to damage (optimality conditions):

$$\min_{\alpha \geq a_t} \mathcal{E}(u,\alpha) \qquad [5.53]$$

Also, concerning the equilibrium equation in [5.5] and the definition of stress [5.45], they are the optimality conditions that characterize a solution of the minimization problem with respect to displacements:

$$\min_{v \in \mathcal{C}} \mathcal{E}(v,a) \qquad [5.54]$$

It should also be noted that minimization under constraint [5.53] can also be expressed by means of an indicator function $I_{\mathbb{R}^+}$ equal to 0 if its argument is positive and $+\infty$ otherwise:

$$\min_{\alpha} \left[\mathcal{E}(u,\alpha) + I_{\mathbb{R}^+}(\alpha - a_t)\right] \qquad [5.55]$$

The link with the expression proposed in [5.17] can thus be achieved by choosing as dissipation potential:

$$\widehat{D}(\dot{a}) \equiv \omega\, \dot{a} + I_{\mathbb{R}^+}(\dot{a}) \quad [5.56]$$

5.3.4.3. *Interpretation in terms of phase field model*

This gradient damage model was examined in Lorentz and Godard (2011) where an analytical solution is given for a simple tension problem[7]. Figure 5.11 describes the shape of the localization band and the global response of the specimen in terms of stress versus opening curve. The simple tension test provides interpretation to the internal parameters of the model ω and c.

Indeed, the fracture energy G_F (the energy consumed to fracture the specimen divided by the cross section area of the specimen) and the width of the localization band w_{FPZ} (using the notation introduced in section 5.2.4) are equal to:

$$G_F = \sqrt{\frac{32}{9}\, c\, \omega}\ ;\ w_{FPZ} = \sqrt{8\,\frac{c}{\omega}} \quad [5.57]$$

The internal parameters of the model ω and c can therefore also be expressed according to G_F and w_{FPZ} :

$$\omega = \frac{3}{2}\frac{G_F}{w_{FPZ}}\ ;\ c = \frac{3}{16} G_F\, w_{FPZ} \quad [5.58]$$

Now introducing [5.58] into the expression of total energy [5.50], we get:

$$\mathcal{E}(u,a) = \int_{\Omega} \left[(1-a)^2\, w(\varepsilon) + \frac{3}{2}\frac{G_F}{w_{FPZ}} \left(a + \frac{w_{FPZ}^2}{16} \nabla a \cdot \nabla a \right) \right] d\Omega - \mathcal{W}_e(u) \quad [5.59]$$

The minimizations [5.53] and [5.54] with respect to the damage and displacement fields as well as the expression of the total energy [5.59] correspond to one of the variants of the phase field formulations as presented in Amor et al. (2009).

---

7 Strictly speaking, the analytical solution does not exactly apply for this model because the width of the localization band decreases during the process, while the solution is built under the assumption that no damaged points unloads. The condition of irreversibility will be ignored here, without affecting the orders of magnitude of the calculated quantities.

The approximation of the problem of LEFM by phase field therefore leads strictly to the same problem to be solved as a non-local formulation with damage gradient. In particular, the mathematical and numerical properties of the formulation are the same.

However, in line with a phase-field regularized model, the scale parameter w_{FPZ} is intended to tend to zero. However, the peak stress f_t (see Figure 5.11) is expressed as follows:

$$f_t = \sqrt{\frac{3}{2}\frac{E\,G_F}{w_{FPZ}}} \qquad [5.60]$$

where E is Young's modulus (bar stiffness). When w_{FPZ} tends to zero, the peak stress tends to infinity. We thus indeed retrieve a Griffith brittle fracture model where G_F is the fracture energy (only parameter in addition to elastic modulus) and for which the stresses are singular at the crack point.

On the other hand, in line with a damage model, the peak stress f_t is a model parameter.

However, relation [5.60] can be reversed as follows:

$$w_{FPZ} = \frac{3}{2}\frac{E\,G_F}{f_t^2} \qquad [5.61]$$

In this elementary model, knowledge of elastic modulus, peak stress and fracture energy sets the scale of non-local interactions: it is no longer intended to tend toward zero but takes a given value.

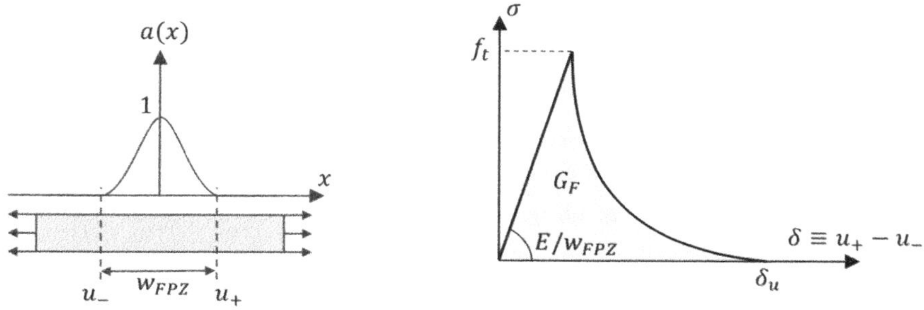

Figure 5.11. *Uniaxial tension: damage gradient model response. For a color version of this figure, see www.iste.co.uk/lebon/mechanics.zip*

5.3.5. Approximating a cohesive model with a gradient model

The quantity that appears in relation [5.61] is related to the length introduced by Hillerborg (1976), which is directly related to the length of the FPZ in a cohesive model. The basic model introduced in section 5.3.4 does not make it possible to take into account both the length of the process zone L_{FPZ} – linked to the peak stress – and the width of the localization band w_{FPZ} (see Figure 5.6). This is not penalizing in the context of the regularization of a Griffith fracture model since the size of the process zone is zero (infinite peak stress): the same small parameter can simultaneously account for the fictitious crack width and the length of the cohesive zone, which is also fictitious in this case. This is also a relevant choice when a Griffith model is suitable, that is, when the size of the FPZ is small compared to the characteristic scales of the structure under study.

When this length is no longer negligible, it is necessary to effectively model the process zone. On the other hand, the scale of non-local interactions, which is reflected in the width of the localization bands, may remain small compared to that of the structure, as explained in section 5.2.4. This is the basic assumption of cohesive models for which the phenomena are concentrated on a surface (zero localization band width). This cohesive model may therefore be regularized using a non-local damage model for which the non-local scale w_{FPZ} is uncorrelated from that resulting from the actual microstructures. In practical terms, it should be possible to independently control the length of the FPZ L_{FPZ} and its width w_{FPZ}, which is intended to tend toward zero[8]. For a given fracture energy G_F and stiffness E, the first depends on the peak stress f_t while the second depends on the volume energy consumed by damage ω.

In the model in section 5.3.4, the stiffness function $\rho(a)$, which affects the elasticity models and is defined in [5.44], is not rich enough to independently parameterize its behavior in the vicinity of 0 (which sets the peak stress) and the rest of its evolution (which sets ω). This is why Lorentz (2017) proposed an extended form of expression [5.44], consistent with the uniaxial tension analytical solution:

$$\rho(a) \equiv \frac{(1-a)^2}{(1-a)^2 + 3\dfrac{E\,G_F}{f_t^2\,w_{FPZ}}a(1+p\,a)} \quad [5.62]$$

Here, the width w_{FPZ} (the small parameter) can be chosen independently of the physical characteristics of the model, in particular the peak stress. When w_{FPZ} tends

8 In practice, a value should be chosen to be small relatively to the scale of geometric details of the structure and the load that ought to be taken into account.

to zero, a cohesive model is obtained whose response is illustrated in Figure 5.12. In particular, the role of the parameter p is clearly visible which acts on the softening slope at the origin as well as on the opening at final fracture[9]. Issues of internal stability of the model (convexity, which bounds the attainable volume energy values ω) reduces the range of possible values for the band width:

$$w_{FPZ} \leq \left(\frac{3}{p+2}\right)\frac{E\,G_F}{f_t^2} \qquad [5.63]$$

The band width is bounded by a term related to the length of the FPZ. The existence of this upper bound is not surprising in itself; it has already been mentioned in section 5.2.4. For practical reasons (mesh size), it may be desirable to achieve higher band widths. This means that we could ignore the effects of the structure at the length scale of the FPZ. It is therefore implicitly considered that the size of the FPZ itself is negligible compared to the scale of the structure. In this case, we might as well adopt Griffith's model.

It can also be noted that the stiffness function [5.44], well adapted to the Griffith model, corresponds to expression [5.62] for the cohesive model, provided that relation [5.61] is imposed between f_t and w_{FPZ} and that $p = -1/2$ be set. The "cohesive" characteristics of a Griffith model then once again come across: infinite peak stress and zero fracture opening. Also, it can be observed that the constraint in [5.63] is always verified, given the relationship imposed between f_t and w_{FPZ}.

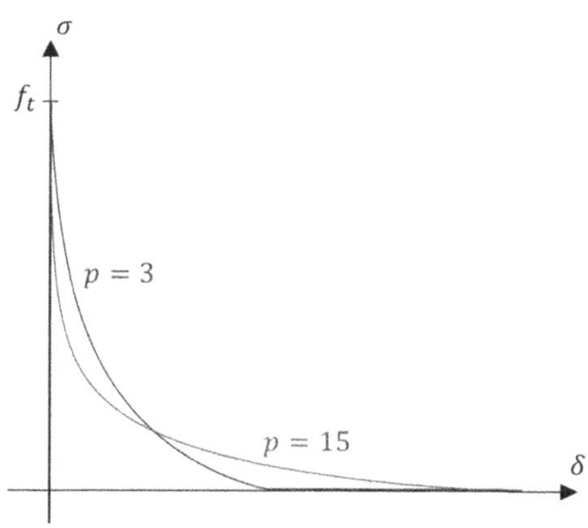

Figure 5.12. *Cohesive model asymptotic response when $w_{FPZ} \to 0$. For a color version of this figure, see www.iste.co.uk/lebon/mechanics.zip*

9 An even richer version of the stiffness function $\rho(a)$ makes it possible to distinguish these two quantities, but at the cost of an additional parameter to be identified.

This analysis is conducted within a uniaxial framework; nevertheless, the results can be extended to 2D and 3D configurations, at least as long as a mode I (opening) load is maintained. Figure 5.13 thus represents a detailed illustration in Lorentz (2017). This is a purely numerical test, without experimental existence. It models a pressurized cavity of asymmetrical and eccentric shape with thick concrete walls; see the inset between the two graphics in Figure 5.13. The pressure generates a crack on the right-hand side of the cavity and then on its left side. Two simulations are carried out: the first using the damage gradient model whose stiffness function is given in [5.62] and whose scale w_{FPZ} is set to $1/10$ the height of the cavity. The iso-values of the damage field can be discerned in Figure 5.13 in the inset between the two graphs, in the form of "spread cracks" on either side of the cavity. The second simulation is conducted with the cohesive model which corresponds to the limit of the non-local model when $w_{FPZ} \rightarrow 0$. It can be seen that the overall pressure–volume response as well as the opening profile of the crack actually match.

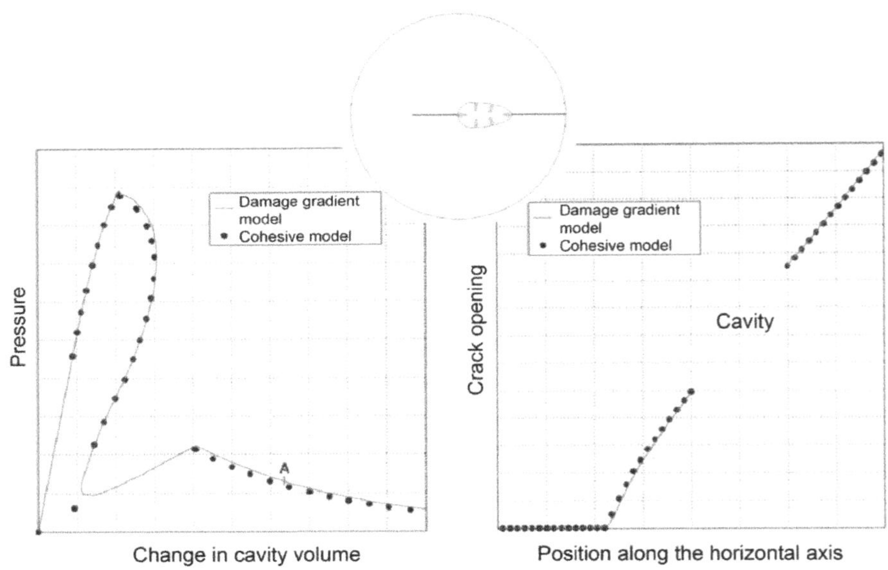

Figure 5.13. *Virtual pressurized cavity: damage gradient model versus cohesive model. For a color version of this figure, see www.iste.co.uk/lebon/mechanics.zip*

5.4. Phenomenological aspects of quasi-brittle behavior

So far, only the non-local nature of brittle damage models has been highlighted. Still, other phenomenological specificities also deserve some attention. Given the great diversity of brittle materials, claiming to be comprehensive here would be inappropriate. That is the reason why it is preferable to address three generic questions: damage-induced anisotropy, the unilateral character that results from the asymmetry between opening and closing the microcracks and, finally, the ability of the model to describe a cracked macroscopic state at the end of the damage; this last point is intrinsically linked to taking non-locality into account.

5.4.1. *Isotropy or anisotropy?*

Macroscopic-scale brittle damage often results from a concentration of microcracks within the microstructure. Their orientation a priori depends on loading, typically perpendicular to the main tensile direction. This privileged orientation induces a microstructural polarization and therefore an anisotropy of its average behavior.

This phenomenological interpretation is at the origin of smeared crack models that have long been established (for a review, see Rots (1988)). They introduce an irreversible strain to represent the effects of damage, based on this preferred orientation. It can be set once and for all, at the time of onset: it is then referred to as fixed crack type models. They are however ill-suited when the loading directions rotate over time because they then underestimate the effect of loading on damage (the load is projected on an "outdated" direction). A second family of smeared crack type of models, known as rotating crack, adapts the preferred direction during loading and associates a level of damage to each direction (the component of irreversible strain in this direction). This formulation raises the issue of the effects of history when directions rotate: how does damage in the current direction "inherit" damage in a different previous direction? This physical problem is accompanied by a numerical problem when the definition of the preferred directions is based on the eigendirections of strain. Indeed, they are not uniquely defined[10]. Therefore, how can the set of eigendirections at a given moment be chosen to ensure consistency with the choices of past directions? This question, which does not seem to be discussed in the literature, is illustrated in Figure 5.14 in 2D. The eigen-coordinate system gradually rotates between times t_n and t_{n+3}. In addition, if the eigendirections are associated with the eigenvalues ordered in descending order (blue for the large one, red for the small one), a sudden permutation between t_{n+1} and t_{n+2} can be observed, symptomatic of eigenvalues "crossing" (there is an instant

10 When the eigenstrains are distinct, there exists a finite number of sets of eigendirections (any permutation of the eigendirections between them is also a set of eigendirections). Also, when two principal strains are equal, there exists an infinite number of sets of eigendirections.

between t_{n+1} and t_{n+2} where the two eigenvalues are equal). The classification of the eigendirections according to the order of the associated eigenvalues is therefore not relevant: the damage carried by the blue direction for t_{n+1} is a priori unrelated to that carried by the blue direction for t_{n+2}. Concerning the memory of past damage, a better criterion seems to us rather associated with a form of time continuity of the eigendirections: at a time t, it is preferable to choose the closest set of eigendirections (in a norm to be defined) to the set retained at the previous instant. Far from being trivial, these issues appear to reflect a serious problem of rotating crack models.

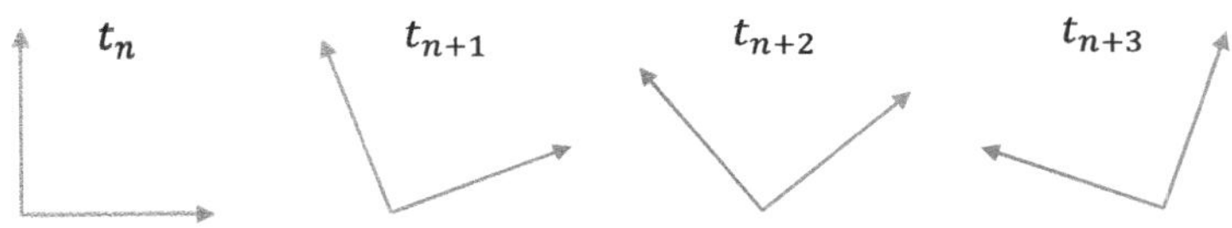

Figure 5.14. *Problem of tracking eigendirections over time (in blue, eigendirection attached to the largest principal strain, in red, the one attached to the smallest). For a color version of this figure, see www.iste.co.uk/lebon/mechanics.zip*

Subsequently, damage-induced anisotropy was modeled by the introduction of (second-order or higher) tensor damage variables with effect on material stiffness. The literature is very abundant on this topic and we will cite only three representative examples (Halm and Dragon 1998; Badel et al. 2007; Desmorat et al. 2007). When the behavior law is based on the eigen directions of the damage tensor, some level of attention is again required with regard to the objectivity of the model, namely its independence to the choice of the set of eigendirections. Therefore, by denoting by D the second-order damage tensor, and by d_i its eigenvalues and n_i the associated eigendirections, it is recalled that the following expression is objective, where f designates any function:

$$D \mapsto \sum_i \mathrm{f}(d_i)\ n_i \otimes n_i \qquad [5.64]$$

It is indeed a function of the tensor D. Typically, when the function f corresponds to the positive (resp. negative) part of a scalar[11], expression [5.64] refers to the positive $\langle D \rangle_+$ (resp. negative $\langle D \rangle_-$) part of the tensor D. On the other hand,

11 The notation $\langle x \rangle$ was already introduced to designate the positive part of a scalar x. When it is necessary to clearly distinguish the positive and negative parts of x, they should be, respectively, denoted by $\langle x \rangle_+$ and $\langle x \rangle_-$.

expression [5.65] which also depends on another tensor, for example, strains ε, is not objective, although used in some models:

$$\sum_i (n_i \cdot \varepsilon \cdot n_i)\, n_i \otimes n_i \qquad [5.65]$$

When two eigenvalues of the damage tensor are equal, the result depends on the choice of associated eigendirections (in infinite number). Also, although it may be assumed that this difficulty appears only in the presence of multiple eigenvalues, issues of numerical accuracy in the computation of eigendirections also arise in the vicinity of these situations, resulting in an improper evaluation of [5.65]. Therefore, constitutive models in which the damage tensor is explicitly involved and not its eigendirections will be preferable.

In general, we should keep in mind that taking into account damage-induced anisotropy results in complicated models. They can potentially lead to undesirable and difficult to anticipate responses, given the wide variety of possible loading paths. They are also challenging to implement numerically, especially since their properties in this respect are often not optimal (e.g. to guarantee the existence and uniqueness of the solution to the problem of constitutive integration). It is therefore legitimate to question the added value of these models compared to isotropic variants.

In practice, it all depends on the extent of the damaged areas and the scale at which they are examined. Consider the example of the structure described in Figure 5.15. A first, horizontal load direction is at the origin of the occurrence of vertical microcracks (softening in the horizontal direction), according to Figure 5.15(a). They are organized in a narrow band, as a consequence of the phenomenon of localization. Since we are not interested in the fine distribution of mechanical fields in this area but rather in their consequences on the structure, the locally anisotropic character of microcracks does not matter. Indeed, an additional load in another direction, namely vertical, would only engage the anisotropy of the model inside the damaged area (which retained its stiffness in the vertical direction), according to Figure 5.15(b). Since its thickness is small, the overall structural response would show little or no sensitivity to the anisotropic character of the damaged material (if isotropic, too flexible in the vertical direction), according to Figure 5.15(c). Also, it is indeed the geometric distribution of damage, distributed in the form of a vertical macrocrack, which is then at the origin of the anisotropy at the structural scale. Consequently, it seems that an isotropic constitutive relation suffices in the presence of a localized damage distribution, and when merely seeking to predict its effect at the structural scale. This is also the purpose of phase field models which regularize a macro-crack isotropically (the phase field is scalar). On the other hand, the response is potentially different in the presence of a diffuse damage distribution, as discussed in section 5.2.1; in this case, in fact, the volume of the damaged area is no longer negligible and mobilizing the anisotropy of the material behavior in this area can have a visible effect on the structural response.

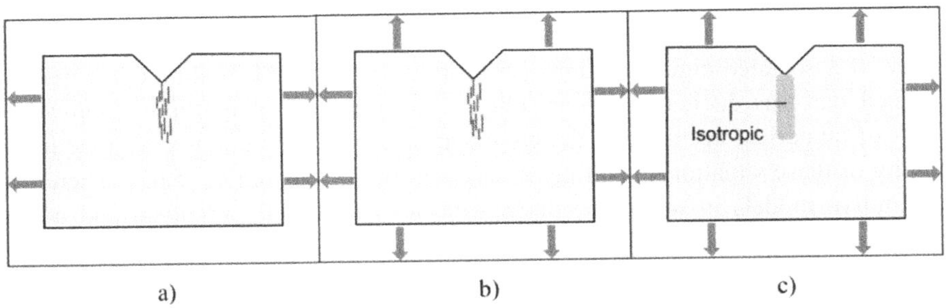

Figure 5.15. *Limited effect of anisotropy in the presence of damage localization. For a color version of this figure, see www.iste.co.uk/lebon/mechanics.zip*

5.4.2. *Unilateral nature*

Even if we merely consider an isotropic constitutive relation as suggested in section 5.4.1, it is possible to distinguish a tensile mechanical state from a compressive one, taking into account the sign of the principal strains or stresses. This is often important to account for residual compressive stiffness in damaged areas. It is essential, for example, in the presence of a volume load of self-weight type: thus, in Figure 5.15(c), the damaged area could not withstand self-weight loading with no residual stiffness since zero stresses cannot balance the weight. Another typical case is that of the application of a surface load, pressure for example, on a damaged area (see the diagram in Figure 5.16 which will be analyzed below). Again, only residual compressive stiffness allows withstanding wall pressure loading. On the other hand, totally damaged areas should no longer present tensile stiffness, otherwise there is a risk of introducing artificial cohesion forces on both sides of the damaged area. Therefore, a totally damaged point should:

– retain residual compressive stiffness, to withstand states of compressive stresses;

– have zero stress for extension strain states to allow crack opening without residual cohesive efforts.

This progressive distinction between tensile and compressive states during the course of damage is often referred to as the "unilateral effect". This is also sometimes referred to as "crack (re)closure" but this term seems less appropriate because it implies an opening-closing cycle, whereas the unilateral effect also manifests itself under monotonic loading (example of shear stress which combines tensile and compression directions). The term "tension-compression asymmetry" is also sometimes used, but it can be confusing because it is also used to refer to an asymmetric damage threshold (when damage occurs at lower tensile stress levels than compressive ones).

One difficulty is to effectively define compressive and tensile states, which is not as straightforward as it seems outside of one-dimensional situations. A proposition by Fichant et al. (1999) involves the sign of the principal stresses and defines the stress–strain relation of the broken material as follows:

$$\sigma = \langle \mathbb{E} : \varepsilon \rangle_{-} \qquad [5.66]$$

where $\mathbb{E}$ denotes the Hooke elasticity tensor. It has however the disadvantage of leading to an elastic law that no longer derives from a strain energy (elastic cycles can produce energy).

It is preferable to define the compressive and tensile states directly from the Helmholtz free energy. We reconsider here a definition proposed in Lorentz (2017). We denote by w_c the local contribution of a broken point to the free energy; using the notations of section 5.3, $w_c(\varepsilon) \equiv \Phi(\varepsilon, a_c)$, where a_c designates the damage level corresponding to total failure (in many models, $a_c = 1$). The potential w_c is defined up to a constant; it is chosen so that $w_c(0) = 0$. It is further assumed that w_c is positively homogeneous of degree 2 (linear response in each strain direction) and that $\sigma = \partial w_c / \partial \varepsilon$, which excludes, for example, non-local strain gradient models (we will later see what can be said of them). The tensile states $\mathcal{U}_\varepsilon$ are then characterized as the *strains* for which the strain energy w_c is zero and compression states $\mathcal{U}_\sigma$ such as *stresses* for which the complementary energy w_c^* (Legendre transform) is finite, that is, stresses that the broken material can withstand. It is shown in Lorentz (2017) that these two sets are cones orthogonal to one another:

$$\mathcal{U}_\varepsilon \equiv \left\{ \varepsilon \,;\, w_c(\varepsilon) = 0 \right\} \,;\, \mathcal{U}_\sigma \equiv \left\{ \sigma \,;\, w_c^*(\sigma) \neq \infty \right\} \,; \qquad [5.67]$$
$$\forall\, (\varepsilon, \sigma) \in \mathcal{U}_\varepsilon \times \mathcal{U}_\sigma \quad \varepsilon : \sigma = 0$$

In particular, it shows that the larger one of these sets, the smaller the other. Therefore, the choice of considering that all positive-trace strain states are tensile states (Comi and Perego 2001; Amor et al. 2009) implies that only negative-trace spherical stresses are withstood:

$$\mathcal{U}_\varepsilon = \left\{\varepsilon\, ;\, \mathrm{tr}(\varepsilon) \geq 0\right\} \Leftrightarrow \mathcal{U}_\sigma = \left\{\sigma\, = p\,\mathrm{Id}\, ;\, p \leq 0\right\} \qquad [5.68]$$

Under this assumption, the broken material behaves like a fluid, which may prove too restrictive as discussed hereafter. That is why Lorentz (2017) proposed another choice. Intuitively, it is assumed that any strain whose three principal strains ε_i are positive is a tensile state and that any stress whose three principal stresses σ_i are negative is a compressive state. Then, the orthogonality property in [5.67] shows that, under these two conditions, these sets cannot be larger:

$$\mathcal{U}_\varepsilon = \left\{\varepsilon\, ;\, \forall\, i\ \ \varepsilon_i \geq 0\right\} \Leftrightarrow \mathcal{U}_\sigma = \left\{\sigma\, ;\, \forall\, i\ \ \sigma_i \leq 0\right\} \qquad [5.69]$$

Now, we still have to propose a form of free energy in accordance with [5.69]. A common choice is based on a gradual transition of the "law of mixtures" type between the strain energy of the sound material denoted w_e and that of the totally damaged material w_c :

$$\Phi(\varepsilon, a) = \rho(a) w_e(\varepsilon) + [1 - \rho(a)] w_c(\varepsilon) \qquad [5.70]$$

where $\rho(a)$ is a stiffness attenuation function that decreases from 1 (for the sound material) to 0 (for the broken material). The sound material is assumed to be linear elastic; by denoting λ and μ its Lamé coefficients, its strain energy is written simply as:

$$w_e(\varepsilon) = \frac{1}{2}\lambda\,(\mathrm{tr}\,\varepsilon)^2 + \frac{1}{2}2\mu \sum_i \varepsilon_i^2 \qquad [5.71]$$

Also, concerning the totally damaged material, the following form is proposed, in agreement with the equivalence in [5.69]:

$$w_c(\varepsilon) = \frac{1}{2}\lambda \langle \mathrm{tr}\,\varepsilon\rangle_-^2 + \frac{1}{2}2\mu \sum_i \langle \varepsilon_i \rangle_-^2 \qquad [5.72]$$

It is indeed zero if and only if all the principal strains are positive. We retrieve the form of free (local) energy proposed in particular in Badel et al. (2007).

It should be noted that despite its intuitive nature, this choice of energy for the broken material faces a problem as detailed in Lorentz (2017). A stress that combines an opening and a slip (mode II or III) of the macrocrack necessarily results in a strain in the broken material of which one of the eigenvalues is negative, thus giving rise to non-zero strain energy and therefore stiffness. As long as the opening remains confined, this parasitic stiffness is compatible with an interpretation in terms of interlocking of crack lips, at least qualitatively (Jefferson and Mihai 2015). However, for larger openings (greater than the size of asperities), this interpretation is no longer conceivable, which then prompts a transition toward crack models (see section 5.2.4).

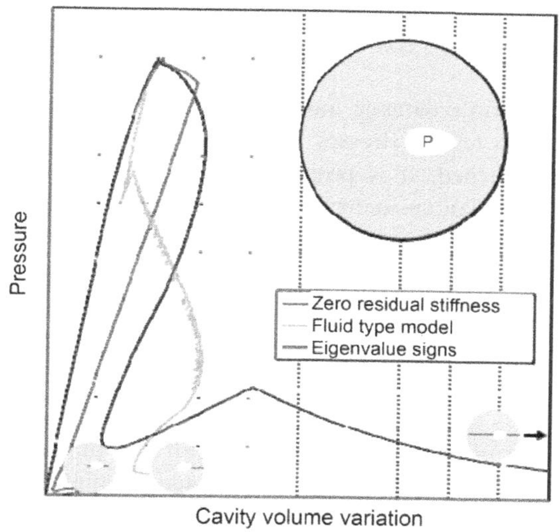

Figure 5.16. *Effect of the choice of unilateral model (from three variants) on the overall response of a virtual cavity under pressure. For a color version of this figure, see www.iste.co.uk/lebon/mechanics.zip*

To illustrate the consequences of the unilateral character, it is proposed to re-examine the response of the virtual concrete enclosure whose inner cavity is under pressure (see Figure 5.16). The enclosure responses in terms of cavity volume versus applied pressure are reported for three unilateral models. The first exhibits no residual stiffness, that is, $w_c = 0$ and $\mathcal{U}_\sigma = \{0\}$. The second corresponds to the fluid model previously mentioned and in which the tensile strains are those whose trace is positive. The third model corresponds to the residual strain energy in equation [5.72]. Reference is made to Lorentz (2017) for more information on the formulation that was used; we simply mention that this is a non-local model with damage gradient. The expected effects can be correctly observed. Actually, for the model without residual stiffness, the computation ceases as soon as a point of the inner wall is totally damaged: it cannot withstand any stress, and there no longer exists a non-zero pressure solution. For the fluid model, the computation continues further, until the crack reaches the outer wall. At that point, there no longer exists a non-zero pressure solution because the damaged material would flow outward like a fluid. Finally, the third unilateral model makes it possible to complete the calculation: a first crack is initiated and then penetrates the wall, a second crack then develops on the other side, until the enclosure is separated into two parts.

5.4.3. *Asymptotic fracture behavior*

As seen above, a brittle damage model is expected to reflect fracture behavior and, in particular, zero tensile stresses through the damaged area when a certain level of damage is reached. It is proposed to examine what the conditions are relating to constitutive relations under which a non-local model can account for this phenomenon. For this purpose, a strain gradient model (see section 5.3.2.2) and a damage gradient model (see section 5.3.2.3) are considered alternately. The typical application case is again the uniaxial solicitation of a bar as shown in Figure 5.1.

5.4.3.1. *Fracture response of a strain gradient model*

The purpose here is to determine the level of residual tension applied to the bar at the time when the ultimate damage is reached at a point. The homogeneous response (i.e. without gradient effect) of a material point is considered to be represented via a curve (ε, σ) (see Figure 5.17(a)). We call $\Pi^H(\varepsilon)$ the strain potential, where the letter H refers to the fact that this potential depends on history effects (load/unloading in the history of the strain). It is a function C^1 from which the strain is derived, by definition:

$$\sigma = \frac{\mathrm{d}\Pi^{\mathrm{H}}}{\mathrm{d}\varepsilon}(\varepsilon) \qquad [5.73]$$

The value of Π^H is interpreted as the area under the curve (ε, σ) followed at the material point under consideration. The potential $\Pi(\varepsilon)$ is also introduced which corresponds to a loading path without elastic unloading.

If $\hat{\sigma}(\varepsilon)$ denotes the curve obtained with no elastic unloading (increasing ε), it is written as:

$$\Pi(\varepsilon) \equiv \int_0^{\varepsilon} \hat{\sigma}(e)\, de \qquad [5.74]$$

This is an increasing function, due to the positivity of $\hat{\sigma}$. In addition, $\Pi(\varepsilon) \geq \Pi^H(\varepsilon)$ regardless of the strain and history of loads/unloads and $\Pi(\varepsilon) = \Pi^H(\varepsilon)$ for a point in a current loading state with damage (see the interpretation in terms of area under the curve).

The equilibrium equation in the bar is written according to [5.28] as follows, denoting T the (tensile) force applied at the ends of the bar:

$$\sigma - c\,\nabla^2\varepsilon = T\,;\ \sigma = \frac{\mathrm{d}\Pi^{\mathrm{H}}}{\mathrm{d}\varepsilon}(\varepsilon) \tag{5.75}$$

where the parameter c reflecting the weight of non-local interactions has been considered to be spatially uniform. This differential equation is equivalent to the following first integral:

$$\frac{d}{dx}\left[\Pi^{\mathrm{H}}(\varepsilon) - \frac{1}{2}c\,(\nabla\varepsilon)^2 - T\,\varepsilon\right] = 0 \Rightarrow \Pi^{\mathrm{H}}(\varepsilon) - \frac{1}{2}c\,(\nabla\varepsilon)^2 - T\,\varepsilon = \Pi^0 \tag{5.76}$$

where Π^0 is the integration constant. To set the latter, it can be noticed that the smoothness of the strain field is $\varepsilon \in H^1$ (otherwise infinite free energy), resulting here in the continuity of ε. The interface terms of the integration by parts that led to equation [5.28] also impose the continuity of $c\,\nabla\varepsilon$, and therefore of $\nabla\varepsilon$. In this 1D problem, strain is continuously differentiable. We denote by ε^M the maximum strain in the bar for a given loading condition. By driving the solution through increasing maximal strain, the point where ε^M is reached is necessarily a loading point, so that $\Pi^H(\varepsilon^M) = \Pi(\varepsilon^M)$. In addition, the gradient of ε is zero (characterization of the maximum of a field C^1), so that the integration constant is equal to:

$$\Pi^0 = \Pi(\varepsilon^M) - T\,\varepsilon^M \tag{5.77}$$

Also, the first integral is finally written as:

$$T = \frac{1}{\varepsilon^M - \varepsilon}\left[\frac{1}{2}c\,(\nabla\varepsilon)^2 + \Pi(\varepsilon^M) - \Pi^{\mathrm{H}}(\varepsilon)\right] \tag{5.78}$$

To analyze the situation when the bar breaks, we start by getting rid of the unloading problem, by replacing Π^H with Π. The first integral is then only an inequality, but without history effect:

$$T \geq \frac{1}{\varepsilon^M - \varepsilon}\left[\frac{1}{2}c\,(\nabla\varepsilon)^2 + \Pi(\varepsilon^M) - \Pi(\varepsilon)\right] \tag{5.79}$$

Since Π is an increasing function, we have $\Pi(\varepsilon^M) \geq \Pi(\varepsilon)$, which provides a lower bound for the last two terms of [5.79]:

$$T \geq \frac{1}{\varepsilon^M - \varepsilon}\left[\frac{1}{2} c\ (\nabla\varepsilon)^2\right] \tag{5.80}$$

The term in $\nabla\varepsilon$ can also be eliminated. Actually, by denoting by ε^m the minimum of the current strain field in the bar and L the length of the bar, there indeed exists a point where:

$$|\nabla\varepsilon| \geq \frac{\varepsilon^M - \varepsilon^m}{L} \tag{5.81}$$

Moreover, at this point, we have $\varepsilon^m \leq \varepsilon \leq \varepsilon^M$ of course, the following inequality is obtained, less restrictive but sufficient for our analysis:

$$T \geq \frac{c}{2\,L^2}(\varepsilon^M - \varepsilon^m) \tag{5.82}$$

This inequality allows us to identify an important property: as long as there is a strain gradient in the bar, typically in the presence of a localized solution, the applied force T is necessarily non-zero. This is also true when the ultimate damage is reached. It therefore appears that with a gradient strain model, there is a non-zero residual tensile force when the ultimate damage state is reached, regardless of the shape of the homogeneous response (σ, ε). The zero ultimate stress property in the homogeneous constitutive is not transferred in terms of force into the bar in the presence of a localized solution, as illustrated in Figure 5.17(b).

A second property stems from the control of $\nabla\varepsilon$ in free energy. By denoting x^M and x^m the points where the *extrema* ε^M and ε^m are reached, we have the following increase:

$$\frac{c}{2\,L}(\varepsilon^M - \varepsilon^m)^2 = \frac{c}{2\,L}\left(\int_{x^m}^{x^M} \nabla\varepsilon\, dx\right)^2 \leq \frac{c}{2\,L}\left(\int_0^L |\nabla\varepsilon|\, dx\right)^2 \leq \int_0^L \frac{1}{2} c\,\nabla\varepsilon^2\, dx \tag{5.83}$$

Since the free Helmholtz energy remains finite, the difference $\varepsilon^M - \varepsilon^m$ is bounded. If the fracture strain of the local model is infinite (i.e. the stress σ asymptotically tends to zero), the ultimate state of the bar is reached for $\varepsilon^M \to \infty$

and therefore also $\varepsilon^m \to \infty$. The entire bar tends toward a state of infinite strain and there is no elastic unloading zone (the localization band has extended to the entire bar).

This is the parasitic widening of the band already mentioned in section 5.3.3 for an unbounded variable.

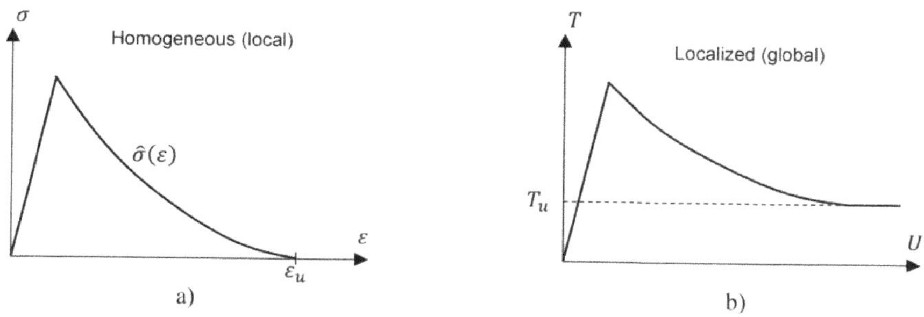

Figure 5.17. *Uniaxial tension: homogeneous and localized responses with a strain gradient model*

5.4.3.2. *Gradient damage model failure response*

The same bar is now examined assuming that the structure response is governed by a gradient damage model similar to that introduced in section 5.3.2.3. A free energy and a threshold function are considered therein as follows:

$$\Phi(\varepsilon, a) \equiv \rho(a)\, w(\varepsilon)\;;\; g(A) \equiv A - \omega \qquad [5.84]$$

The function of loss of stiffness $\rho(a)$ is continuously differentiable and cancels out for the ultimate damage $a = 1$ in power law $(1 - a)^q$, which allows us to examine the influence of the exponent q:

$$\rho(a) \equiv \rho_0(a)\,(1 - a)^q\;;\; \rho_0(1) \neq 0\;;\; q > 0 \qquad [5.85]$$

For a homogeneous response, the evolution of the strain when damage develops is governed near $a = 1$ by:

$$w(\varepsilon) = -\omega \left(\frac{\mathrm{d}\rho}{\mathrm{d}a}\right)^{-1} \approx \frac{\omega}{q\, \rho_0(a)}(1 - a)^{1-q} \qquad [5.86]$$

Assuming, for example, that w is a quadratic function (linear elasticity), it is deduced that the ultimate strain ε^u corresponding to $a = 1$ is zero if $q < 1$, finite and non-zero if $q = 1$ and infinite if $q > 1$. These three schemes are illustrated in Figure 5.18.

In the presence of a localized response, gradient terms are activated. An analytical solution was provided in Lorentz and Godard (2011). It is expressed in terms of stress σ in the bar and opening displacement of the localized area, denoted by δ as in Figure 5.11. Due to the equilibrium equation, σ is uniform in the bar; the stress corresponds to the tensile force (per unit area). It is zero when the ultimate damage is reached in $x = 0$, conversely to the example of the strain gradient model. As for the ultimate opening δ^u, it is zero if $q < 2$, finite and non-zero if $q = 2$ and infinite if $q > 2$. These schemes are also illustrated in Figure 5.18.

By comparing the results for a homogeneous response and a localized response, it can be seen that the exponents corresponding respectively to a finite ultimate strain ε^u $(q = 1)$ and a finite ultimate opening δ^u $(q = 2)$ are not identical. In addition, for a finite (non-zero) ultimate opening, the ultimate strain is infinite. This is explained by the stress–strain relation:

$$\sigma = \rho(a)\frac{\mathrm{d}w}{\mathrm{d}\varepsilon}(\varepsilon) \tag{5.87}$$

At any point $x \neq 0$, the damage has not reached its ultimate value (because the solution is localized), that is $a < 1$, so that the stiffness function ρ defined in [5.85] is non-zero. Also, since the stress is zero (ultimate state and equilibrium equation), the strain is zero. In the ultimate state, the ultimate strain is zero everywhere except for $x = 0$. The ultimate opening is finite and non-zero only if the deformation is a Dirac for $x = 0$, therefore, effectively infinite.

Again, it appears that the properties of the homogeneous model with respect to the ultimate state (zero, finite or infinite strain) do not transfer to the non-local model (zero, finite or infinite opening). It is also interesting to note that applying a local *crack band* type model accompanied by a control of the fracture energy by a Hillerborg method (see section 5.2.2), would lead to a finite opening for $q = 1$ since it is the product of the ultimate strain by the mesh size.

These remarks, as well as those in section 5.4.3.1, lead us to an important caveat. A non-local model cannot be seen as the combination of a local model and a spatial smoothing technique. In fact, the properties of the local model can be qualitatively altered significantly. Typically, in the previous example, constructing a model for which the fracture opening is finite requires choosing $q = 2$ and not $q = 1$ as the local model would suggest. Therefore, regarding the localized regime, there is no need to consider the local (or homogeneous) response when developing the model; only the non-local response makes sense.

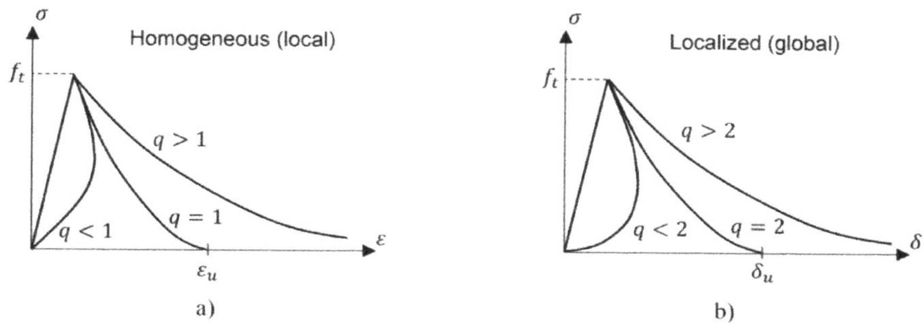

Figure 5.18. *Uniaxial tension: homogeneous and localized responses with a damage gradient model*

5.5. Numerical solving methods

In the previous sections, it has been observed that modeling quasi-brittle damage raises several very specific questions. Furthermore, not surprisingly, the responses that have been made at the level of the continuous medium have implications on the numerical implementation. It is proposed to first examine numerical formulations capable of taking into account non-locality without altering the traditional architecture of finite element codes, namely a separation between "linear algebra at the scale of the global discrete problem" and "nonlinear solution at the local scale of the material point".

The first refers to the solution of equilibrium equations, the second to the integration of constitutive relations. To this end, it is chosen to focus on the gradient formulations in sections 5.3.2.2 and 5.3.2.3, well suited to finite element treatment, after recalling the usual solving method in the presence of local nonlinear behavior. In a second step, the practical difficulties that make solving damage problems so challenging are examined and a few countermeasures drawn from the author's feedback are proposed, without claiming to be comprehensive nor any preeminence over other techniques.

5.5.1. *Impact of non-locality*

5.5.1.1. *Local formulations*

In order to illustrate the global solving approach and introduce the corresponding notations, the simple case of a local constitutive model is first examined, described by equations [5.6], [5.9], [5.13] and [5.14], which are recalled here for ease of reading.

REMINDER.–

$$\forall\, \delta u \in \mathcal{C}^0 \quad \int_\Omega \sigma : \hat{\varepsilon}(\delta u)\, d\Omega = \frac{\partial \mathcal{W}_e}{\partial u}(u) \star \delta u \; ; \; \sigma \equiv \frac{\partial \Phi}{\partial \varepsilon}(\varepsilon, a) \qquad [5.6]$$

$$\mathcal{D} = \int_\Omega A\, \dot{a}\, d\Omega \; ; \; A \equiv -\frac{\partial \Phi}{\partial a}(\varepsilon, a) \qquad [5.9]$$

$$\dot{a} = \lambda\, N(A\, ; \varepsilon, a) \qquad [5.13]$$

$$g(A;\, \varepsilon, a) - v(\lambda) \leq 0\, ; \, \lambda \geq 0\, ; \, \lambda\big[g(A;\, \varepsilon, a) - v(\lambda)\big] = 0 \qquad [5.14]$$

In general, this problem is discretized in time by means of a first-order implicit scheme, for its (relative) simplicity and stability qualities. The displacement field which is the main unknown is discretized in space by finite elements. The integrals are evaluated by means of an adequate integration scheme. We now explain each of these steps.

The displacement field u as well as its variation δu are discretized on a (same) basis of shape functions that are denoted N_n^u (the letter u is a reminder that these are shape functions for displacements):

$$u(x) = \sum_n N_n^u(x)\, U_n = [N^u(x)] \cdot [U]\, ; \; \delta u(x) = \sum_n N_n^u(x)\, \delta U_n = [N^u(x)] \cdot [\delta U] \qquad [5.88]$$

where a compact bracket-based notation is used to designates the sum on n. Here, $[U]$ gathers all the degrees of freedom of the problem. They are required to satisfy the kinematic conditions:

$$[U] \in C\, ; \; [\delta U] \in C^0 \qquad [5.89]$$

For the sake of simplicity, C is considered to be an affine manifold (which may depend on time) and C^0 the associated vector subspace. The strain operator $\hat{\varepsilon}$ can be discretized in turn. While still in the framework of small strain, $\hat{\varepsilon}$ is a linear operator.

Applied to a displacement field with a form as in [5.88], it is written as:

$$\varepsilon(x) = \hat{\varepsilon}(u)(x) = \sum_{n} \hat{\varepsilon}(N_n^u)(x)\, U_n = [B^u(x)] \cdot [U] \qquad [5.90]$$

The discretized strain operator, denoted here by $[B^u]$, depends on the derivatives of the shape functions. Now, [5.88]–[5.90] can be injected into [5.6] to obtain the discretized equilibrium equation:

$$\forall\, [\delta U] \in C^0 \quad [\delta U] \cdot \int_{\Omega} \sigma(x) \colon [B^u(x)]\, d\Omega = [\delta U] \cdot \left(\frac{\partial W_e}{\partial u} \star [N^u]\right) \qquad [5.91]$$

We denote by F_e^u the vector of external forces:

$$[F_e^u] \equiv \frac{\partial W_e}{\partial u} \star [N^u] \qquad [5.92]$$

and F_i^u the vector of internal forces:

$$[F_i^u] \equiv \int_{\Omega} \sigma(x) \colon [B^u(x)]\, d\Omega \qquad [5.93]$$

Finally, the integral is approximated by a quadrature formula (usually Gaussian quadratures for hexahedra and Hammer quadratures for tetrahedra) where x_g are the sampling points and w_g the associated weights:

$$[F_i^u] \approx \sum_{g} w_g\, \sigma(x_g) \colon [B^u(x_g)] \qquad [5.94]$$

The discretized equilibrium equation is finally written as:

$$\forall\, [\delta U] \in C^0\ [\delta U] \cdot ([F_i^u] - [F_e^u]) = 0 \Leftrightarrow [R^u] \equiv \operatorname*{proj}_{C^0}\left([F_i^u] - [F_e^u]\right) = [0] \qquad [5.95]$$

where the residual $[R^u]$ has been introduced as the orthogonal projection of $[F_i^u] - [F_e^u]$ onto C^0.

To evaluate the residual according to $[U]$, it is now necessary to calculate the stresses at each point x_g knowing the history of the strains and the internal variables at this point. For this purpose, a discretization in time is performed, denoting by t_n the sequence of calculation instants. We are interested in a given point x_g and a time step $[t_n\,;\,t]$ in which the index $n+1$ was omitted to simplify the notation.

It is considered that the different quantities of the problem ε_n, a_n and σ_n are known at time t_n and their values is looked for at time t according to the main unknown of the problem $[U]$ at time t. The strain ε is expressed with respect to $[U]$ via [5.90]; the stress is function of (ε, a) via [5.6]. Lastly, a has to be determined. According to [5.9], the driving force A is a function of (ε, a). Equations [5.13]–[5.14] are then discretized in time via an implicit Euler scheme:

$$\frac{a-a_n}{t-t_n} = \lambda\, N(A(\varepsilon, a)\,;\varepsilon, a) \qquad [5.96]$$

$$g(A(\varepsilon, a);\ \varepsilon, a) - v(\lambda) \leq 0\,;\ \lambda \geq 0\,;\ \lambda\big[g(A(\varepsilon, a);\ \varepsilon, a) - v(\lambda)\big] = 0 \qquad [5.97]$$

In favorable problems where there exists a solution[12], the two equations lead to expressing λ and especially a with respect to ε (and a_n), which is denoted as:

$$a = \hat{a}(\varepsilon)\,;\ \sigma = \hat{\sigma}(\varepsilon) \equiv \frac{\partial \Phi}{\partial a}\big(\varepsilon, \hat{a}(\varepsilon)\big) \qquad [5.98]$$

It is therefore known how $[F_i^u]$ is calculated with respect to $[U]$. The discretized problem in time and space ultimately amounts to the following nonlinear system, often solved by a Newton–Raphson method or one of its variants:

$$U \in C\,;\ [R^u]([U]) = [0] \qquad [5.99]$$

5.5.1.2. *Strain gradient formulations*

It is proposed here to examine the numerical implementation of the strain gradient formulation as introduced in section 5.3.2.2. The starting point is the variational formation of [5.5] and [5.27] of the equilibrium equations, well suited to be treated by finite elements. Initially, a time discretization is performed and the

12 Ideally, the solution is unique. If there exists several solutions a, the closest to a_n should be retained by consistency with the time discretization, which may prove to be difficult on a numerical level, even if the number of unknowns is very small at this stage.

problem is examined with a time step $[t_n\,;\,t]$. As the evolution of the internal variables is governed by a local process, the expression obtained in section 5.1.1 is found: knowing the strains and the internal variables at the beginning of the time step, the values of the latter at the end of the time step are formally written as $a = \hat{a}(\varepsilon)$, where the same notations as in section 5.1.1 are maintained while again omitting to mention the dependency on the values of the variables at the beginning of the time step.

The stress expression σ can then be recovered at the end of the time step:

$$\sigma \equiv \frac{\partial \Phi}{\partial \varepsilon}(\varepsilon, a) \quad \Rightarrow \quad \sigma = \hat{\sigma}(\varepsilon) \equiv \frac{\partial \Phi}{\partial \varepsilon}\big(\varepsilon, \hat{a}(\varepsilon)\big) \qquad [5.100]$$

At the scale of the time step, this is tantamount to solving a nonlinear elastic problem with strain gradient.

The point is actually to find the solution displacement field $u \in \mathcal{C}$ such that:

$$\forall\, \delta u \in \mathcal{C}^0 \int_\Omega \left[\hat{\sigma}(\varepsilon) : \delta\varepsilon + c\,\nabla\varepsilon : \nabla\delta\varepsilon\right] d\Omega = \frac{\partial \mathcal{W}_e}{\partial u} \star \delta u \text{ where } \varepsilon = \hat{\varepsilon}(u) \qquad [5.101]$$

A direct elementary approach consists of discretizing by finite elements the displacement field and its variation in [5.101].

This approach however presents a major drawback: the discretized displacement field must have a smoothness $\mathcal{C}^1$ (and $\mathcal{C}^2$ piecewise) to be able to calculate its second derivative. Such finite elements are constraining.

This is why the implementation of a relaxed formulation is preferred. An auxiliary field e is introduced intended to be equal to the strain ε. A Lagrange multiplier τ ensures this equality constraint.

The Lagrangian corresponding to the relaxed potential energy is then written as:

$$\mathcal{L}(u, e, \tau\,; a)$$

$$\equiv \int_\Omega \left[\Phi(\varepsilon, a) + \frac{c}{2}\,\nabla e : \nabla e + \tau : (e - \varepsilon)\right] d\Omega - \mathcal{W}_e(u) \text{ where } \varepsilon = \hat{\varepsilon}(u) \qquad [5.102]$$

Also, to stabilize the formulation, the use of an augmented Lagrangian is even recommended:

$$\mathcal{L}_r(u, e, \tau\,; a) \equiv \mathcal{L}(u, e, \tau\,; a) + \int_\Omega \frac{\mathrm{r}}{2}(e - \varepsilon)^2\, d\Omega \qquad [5.103]$$

In such a form, a connection can be observed with the micromorphic models of section 5.3.2.4 where the auxiliary strain e acts as the kinematic variable χ. However, unlike micromorphic models, the quadratic augmentation term does not influence the solution of the continuous problem since the Lagrange multiplier ensures the constraint $e = \varepsilon$. The value of the augmentation coefficient $r \geq 0$ is independently chosen of the constraint $e = \varepsilon$.

This Lagrangian augmentation term should therefore not be taken for a penalization method in which the penalization coefficient must be high enough to ensure the constraint in an approximate way, with the resulting ill-conditioning and its impact on the convergence of solving algorithms. The interest of this augmentation term will be later examined.

The solution (u, e, τ) to the problem is an augmented Lagrangian saddle point. It is characterized by the following three stationarity conditions:

$$\begin{aligned} &\forall\, \delta u \in \mathcal{C}^0 \quad 0 = \frac{\partial \mathcal{L}}{\partial u}\big(u, e, \tau\,; \hat{a}(\varepsilon)\big) \star \delta u \\ &= \int_\Omega \Big[\hat{\sigma}(\varepsilon) - \tau + r(\varepsilon - e)\Big] : \delta\varepsilon\, d\Omega - \frac{\partial W_e}{\partial u}(u) \star \delta u \end{aligned} \qquad [5.104]$$

$$\begin{aligned} &\forall\, \delta e \quad 0 = \frac{\partial \mathcal{L}}{\partial e}\big(u, e, \tau\,; \hat{a}(\varepsilon)\big) \star \delta e \\ &= \int_\Omega \Big[c\, \nabla e \,\vdots\, \nabla \delta e + \big(\tau - r(\varepsilon - e)\big) : \delta e\Big]\, d\Omega \end{aligned} \qquad [5.105]$$

$$\forall\, \delta\tau \quad 0 = \frac{\partial \mathcal{L}}{\partial \tau}\big(u, e, \tau\,; \hat{a}(\varepsilon)\big) \star \delta\tau = \int_\Omega (e - \varepsilon) : \delta\tau\, d\Omega \qquad [5.106]$$

The fields (u, e, τ) and their variations by finite elements can then be discretized. Reusing the notations of section 5.1.1, which are completed with $[N^e]$ and $[B^e]$ the shape functions to discretize the field e as well as their gradient and $[N^\tau]$ the shape functions to discretize the field τ. The discretized system corresponding to [5.104]–[5.106] is then written (without specifying the quadrature formula, identical for the three equations) as:

$$[R^u] \equiv \underset{C^0}{\mathrm{proj}}\left([F_i^u] - [F_e^u]\right) = [0] \; ; \; [F_i^u] \equiv \int_{\Omega} \left[\hat{\sigma}(\varepsilon) - \tau + r(\varepsilon - e)\right] : [B^u] \, d\Omega \qquad [5.107]$$

$$[R^e] \equiv \int_{\Omega} \left[c \, \nabla e \, \vdots \, [\mathrm{B}^e(x)] + \left(\tau - r(\varepsilon - e)\right) : [N^e(x)]\right] d\Omega = [0] \qquad [5.108]$$

$$[R^\tau] \equiv \int_{\Omega} (e - \varepsilon) : [N^\tau(x)] \, d\Omega = [0] \qquad [5.109]$$

where the definition of $[F_e^u]$ is identical to that given in [5.92] and:

$$\varepsilon = [B^u] \cdot [U] \; ; \; e = [N^e] \cdot [E] \; ; \; \nabla e = [B^e] \cdot [E] \; ; \; \tau = [N^\tau] \cdot [T] \qquad [5.110]$$

This nonlinear algebraic system is a function of the degrees of freedom $[U]$, $[E]$ and $[T]$; it can be solved, for example, by a Newton–Raphson method (monolithic solution). Only derivatives of at most the first-order can be seen in the stationarity conditions, so that continuous finite elements (and C^1 piecewise) should be adequate. Quadratic shape functions are recommended for displacement and linear functions for the auxiliary strain e and the Lagrange multiplier τ. It is actually preferable to have an order of difference between the discretization of u and that of e to reduce the locking risks between ε and e which appear at the same level in the equations (as in thermo-mechanics, where an order of difference between the discretization of displacement and temperature is recommended). The chosen discretization order of τ is the same as that of e to verify the LBB condition (see, for example, Ern and Germond 2004; Boffi et al. 2013), while sufficiently constraining the auxiliary strain: $[N^e] = [N^\tau]$.

By comparing [5.28] to [5.104], an interpretation can be given to the Lagrange multiplier τ. After spatial convergence, it indeed acts as $\operatorname{div} \mathrm{S}$. In addition, the role of the augmentation term can also be reconsidered. The softening behavior is present in the function $\hat{\sigma}(\varepsilon)$ while the non-local regularization stems from ∇e. Despite that the fields e and ε are intended to be equal, this is only approximately achieved after spatial discretization because the discretization space of $\hat{\varepsilon}(u)$ is not included in that of τ. For this numerical reason, control over ∇e may not be sufficient to smooth the problem in u, leading to parasitic localizations of the strain. This is where the augmentation term comes in: it is interpreted in [5.104] as a spring of stiffness r with respect to the strain ε. If r is larger than the absolute value of the softening modulus (more precisely, the smallest eigenvalue of the acoustic tensor), then the behavior with respect to ε and for given (e, τ) is no longer softening: the parasitic oscillations disappear. In addition, it should be remembered that the influence of r on the discretized solution decreases with spatial refinement (r does not play any role after spatial convergence).

5.5.1.3. *Gradient formulation of internal variables*

Solutions for the numerical implementation of damage gradient models presented in section 5.3.2.3 are examined here. Again, the focus is on solving the semi-discretized problem in time, over a time step $[t_n \,;\, t]$ for which all the variables of the problem are known at time t_n.

A first example concerns generalized standard materials, for which the time semi-discretized problem can be interpreted as the minimization of the following total energy (see section 5.3.1.3):

$$\begin{aligned} &\min_{u,a} \mathcal{E}_{tot}(u,a) \qquad \text{where} \quad \mathcal{E}_{tot}(u,a) \\ &\equiv \mathcal{F}(u,a) + \Delta t\, \widehat{\mathcal{D}}\left(\frac{a-a_n}{\Delta t}\right) - \mathcal{W}_e(u) \end{aligned} \qquad [5.111]$$

where $\widehat{\mathcal{D}}$ denotes the dissipation potential, as defined in section 5.3.1.3. By achieving simultaneous minimization with respect to the pair (u, a), only stable states are taken into account. The class of phase field models presented in section 5.3.4 typically falls within this framework, since the total energy is its basis. The two variables of the problem, u and a, appear in [5.111] through their values and gradients.

A finite element Lagrange discretization is therefore possible:

$$u = [N^u] \cdot [U] \,;\, \varepsilon = [B^u] \cdot [U] \,;\, a = [N^a] \cdot [A] \,;\, \nabla a = [B^a] \cdot [A] \qquad [5.112]$$

The problem is then reduced to solving a discrete minimization. In the literature, linear shape functions are usually employed to discretize both damage and displacement. To minimize the risks already mentioned in section 5.5.1.2 of strain/damage locking (as in thermo-mechanics), a combination of linear shape functions for damage and quadratic for displacement may be preferred; this is a choice which in addition leads to more accurate results for a given mesh but which also incurs higher costs.

A common practice in the phase field model community consists of achieving the minimization of the discretized total energy by a succession of alternating minimizations with respect to $[U]$ (elastic problem with given damage) then with respect to $[A]$ (damage evolution for given strains), and until convergence. An alternative consists even in not iterating until convergence but to progress directly to the next time step when the first iteration completes. In this case, this is referred to as a staggered scheme but it requires potentially very fine time steps to achieve time convergence; otherwise, the predicted crack paths may be wrong (Gerasimov and De Lorenzis 2016; Seles et al. 2019). The performance gain of the staggered scheme is not demonstrated and the approach seems risky for lack of convergence indicators: it is preferable to iterate until the minimum is reached.

The choice of alternating minimization was originally motivated, for the simplest models, by the quadratic nature of the energy with respect to u and with respect to a independently, thus by solving two linear problems at each iteration, usually with the guarantee of the (potentially slow) convergence of the process. For richer models, for example, when the elasticity is no longer linear due to the unilateral stiffness/damage coupling seen in section 5.4.2 or the presence of other dissipative processes such as plasticity, the alternating scheme may lose its simplicity and efficiency. A simultaneous minimization with respect to $[U]$ and $[A]$ is then preferable: this is referred to as monolithic problem solution. However, the guarantees of convergence of the iterative minimization process are reduced, in particular because of the lack of convexity of the total energy with respect to the pair (u, a).

This numerical approach by minimization of the total energy has its interest, especially because of the availability of well-known optimization algorithms. Nonetheless, two main issues have to be dealt with. On the one hand, it is based on the existence of a total energy that does not exist when leaving the framework of generalized standard materials and which, even in the best case, can be difficult to express (Legendre transform of a threshold function to construct the dissipation potential). On the other hand, the total energy is generally not a smooth function: it is not necessarily differentiable (this is often the case when internal variables are not restricted to a single scalar) and contains terms that refer to constraints on variables, typically an irreversibility constraint $a \geq a_n$ (depending on the model, these constraints can be non-linear). However, an optimization problem under inequality constraints is (much) more challenging to deal with, despite the fact that there are of course dedicated algorithms.

Given these issues, another approach, based on Lagrangian relaxation, may be preferable, as in section 5.5.1.2, which is also named the decomposition-coordination method for solving optimization problems (Fortin and Glowinski 1983).

Since a direct approach based on [5.35] would be very difficult to implement because of the second spatial derivative of a, an auxiliary damage variable b is introduced intended to be equal to a by virtue of the introduction of a Lagrange multiplier λ. For stability reasons, a quadratic augmentation term is also introduced whose coefficient is denoted by r.

The (augmented) Lagrangian corresponding to the relaxed potential energy is written as follows:

$$\mathcal{L}_r(u,a,\lambda,b) \equiv \int_\Omega \left[\Phi(\varepsilon,b) + \frac{c}{2}\,\nabla a\cdot\nabla a + \lambda\,(a-b) + \frac{r}{2}(a-b)^2\right] d\Omega - \mathcal{W}_e(u) \qquad [5.113]$$

It can be noted that a acts as damage concerning non-local interactions, that is, at the global structural scale, while b is involved at the level of strong local constitutive nonlinearities; this partition is motivated by a separation between "global quadratic *versus* local non-linear". Dissipation is thus carried by b while a and λ do not contribute to it.

Following the approach in [5.8], the driving force of damage is defined as:

$$\mathcal{B}\star\delta b \equiv -\frac{\partial\mathcal{L}_r}{\partial b}(u,a,\lambda,b)\star\delta b = \int_\Omega \tilde{B}\,\delta b\ d\Omega \text{ where } \tilde{B} = -\frac{\partial\Phi}{\partial \mathrm{b}}(\varepsilon,b) + \lambda\ + r\,(a-b) \qquad [5.114]$$

Also, the evolution of b is governed by the same local relation as in [5.12]:

$$\dot{b}(x) \leftarrow \Gamma\left(\tilde{B}(x)\,;\ \varepsilon(x), b(x)\right) \qquad [5.115]$$

The time discretization scheme makes it possible to express b at the end of the time steps by integrating the constitutive law, which can be formally written as $= \hat{b}$

$(\varepsilon, a, \lambda)$. The stationarity conditions of the Lagrangian with respect to a and λ subsequently lead to the following variational equations:

$$\forall\, \delta a \quad 0 = \frac{\partial \mathcal{L}_r}{\partial a}(u, a, \lambda, b) \star \delta a$$
$$= \int_\Omega \left[c\, \nabla a \cdot \nabla \delta a + \lambda\, \delta a + r\,(a - b)\, \delta a \right] d\Omega \qquad [5.116]$$

$$\forall\, \delta\lambda \quad 0 = \frac{\partial \mathcal{L}_r}{\partial \lambda}(u, a, \lambda, b) \star \delta\lambda = \int_\Omega (a - b)\, \delta\lambda\, d\Omega \qquad [5.117]$$

The definition of internal forces and the equilibrium equations remain unchanged:

$$\forall\, \delta u \in \mathcal{C}^0 \quad 0 = \frac{\partial \mathcal{L}_r}{\partial u}(u, a, \lambda, b) \star \delta u$$
$$= \int_\Omega \sigma : \delta\varepsilon\, d\Omega - \frac{\partial \mathcal{W}_e}{\partial u} \star \delta u \text{ where } \sigma \equiv \frac{\partial \Phi}{\partial \varepsilon}(\varepsilon, b) \qquad [5.118]$$

where the stresses can also be formally denoted $\sigma = \hat{\sigma}(\varepsilon, a, \lambda)$.

It can easily be shown that this relaxed formulation is equivalent to the original formulation [5.31]–[5.35].

Actually, $a = b$ as foreseen according to [5.117]. In this case, the stresses and the equilibrium equation in [5.118] coincide with [5.31]. By integration by parts of [5.116], it is obtained that $\nabla a \cdot n = 0$ at the edge of the structure and the interpretation of the Lagrange multiplier is provided as: $\lambda = \mathrm{div}(c\, \nabla a)$. It implies that the thermodynamic force $\tilde{A}$ in [5.33] is equal to $\tilde{B}$ in [5.115] and therefore that the evolution equations of [5.35] and [5.115] coincide, which finalizes our demonstration of the agreement of the two formulations for the continuous problem in space.

This relaxed formulation is well suited for a finite element discretization. Indeed, the fields u, a and λ are discretized using shape functions $[N^u]$, $[N^a]$ and $[N^\lambda]$ while ε, σ and b are treated at the level of integration points, classically since they are quantities specific to the constitutive law.

The discretized system is written (without explicitly expressing the quadrature formula, identical for the three equations) as:

$$[R^u] \equiv \underset{C^0}{\text{proj}}\left([F_i^u] - [F_e^u]\right) = [0]\ ;\ [F_i^u] \equiv \int_{\Omega} \sigma : [B^u]\, d\Omega \qquad [5.119]$$

$$\int_{\Omega} \Big[c\, \nabla a \cdot [B^a(x)] + \lambda\, [N^a(x)] + r\,(a-b)\, [N^a(x)]\Big]\, d\Omega = [0] \qquad [5.120]$$

$$\int_{\Omega} (a-b)\, \big[N^\lambda\big]\, d\Omega = [0] \qquad [5.121]$$

where the displacement equation is the same as [5.95] and:

$$\begin{aligned} &\varepsilon = [B^u] \cdot [U]\ ;\ a = [N^a] \cdot [A]\ ;\ \nabla a = [B^a] \cdot [A]\ ;\ \lambda = \big[N^\lambda\big] \cdot [L] \\ &b = \hat{b}(\varepsilon, a, \lambda)\ ;\ \sigma = \hat{\sigma}(\varepsilon, a, \lambda) \end{aligned} \qquad [5.122]$$

For the same reasons as in section 5.5.1.2, quadratic shape functions are recommended for discretizing u and linear for a and λ, that is, $[N^a] = \big[N^\lambda\big]$.

Finally, the augmentation coefficient r (not to be mistaken for a penalization) is chosen to avoid parasitic oscillations, again as in section 5.5.1.2. As a matter of fact, non-local regularization stems from the term in ∇a and the softening results from (u, b). Since the Lagrange multiplier λ does not exactly guarantee $a = b$ in the discretized problem localization patterns are possible.

The quadratic term proportional to r provides a form of positive strain hardening with respect to b which opposes parasitic localization: r is interpreted as a strain hardening modulus. If it is greater than the absolute value of the actual softening modulus, the problem in (u, b) becomes well-posed with given (a, λ), ending any parasitic localization. A numerical confirmation of this property can be found in a strain softening gradient plasticity model (Scherer 2020).

5.5.2. *Difficult to perform computations*

Numerical simulations in the presence of quasi-brittle damage are generally cumbersome, because they consist of large-sized problems, and they are difficult to carry out; the solving algorithms are met with some challenges in terms of convergence: according to the chosen solving method, oscillations of residuals $[R]$ are observed, as well as very slow convergence, cycles, even divergence during iterations.

5.5.2.1. *Large-sized problems*

To obtain a reasonably accurate solution in a finite element computation, a fine mesh is conventionally recommended in areas where geometric details or loading gradients cause a fast spatial variation of the solution (excluding singularities). This still holds for damage problems but with specificities that considerably increase the number of unknowns (nodal degrees of freedom) of the problem:

– The elements must no longer be fine relatively to the characteristic scales of the geometry and loading but relatively to the non-local characteristic scale w_{FPZ} (introduced in section 5.2.4), which itself ought to be small compared to the geometry and loading scales. The elements are therefore finer than in a local nonlinear calculation due to this embedded scale. In practice, four elements seem a minimum to describe a damage profile as in Figure 5.11. They are therefore at least four times smaller than in a conventional nonlinear calculation, which leads in 3D to 64 times more elements. It should also be remembered that even when there are no structural details, an artificially high scale w_{FPZ} cannot be chosen because it is upper bounded by a length derived from the characteristics of the material (of about the length of the FPZ). To overcome this upper bound, a Griffith-type fracture model has to be considered, as in section 5.2.4.

– The areas where damage develops are not necessarily a priori known; moreover, they evolve during loading (propagation). With fixed mesh, this requires anticipating large portions of the finely meshed structure. If the essential of the structure nevertheless remains linear elastic, static condensation techniques make it possible to reduce the cost of the global linear algebra phase since the matrices remain identical from one iteration to another (Llau et al. 2015). Another potentially complementary approach consists of adapting the mesh during the evolution of the damage, such that only the damaged areas and their immediate periphery are finely discretized (see an example in Figure 5.19). This solution has been studied in the literature (Rodriguez-Ferran and Huerta 2000; Patzak and Jirasek 2004; Badnava et al. 2018; Nagaraja et al. 2018), but more often for ductile damage; it is also covered in Chapter 7. Although attractive, it presents two kinds of drawbacks: on the one hand, the projection of mechanical fields from one mesh onto another, and on the other hand, the correction of fields after projection to find a state that satisfies the discretized equations of the problem (equilibrium and non-locality). This last step can be challenging because of the problems of convergence raised by damage phenomena, as will be seen hereafter.

– Given the Lagrangian formulations introduced in section 5.5.1, the finite elements that were used have more degrees of freedom per mesh than conventional formulations. This increases the problem size, typically from 16% for a damage gradient formulation up to 100% for a strain gradient formulation (based on a displacement quadratic hexahedron linear with respect to Lagrangian variables).

The size of the discretized problems has of course a direct impact on the performance of the calculation which may prove incompatible with industrial or computer constraints (for example, memory footprint), even when relying on massive parallelism. Moreover, this size only exacerbates the convergence problems inherent in quasi-brittle damage, as detailed in section 5.5.2.2.

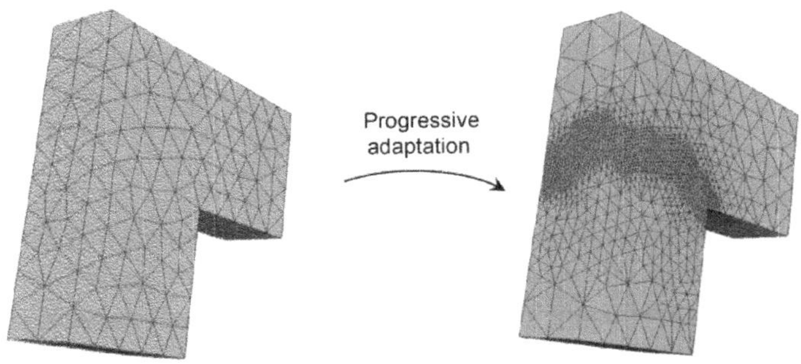

Figure 5.19. *Structure under opening and shear loading: compatible mesh adaptation. For a color version of this figure, see www.iste.co.uk/lebon/mechanics.zip*

5.5.2.2. *Damage pattern cycles*

As seen above, the localization mechanism reflects a "competition" between neighboring material points: the damage in one causes elastic unloading in its neighbors by lowering the stresses withstood. There is then a risk that the solving algorithm constantly switches from one pattern of damaged points to another, hence the oscillations observed in the residuals.

This phenomenon is (somewhat) reduced by non-local interactions because they compel neighboring material points into some form of "co-operation": a point that is damaged causes damage to its neighbors by gradient effect. In contrast, the existence of multiple, potentially unstable solutions worsens this problem.

To constrain these possible oscillations, some solving methods can enforce one damage configuration at the expense of the others:

– An explicit or semi-explicit scheme (Oliver et al. 2008; Miehe et al. 2010) no longer seeks to satisfy the equilibrium equations during the current time step and attempts to correct the lack of equilibrium during the next step.

– With an implicit scheme, the quality of solving the equations can also be temporarily or permanently lowered (by increasing or disregarding the convergence thresholds). However, this practice can hardly be recommended, since it accepts to deviate in an uncontrolled way from the equations of physics, at the risk of introducing significant errors (see Gerasimov and De Lorenzis 2016; Seles et al. 2019). On the contrary, the use of reasonably objective convergence criteria with respect to units (different degrees of freedom may have different units; see the hybrid formulations presented in section 5.5.1) and mesh size (nodal residuals are related to local mesh size) is even recommended (see, for example, Lorentz and Badel 2011).

– Another approach consists of freezing the load and allowing a succession of unbalanced intermediate states until the imposed load is found (Michel et al. 2018): unlike the previous crude solution, the targeted load ends up in equilibrium via a temporarily unbalanced path, which can also be interpreted as an increase of the damage in the constitutive law over a time step.

All these methods have the advantage of reinforcing a damage pattern, but with the risk that the pattern in question is not one of those achievable in a continuous time process: it is the intensity of the allowed momentum unbalance that selects the achievable patterns.

When the problem can be expressed as a minimization of energy, as in [5.111] for example, the cycles between neighboring patterns can be interpreted in terms of the (very strong) non-convexity of the energy in question, of which a very schematic illustration is given in Figure 5.20. It can be seen that it is particularly perturbing for any strategy based on the local energy curvature and the search for a stationary point, such as the Newton–Raphson method.

Methods based on the use of a BFGS-type "mean curvature" (Bonnans et al. 2006) or an artificial but positive definite curvature such as the use of the elasticity operator instead of the tangent matrix can sometimes address it somewhat, but usually at the cost of very slow convergence rate.

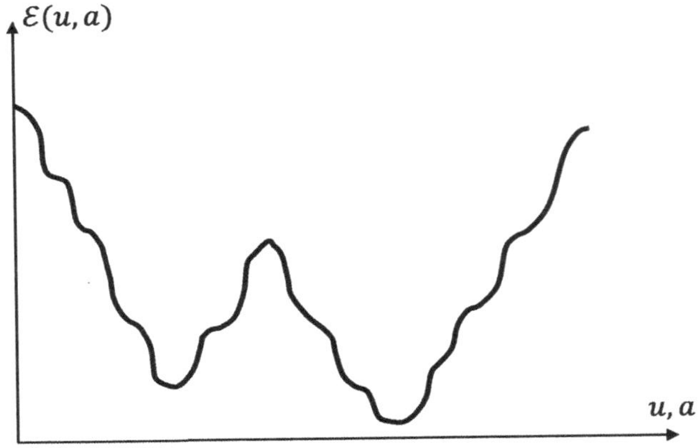

Figure 5.20. *Schematic representation of total energy according to state variables: fast changes in curvature and multiplicity of solutions*

Other solutions, based on the construction of a series of iterations of decreasing energies, may also prohibit these cycles: this is the idea of alternating minimization already mentioned in section 5.5.1.3 that also has the advantage of favoring stable solutions. Linear search methods based on the energy value can also lead thereto (Bonnans et al. 2006). Convergence may however also be slow.

5.5.2.3. *Structural instabilities*

A second set of numerical difficulties stems more directly from the brittle nature of damage, which leads to potential structural instabilities. To understand its origin, we may reconsider the energy balance expressed in equation [5.8]:

$$\mathcal{P}_e = \dot{\mathcal{F}}_{el} + \dot{\mathcal{F}}_d + \mathcal{D} \qquad [5.123]$$

where a distinction has been made in the Helmholtz free energy between elastic energy and damage-related energy. When damage propagates, this is reflected by a cost in terms of energy consumed by the mechanism (stored plus dissipated): $\dot{\mathcal{F}}_d + \mathcal{D} > 0$. On the other hand, the stress reduction that is associated with structural softening and equilibrium releases part of the elastic energy: $\dot{\mathcal{F}}_{el} < 0$. The quasi-brittle nature of damage refers to a moderate energy consumption, while the restoration of elastic energy is all the greater when the unloading area is large. The latter can prevail and lead to $\mathcal{P}_e < 0$. To preserve a quasi-static situation, the external environment must therefore absorb the energy released. If this is not possible, a dynamic evolution begins and a term of kinetic energy variation $\dot{K}$

complements the balance equation [5.123]: the structure becomes unstable. Such a situation is schematically illustrated in terms of the overall response of a damaged structure in Figure 5.21(a): the imposed displacement U and the force F are dual quantities, so that $\mathcal{P}_e = F\,\dot{U}$, the area under the curve is interpreted as the energy brought by the external environment. To allow a quasi-static and continuous evolution in terms of damage evolution (no damage jump), it is necessary to lower the displacement as soon as the critical point C is reached; this is then called a *snap-back* (as in, for example, Figure 5.13).

For a real load U that continues to increase beyond the point C, a rigorous dynamic analysis is required (Dumouchel et al. 2008). However, the corresponding numeric schemes must be adapted to capture wave propagations that result from instability (Zhou et al. 2005; Gravouil et al. 2009; Bourdin et al. 2010; Radovitsky et al. 2011; Doyen et al. 2012; Li et al. 2016; Aduloju and Truster 2019). To achieve quality numerical simulations, a huge challenge has to be overcome, not to mention the use of very small time steps, and it might be preferable to remain within the framework of quasi-static modeling, at the cost of a few trade-offs.

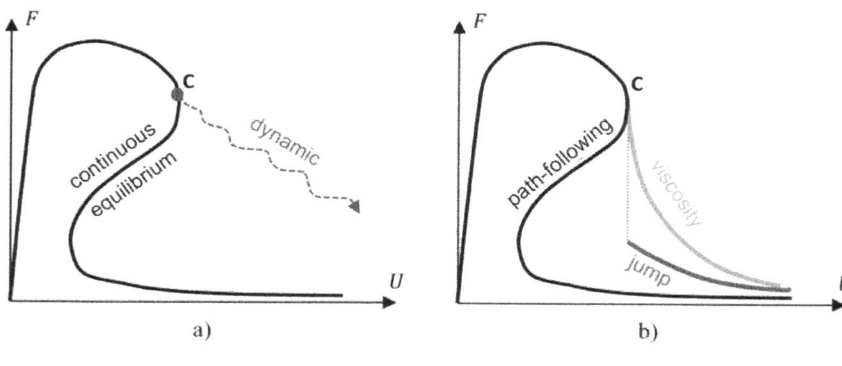

Figure 5.21. *Schematic response of a structure in the presence of instability. For a color version of this figure, see www.iste.co.uk/lebon/mechanics.zip*

– *Solution jumps*: we may be tempted to cross the snap-back abruptly, as illustrated by the blue dotted line in Figure 5.21(b). Since the arrival state is a priori far removed from the mechanical state at the point C, a Newton–Raphson method will generally result in failure (its convergence is limited to a basin of attraction around the solution). On the other hand, more robust – but potentially slow – methods have been mentioned in section 5.5.2.1: unbalanced intermediate state paths, alternating minimization, quasi-Newton method with elastic operator. The energy consumed by the structure is artificially increased by the numerical scheme (kinetic energy production ignored). It can also be noted that the state of arrival does not a priori coincide with that which would be obtained by following equilibrium states and with continuity of damage evolution (black response in Figure 5.21(b), due to the dependency on the loading history). Finally, if the whole line in blue dotted lines is interpreted as a single time step with infinitely small ΔU, the damage increment Δa is finite therein: the incremental constitutive law no longer corresponds to the discretization of a continuous law in time; a set of algebraic equations replaces the physics of the model.

– *Loading control*: an alternative solution consists of actually calculating the black response in Figure 5.21(b). To this end, it is necessary to control the intensity of loading so as to follow this response consisting of infinitely close equilibrium states, that is, for which damage evolution is continuous. This is referred to as path-following methods (Crisfield 1981; de Borst 1987; Hellweg and Crisfield 1998). The loading intensity, denoted η, becomes an additional unknown of the discretized problem during a time step, while an additional equation is introduced for controlling the displacement increment via a path-following function P. In a simple version, it is written as:

$$\begin{cases} [F_i] = \eta\, [F_e] \\ P([\Delta U]) = \Delta t \end{cases} \qquad [5.124]$$

The effectiveness of the method depends on the choice of path-following function P. The functions recommended for overcoming geometric instabilities (buckling), which relate to a norm $\|[\Delta U]\|$ and are therefore called arc-length methods, are not suitable, in particular because they do not distinguish global elastic unloading from unloading with damage. Several choices have been proposed in the literature to avoid this problem: optimized local displacement jump control (Geers 1999), maximal damage increment control (Lorentz and Badel 2004) and dissipation control during the time step (Gutierrez 2004). These are practically proven methods but they require many time steps to follow the black curve; they can therefore also prove to be slow. In addition, they proceed by adjusting the load in a non-monotonic way (η is brought to decrease and then increase again), which generally does not correspond to the reality of the imposed loading. This is especially true when in addition to mechanical loading, thermal loading takes place (or any other phenomenon in which time is physically involved).

– *Viscosity*: the introduction of viscosity into the problem modeling leads to the occurrence of (viscous) forces that guarantee system equilibrium at all times as loading increases. This is the answer shown in orange in Figure 5.21(b). This viscosity can be real or artificial. In the latter case, a simple model might as well be preferred and whose effect can be overlooked beside instabilities. This artificial viscosity can be introduced at the level of damage evolution (such as viscoplasticity which slows down plasticity evolution) via a viscosity function v in the evolution (equation [5.14]). In this case, the continuous rate (tangent) problem is written as:

$$\forall\ \delta u \in \mathcal{C}^0 \quad \dot{\mathcal{S}} \star \delta u = \frac{d}{dt}\left(\frac{\partial \mathcal{W}_e}{\partial u}\right) \star \delta u\ ;\ \dot{\mathcal{S}} = \frac{\partial^2 \mathcal{F}}{\partial u^2} \star \dot{u} + \underbrace{\frac{\partial^2 \mathcal{F}}{\partial u\, \partial a} \star \dot{a}(u,a)}_{\equiv \mathcal{s}(u,a)} \qquad [5.125]$$

It is important to observe that the rate of damage $\dot{a}$ depends only on the current state (u, a) due to viscosity. The tangent problem is therefore interpreted as an elastic problem with prestress. This is a well-posed problem (as long as the elastic operator verifies the required coercivity and convexity properties). A response exists, whatever the load evolution: the unstable phase can be crossed. Also, the viscous effects remain confined to areas that become damaged. On the other hand, at the computing level, the constitutive law is modified and requires additional work.

Viscosity can also be introduced through viscous overstress σ_v (an additional internal variable), as in the rheological model shown in Figure 5.22 (see, for example, Chen 2019). It is based on two parameters: k_v the stiffness of the viscous branch and τ_v the characteristic time of the viscosity.

If $\mathbb{H}$ denotes the tangent bilinear form of the problem without viscosity, that of the problem with viscosity is written (in square brackets) as:

$$\dot{\mathcal{S}} \star \delta u = \left[(\mathbb{H} \star \dot{u}) \star \delta u + \int_\Omega k_v\, \dot{\varepsilon} : \delta\varepsilon\, d\Omega \right] - \int_\Omega \frac{1}{\tau_v} \sigma_v : \delta\varepsilon\, d\Omega \qquad [5.126]$$

Unlike the previous situation, k_v has therefore to be chosen high enough so that the problem is well posed in $\dot{u}$ and makes it possible to respond to any load evolution.

In addition, viscosity is now generalized throughout the structure. On the other hand, in computing terms, the introduction of viscosity here has no impact on the damage constitutive law (superposition of the stresses deriving from the two models).

Independently of the chosen viscosity model, it consumes more energy than the response following the *snap-back* (the area under the curve $F - U$ is larger). This additional amount of energy is consumed by the viscous mechanism and essentially non-recoverable, unlike a dynamic mechanism. However, a solution branch can be constructed along which damage evolution is continuous and therefore each point is in equilibrium (including the viscous effects). As a small characteristic viscosity time is chosen compared to the loading time in order to restrict viscous effects beside instabilities, small time steps are required to follow the viscous branch and therefore, once again, potentially long computational times.

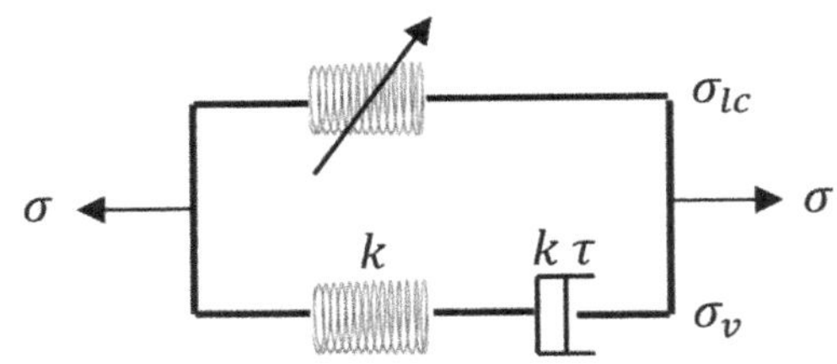

Figure 5.22. *Viscoelastic regularization*

5.5.2.4. *Damage-fracture transition*

For damage models designed to describe behavior up to fracture, totally "broken" points result in zero residual stress and stiffness, at least in some loading directions (see sections 5.4.2 and 5.4.3). This zero stiffness may become a problem with respect to the finite element solution of the equilibrium problem of [5.5]. A typical case is illustrated in Figure 5.23 where a structure is submitted to vertical tension. The optimal energy solution requires only one band of totally broken elements, according to Figure 5.23(a); in this solution, all degrees of freedom remain "attached" to the structure (no artificial rigid body motion). On the other hand, Figure 5.23(b) presents another situation of broken elements for which degrees of freedom (in red) are no longer attached to the structure (they no longer give rise to any strain energy) and the problem arising from the discretization of the equilibrium equation of [5.5] becomes ill-posed, which often results in singular or very ill-conditioned matrices. A (very) special case is that of a symmetric problem for which it might be tempting to represent only half of the structure, according to Figure 5.23(c): the horizontal movements of the red nodes are then no longer controlled (the vertical movements are set by the symmetry condition). These patterns in which the damaged area is too large can occur during equilibrium iterations or even at convergence. Different approaches are used to overcome this issue:

– *Maintain fictitious stiffness and stress at failed points*: this artifice has the inconvenience of introducing cohesive forces through the totally damaged areas, all the higher as the opening is large. Artificial stiffness must therefore remain "small". However, the trade-off between negligible cohesive forces and finding reasonable matrix conditioning is often challenging and would require rigorous systematic parametric analyses.

– *Maintain artificial algorithmic stiffness without effect on constraints*: this is a more attractive solution than the previous one insofar as the constitutive law (and therefore the solution) is not altered. On the other hand, experience feedback shows that this algorithmic gap between tangent matrix and stresses results in a convergence rate that is all the slower for larger damaged area. Unsurprisingly, the problem already mentioned is retrieved when using the elastic operator in the solution algorithm.

– *Remove "broken" elements from modeling*: this is a relatively widely used option (see, for example, Besson and Desmorat 2004). Several challenges have however to be solved. First, it is necessary to determine when an element needs to be removed from modeling. Generally, the constitutive law is evaluated at several integration points per element; however, the absence of stiffness at a single integration point does not imply a rigid mode for the element in question. In addition, loss of stiffness has to be anticipated (otherwise, convergence might fail during the current iteration), but moderately, so as not to perturb the solution or jeopardize convergence because of a stress jump too abrupt. In addition, the – definitive – removal of the finite element means that total loss of stiffness occurs, in contradiction with the proposition of maintaining a unilateral stiffness developed in section 5.4.2. Finally, the problem equations governing the non-local character of behavior would also be altered, wrongly because even a totally damaged point a priori continues to contribute thereto (Geers et al. 1998; Seupel et al. 2018), hence the need to provide specific treatment for the boundary conditions of non-local equations.

– *Maintain a time evanescent artificial stiffness*: this involves introducing an artificial viscosity based on the strain rate (viscoelasticity), as described in section 5.5.2.3. "Broken" elements have an instantaneous stiffness that drops over time. It is a method that has proven itself in ductile fracture (Chen 2019) but that requires properly calibrating the viscosity parameters relatively to the loading rate in order to not cause significant artificial overstresses.

– *Preferentially select solutions without excessive damaged areas*: to give preference to the type of solution in Figure 5.23(a) rather than in Figure 5.23(b), one way to proceed consists of ensuring that damage follows a continuous time evolution and that solution jumps are avoided. To this end, reference is made to the path-following or viscosity techniques introduced in section 5.5.2.3, which allow controlling the damage increment and thus to approximate a continuous time evolution by refining the time discretization.

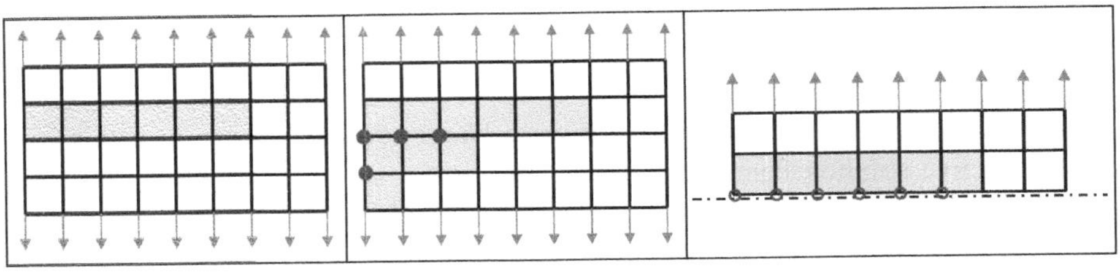

a) b) c)

Figure 5.23. *Different patterns of totally damaged finite elements and free degrees of freedom (in orange, broken elements without stiffness; in white, sound or partially damaged elements). For a color version of this figure, see www.iste.co.uk/lebon/mechanics.zip*

Besides the problems of ill-conditioned or singular matrices, considering a totally damaged state raises other questions. Therefore, we saw in section 5.4.2 a limitation to the unilateral modeling of residual stiffness: sliding movements in the damaged area (mode II or III) lead to the occurrence of spurious stresses that can lead to numerical locking; the crack does not open as much as it should and stresses are transferred through the crack.

Another limitation concerns certain non-local models based on strain control, as in section 5.3.3: the high strains that develop in the cracked zone artificially increase the surrounding strains by non-local interactions, leading to an artificial widening of the localization band as well as spurious dissipation.

The advantage of introducing a real crack in the structure, oriented by the knowledge of the damaged areas, is that of being able to treat these two limits since displacement jumps then no longer have an impact on the constitutive law. This is probably the most satisfactory solution at the physical level, despite remaining to this day numerically challenging because of the various problems to which it refers: geometric identification of the crack path, choice of the time for effectively introducing the crack, crack behavior model (free lips, contact, contact-friction and cohesive model), impact on non-local equations, stress state rebalancing, to name but a few. The interested reader is referred to Chapter 7, in which these questions are addressed. We should however keep in mind that the transition to a crack is all the more stable for the calculation as it takes place in an already broken area and sufficiently behind the FPZ, which militates for the implementation of the other techniques mentioned, at least transiently before the actual insertion of the crack.

5.6. Conclusion

Modeling quasi-brittle fracture using damage models proves to be a relevant approach for describing both the initiation and the onset of cracks as well as the paths and propagation kinetics, until reaching the ultimate failure of the structure. However, it faces several theoretical and numerical difficulties that make its application difficult in practice and requires a certain degree of attention: in the light of current knowledge, the computation of structures in the presence of damage cannot be trivialized. In this respect, it is proposed to repeat here the main messages of the study:

1) The use of damage models should preferably be confined to situations in which the prediction of crack propagation is to be achieved according to a priori unknown paths. Otherwise, simpler formulations are available (onset criteria, cohesive zone models).

2) The damage tends to be located in thin bands due to structural softening and equilibrium. The phenomenon of localization develops at a scale comparable to that of microstructures so that the assumption of scale separation no longer applies. Macroscopic constitutive models, which result from the homogenization of these microstructures, are then necessarily non-local and rely on one (or more) internal lengths.

3) Attempts to ignore the non-local nature lead to ill-posed continuum problems, with no physically acceptable solution. Their discretization by finite elements lead to results that strongly depend on the mesh (size, orientation and mesh nature) and whose objectivity is no longer ensured.

4) The addition of a viscosity does not respond to the lack of scale separation, except perhaps for fast loads for which a high viscosity limits damage gradients.

5) The construction of the "good" non-local model by homogenization is still unfeasible, which explains the number of variants proposed in the literature. They are distinguished by the choice and influence of variables that account for the interactions between neighboring material points. Some of them are not relevant, depending on the type of damage being modeled: uncontrolled localization, lack of solutions in diffuse regime, pathological fracture behavior, etc. If an analysis of their properties is a priori required on a case-by-case basis, several families that introduce the gradient of a state variable in the Helmholtz free energy (strain, internal variable, new kinematic variable) nevertheless maintain common characteristics. This makes it possible to bring forward some systematic properties and recommendations and we recommend in the first instance that these formulations be used.

6) These non-local gradient formulations of state variables are also well suited to a finite element treatment, the state variable under consideration then being added to the nodal degrees of freedom of the discretized problem. A Lagrangian relaxation technique – which introduces a nodal Lagrange multiplier – allows, if necessary, the standard algorithmic architecture that separates the processing of linear algebra over all degrees of freedom and the solution of strongly nonlinear systems related to the constitutive integration at integration points to be preserved.

7) A non-local model is not the combination of a local model and a spatial smoothing technique. As a matter of fact, the quantitative and even qualitative properties of a localized response and a uniform response (without excitation of non-local interaction terms) can be radically different, especially vis-à-vis failure behavior.

8) In the propagation regime, the damage develops in an FPZ whose width (propagation transverse direction) is about the size of the microstructures and reflects the internal length of the non-local model. In order to identify the latter using global experimental quantities, it is necessary to implement tests for which the geometry or loading have characteristic scales small enough to interact with the internal length. However, in the absence of homogeneous microstructures, a high experimental scattering is then expected.

9) Given the challenges deriving from the identification previously mentioned, a choice could be made to neglect the width of the FPZ to turn it into a numerical parameter that ought to tend to zero (therefore small, in practice), while maintaining a constant fracture energy. A non-local model thus makes it possible to approximate the response of a cohesive model while providing access to crack paths (which a cohesive model hardly allows). This is also the underlying idea of phase field models that smooth cracks by numerical thickening, interpreted as a small parameter, and which are formally similar to non-local damage gradient models.

10) In order to represent thin damage bands, it is not necessary in a first stage to make use of a non-local anisotropic model (except in the case of initial anisotropy of the material). On the other hand, many loading situations in which tensile and compression directions overlap require maintaining compressive stiffness in damaged areas. An operational approach consists of expressing the strain energy by distinguishing according to the signs of the principal strains (which remains compatible with an isotropy hypothesis).

11) Numerically, damage problems are hard to solve and the convergence of solution algorithms is often flawed. At least three elements explain this fact: the competition between unloading by damage and elastic unloading (oscillations), the frequent presence of structural instabilities (divergence) and the loss of stiffness of "broken" points (ill-conditioned or even non-invertible tangent matrices).

12) The non-local character leads to large-sized problems because the finite elements must be sufficiently small in comparison to the width of the localization bands. Mesh adaptation techniques can partly address this problem, but at the risk of further weakening the convergence of computations.

13) Relaxing the convergence criteria, even temporarily, to respond to convergence difficulties is not good practice because the quality of the solution is no longer controlled, which can significantly deviate from the actual response in terms of crack depth or even crack pattern.

14) Three strategies emerge, each with their advantages and disadvantages. They are characterized by (i) no longer considering the continuous nature of damage evolution (solution jump); (ii) a non-monotonic time evolution of loading (path-following methods); and (iii) the introduction of artificial viscosity.

15) For physical or numerical reasons, it may be desirable to implement a transition during computations between a localized but nevertheless volume damage distribution and a "real" crack (surface). Although attractive, this choice requires a certain attention in the presence of non-local models and may prove detrimental to the stability of the simulation.

5.7. References

Aduloju, S.C. and Truster, T.J. (2019). A variational multiscale discontinuous Galerkin formulation for both implicit and explicit dynamic modeling of interfacial fracture. *Computer Methods in Applied Mechanics and Engineering*, 343, 602–630.

Alam, S.Y., Saliba, J., Loukili, A. (2014). Fracture examination in concrete through combined digital image correlation and acoustic emission techniques. *Construction and Building Materials*, 69, 232–242.

Allix, O. and Deü, J.-F. (1997). Delayed-damage modelling for fracture prediction of laminated composites under dynamic loading. *Engng. Trans.*, 45, 29–46.

Amor, H., Marigo, J.-J., Maurini, C. (2009). Regularized formulation of the variational brittle fracture with unilateral contact: Numerical experiments. *J. Mech. Phys. Solids*, 57, 1209–1229.

Anand, L., Kim, K., Shawki, T. (1987). Onset of shear localization in viscoplastic solids. *J. Mech. Phys. Solids*, 35, 407–429.

Badel, P., Godard, V., Leblond, J.-B. (2007). Application of some anisotropic damage model to the prediction of the failure of some complex industrial concrete structure. *Int. J. Sol. Struct.*, 44, 5848–5874.

Badnava, H., Msekh, M.A., Etemadi, E., Rabczuk, T. (2018). An h-adaptive thermo-mechanical phase field model for fracture. *Finite Elements in Analysis and Design*, 138, 31–47.

Barenblatt, G.I. (1959). The formation of equilibrium cracks during brittle fracture. General ideas and hypotheses. Axially-symmetric cracks. *J. Appl. Math. Mech.*, 23, 622–636.

Bazant, Z.P. and Oh, B.H. (1983). Crack band theory for fracture of concrete. *Materials and Structures*, 16, 155–177.

Bazant, Z.P. and Pijaudier-Cabot, G. (1988). Nonlocal continuum damage, localization instability and convergence. *J. Appl. Mech.*, 55, 287–293.

Bazant, Z., Le, J., Hoover, C. (2010). Nonlocal boundary layer (NBL) model: Overcoming boundary condition problems in strength statistics and fracture analysis of quasibrittle materials. In *Proc. Fracture Mechanics of Concrete and Concrete Structures – Recent Advances in Fracture Mechanics of Concrete (FraMCoS-7)*, Oh, B.H., Choi O.C., Chung, L. (eds). Korea Concrete Institute, Seoul.

Benallal, A. and Marigo, J.-J. (2006). Bifurcation and stability issues in gradient theories with softening. *Modelling and Simulation in Materials Science and Engineering*, 15, S283–S295.

Benallal, A., Billardon, R., Geymonat, G. (1993). Bifurcation and localization in rate-in-dependent materials. Some general considerations. In *Bifurcation and Stability of Dissipative Systems*, Nguyen, Q.S. (ed.). Springer, Vienna.

Beremin, F.M., Pineau, A., Mudry, F., Devaux, J.-C., D'Escatha, Y., Ledermann, P. (1983). A local criterion for cleavage fracture of a nuclear pressure vessel steel. *Metallurgical Transactions A*, 14, 2277–2287.

Besson, J. and Desmorat, R. (2004). Numerical implementation of constitutive models. In *Local Approach to Fracture*, Besson, J. (ed.). Presses de l'École des Mines de Paris, Paris.

Boffi, D., Brezzi, F., Fortin, M. (2013). *Mixed Finite Element Methods and Applications*. Springer, Berlin.

Bonnans, J., Gilbert, J.C., Lemarechal, C., Sagastizábal, C.A. (2006). *Numerical Optimization: Theoretical and Practical Aspects*. Springer Berlin, Heidelberg.

Borino, G., Failla, B., Parrinello, F. (2003). A symmetric nonlocal damage theory. *Int. J. Solids Struct.*, 40, 3621–3645.

de Borst, R. (1987). Computation of post-bifurcation and post-failure behaviour of strain-softening solids. *Computers and Structures*, 25, 211–224.

de Borst, R. (2003). Numerical aspects of cohesive-zone models. *Eng. Fract. Mech.*, 70, 1743–1757.

de Borst, R. and Sluys, L.J. (1991). Localisation in a Cosserat continuum under static and dynamic loading conditions. *Comput. Methods Appl. Mech. Eng.*, 90, 805–827.

Bourdin, B., Francfort, G.A., Marigo, J.-J. (2000). Numerical experiments in revisited brittle fracture. *J. Mech. Phys. Solids*, 48, 797–826.

Bourdin, B., Francfort, G.A., Marigo, J.-J. (2008). The variational approach to fracture. *J. Elasticity*, 91, 5–148.

Bourdin, B., Larsen, C.J., Richardson, C.L. (2010). A time-discrete model for dynamic fracture based on crack regularization. *International Journal of Fracture*, 168, 133–143.

Chaboche, J.L. and Cailletaud, G. (1996). Integration methods for complex plastic constitutive equations. *Comput. Meth. Appl. Mech. Engrg.*, 133, 125–155.

Charlotte, M., Laverne, J., Marigo, J.-J. (2006). Initiation of cracks with cohesive force models: A variational approach. *Eur. J. Mech. A/Solids*, 25, 649–669.

Chen, Y. (2019). Modeling of ductile fracture using local approach: Reliable simulation of crack extension. Report, Mines ParisTech, Université PSL, Paris.

Cherepanov, G.P. (1967). Crack propagation in continuous media. *J. Appl. Math. Mech.*, 31, 503–512.

Comi, C. (1999). Computational modelling of gradient-enhanced damage in quasi-brittle materials. *Mech. Cohes. Frict. Mater.*, 4, 17–36.

Comi, C. and Perego, U. (2001). Fracture energy based bi-dissipative damage model for concrete. *Int. J. Solids Struct.*, 38, 6427–6454.

Cowan, H.J. (1952). Strength of reinforced concrete under the action of combined stresses, and the representation of the criterion of failure by a space model. *Nature*, 169(4303), 663–663.

Crisfield, M.A. (1981). A fast incremental-iterative solution procedure that handles snap-through. *Computers and Structures*, 13, 55–62.

Desmorat, R., Gatuingt, F., Ragueneau, F. (2007). Nonlocal anisotropic damage model and related computational aspects for quasi-brittle materials. *Eng. Fract. Mech.*, 74, 1539–1560.

Doyen, D., Ern, A., Piperno, S. (2012). Quasi-explicit time-integration schemes for dynamic fracture with set-valued cohesive zone models. *Computational Mechanics*, 52, 401–416.

Drugan, W.J. and Willis, J.R. (1996). A micromechanics-based nonlocal constitutive equation and estimates of representative volume element size for elastic composites. *J. Mech. Phys. Solids*, 44, 497–524.

Dugdale, D.S. (1960). Yielding of steel sheets containing slits. *J. Mech. Phys. Solids*, 8, 100–104.

Dumouchel, P.-E., Marigo, J.-J., Charlotte, M. (2008). Dynamic fracture: An example of convergence towards a discontinuous quasistatic solution. *Continuum Mechanics and Thermodynamics*, 20, 1–19.

Ekeland, I. and Temam, R. (1974). *Analyse convexe et problèmes variationnels*. Dunod/ Gauthier-Villars, Paris.

Elices, M., Guinea, G.V., Gómez, J., Planas, J. (2002). The cohesive zone model: Advantages, limitations and challenges. *Eng. Fract. Mech.*, 69, 137–163.

Enakoutsa, K. and Leblond, J.-B. (2009). Numerical implementation and assessment of the GLPD micromorphic model of ductile rupture. *Eur. J. Mech. A/Solids*, 28, 445–460.

Engelen, R.A.B., Geers, M.G.D., Baaijens, F.P.T. (2003). Nonlocal implicit gradient-enhanced elastoplasticity for the modelling of softening behaviour. *Int. J. Plast.*, 19, 403–433.

Ern, A. and Guermond, J. (2004). *Theory and Practice of Finite Elements*. Springer, New York.

Fédération Internationale du Béton (FIB) (2013). *FIB Model Code for Concrete Structures*. Ernst & Sohn, Berlin.

Fernandes, R., Chavant, C., Chambon, R. (2008). A simplified second gradient model for dilatant materials: Theory and numerical implementation. *Int. J. Solids Struct.*, 45, 5289–5307.

Ferté, G., Massin, P., Moës, N. (2016). 3D crack propagation with cohesive elements in the extended finite element method. *Comput. Meth. Appl. Mech. Engrg.*, 300, 347–374.

Fichant, S., La Borderie, C., Pijaudier-Cabot, G. (1999). Isotropic and anisotropic descriptions of damage in concrete structures. *Mech. Cohes. -Frict. Mater.*, 4, 339–359.

Forest, S. (2009). Micromorphic approach for gradient elasticity, viscoplasticity, and damage. *J. Eng. Mech.*, 135, 117–131.

Forest, S. and Lorentz, E. (2004). Localization phenomena and regularization methods. In *Local Approach to Fracture*, Besson, J. (ed.). Presses de l'École des Mines de Paris, Paris.

Fortin, M. and Glowinski, R. (1983). *Augmented Lagrangian Methods: Applications to the Numerical Solution of Boundary-Value Problems*. North-Holland, Amsterdam.

Francfort, G.A. and Marigo, J.-J. (1998). Revisiting brittle fracture as an energy minimization problem. *J. Mech. Phys. Solids*, 46, 1319–1342.

Francfort, G.A. and Mielke, A. (2006). Existence results for a class of rate-independent material models with nonconvex elastic energies. *J. Reine Angew.*, 595, 55–91.

Fraternali, F. (2007). Free discontinuity finite element models in two-dimensions for in-plane crack problems. *Theor. Appl. Fract. Mec.*, 47, 274–282.

Freddi, F. and Iurlano, F. (2017). Numerical insight of a variational smeared approach to cohesive fracture. *J. Mech. Phys. Solids*, 98, 156–171.

Frémond, M. and Nedjar, B. (1996). Damage, gradient of damage and principle of virtual power. *Int. J. Solids Struct.*, 33, 1083–1103.

Fressengeas, C. and Molinari, A. (1985). Inertia and thermal effects on the localization of plastic flow. *Acta Metall.*, 33, 387–396.

Geers, M.G.D. (1999). Enhanced solution control for physically and geometrically non-linear problems. *Int. J. Numer. Meth. Engng.*, 46, 177–230.

Geers, M.G.D., de Borst, R., Brekelmans, W.A.M., Peerlings, R.H.J. (1998). Strain-based transient-gradient damage model for failure analyses. *Computer Methods in Applied Mechanics and Engineering*, 160, 133–153.

Gerasimov, T. and De Lorenzis, L. (2016). A line search assisted monolithic approach for phase-field computing of brittle fracture. *Computer Methods in Applied Mechanics and Engineering*, 312, 276–303.

Germain, P., Nguyen, Q.S., Suquet, P. (1983). Continuum thermodynamics. *J. Appl. Mech.*, 50, 1010–1020.

Gologanu, M., Leblond, J.-B., Perrin, G., Devaux, J. (1997). Recent extensions of Gurson's model for porous ductile metals. In *Continuum Micromechanics*, Suquet, P. (ed.). Springer, Vienna.

Gravouil, A., Elguedj, T., Maigre, H. (2009). An explicit dynamics extended finite element method. Part 2: Element-by-element stable-explicit/explicit dynamic scheme. *Comput. Meth. Appl. Mech. Engrg.*, 198, 2318–2328.

Griffith, A.A. (1921). The phenomena of rupture and flow in solids. *Philosophical Transactions Roy. Soc. London A*, 221, 163–198.

Gutierrez, M.A. (2004). Energy release control for numerical simulations of failure in quasi-brittle solids. *Commun. Numer. Meth. Engng.*, 20, 19–29.

Hadjab, H.S., Thimus, J.-F., Chabaat, M. (2007). The use of acoustic emission to investigate fracture process zone in notched concrete beams. *Current Science*, 93(5), 648–653 [Online]. Available at: http://hdl.handle.net/2078.1/37341.

Halm, D. and Dragon, A. (1998). An anisotropic model of damage and frictional sliding for brittle materials. *Eur. J. Mech. A/Solids*, 17, 439–460.

Halphen, B. and Nguyen Q.S. (1975) Sur les matériaux standard généralisés. *J. Mécanique*, 14, 39–63.

Hellweg, H.B. and Crisfield, M.A. (1998). A new arc-length method for handling sharp snap-backs. *Computer and Structures*, 66, 704–709.

Hillerborg, A., Modéer, M., Petersson, P.-E. (1976). Analysis of crack formation and crack growth in concrete by means of fracture mechanics and finite elements. *Cem. Concr. Res*, 6, 773–781.

Hoover, C.G. and Bažant, Z.P. (2014). Cohesive crack, size effect, crack band and work-of-fracture models compared to comprehensive concrete fracture tests. *Int. J. Fract.*, 187, 133–143.

Irwin, G.R. (1957). Analysis of stresses and strains near the end of a crack traversing a plate. *Journal of Applied Mechanics*, 24, 361–364.

Jäger, P., Steinmann, P., Kuhl, E. (2008). Modeling three-dimensional crack propagation – A comparison of crack path tracking strategies. *Int. J. Num. Meth. Engng.*, 76, 1328–1352.

Jefferson, A.D. and Mihai, I.C. (2015). The simulation of crack opening-closing and aggregate interlock behaviour in finite element concrete models. *Int. J. Numer. Meth. Engng.*, 104, 48–78.

Jirasek, M. and Rolshoven, S. (2003). Comparison of integral-type nonlocal plasticity models for strain-softening materials. *Int. J. Eng. Sci.*, 41, 1553–1602.

Kouznetsova, V., Geers, M.G.D., Brekelmans, W.A.M. (2002). Multi-scale constitutive modelling of heterogeneous materials with a gradient-enhanced computational homogenization scheme. *Int. J. Numer. Meth. Engng.*, 54, 1235–1260.

Leblond, J.-B. (2003). *Mécanique de la rupture fragile et ductile*. Hermes-Lavoisier, Paris.

Leguillon, D. (2002). Strength or toughness? A criterion for crack onset at a notch. *Eur. J. Mech A/Solids*, 21, 61–72.

Lemaitre, J., Chaboche, J., Benallal, A., Desmorat, R. (2009). *Mécanique des matériaux solides*, 3rd edition. Dunod, Paris.

Li, T., Marigo, J.-J., Guilbaud, D., Potapov, S. (2016). Gradient damage modeling of brittle fracture in an explicit dynamics context. *International Journal for Numerical Methods in Engineering*, 108, 1381–1405.

Llau, A., Jason, L., Dufour, F., Baroth, J. (2015). Adaptive zooming method for the analysis of large structures with localized nonlinearities. *Finite Elements in Analysis and Design*, 106, 73–84.

Lorentz, E. (2017). A nonlocal damage model for plane concrete consistent with cohesive fracture. *Int. J. Fract.*, 207, 123–159.

Lorentz, E. and Andrieux, S. (1999). A variational formulation for nonlocal damage models. *Int. J. Plast*, 15, 119–138.

Lorentz, E. and Andrieux, S. (2003). Analysis of non-local models through energetic formulations. *Int. J. Solids Struct.*, 40, 2905–2936.

Lorentz, E. and Badel, P. (2004). A new path-following constraint for strain-softening finite element simulations. *Int. J. Numer. Meth. Engng.*, 60, 499–526.

Lorentz, E. and Badel, P. (2011). Critères de convergence en mécanique des solides. In *10e colloque national en calcul des structures*, Giens.

Lorentz, E. and Godard, V. (2011). Gradient damage models: Toward full-scale computations. *Comput. Meth. Appl. Mech. Engrg.*, 200, 1927–1944.

Lorentz, E., Besson, J., Cano, V. (2008). Numerical simulation of ductile fracture with the Rousselier constitutive law. *Comput. Meth. Appl. Mech. Engrg.*, 197, 1965–1982.

Lorentz, E., Cuvilliez, S., Kazymyrenko, K. (2011). Convergence of a gradient damage model toward a cohesive zone model. *C.R. Mécanique*, 339, 20–26.

Lorentz, E., Cuvilliez, S., Kazymyrenko, K. (2012). Modelling large crack propagation: From gradient damage to cohesive zone models. *Int. J. Fract.*, 178, 85–95.

Marigo, J.-J. (1981). Formulation d'une loi d'endommagement d'un matériau élastique. *CR. Acad. Sci. Paris II*, 292, 1309–1312.

Marigo, J.-J. (2014). Gradient damage models: Construction and fundamental properties. In 3ème école d'été de méchanique théorique de Quiberon, 22–27 September.

Mazars, J. (1986). A description of micro- and macroscale damage of concrete structures. *Eng. Fract. Mech.*, 25, 729–737.

Mialon, P. (1986). Eléments d'analyse et de résolution numérique des relations de l'élastoplasticité. *EDF bulletin de la DER série C mathématiques informatique*, 3, 57–89.

Michel, B., Helfer, T., Ramière, I., Esnoul, C. (2018). A new numerical methodology for simulation of unstable crack growth in time-independent brittle materials. *Eng. Fract. Mech.*, 188, 126–150.

Miehe, C., Hofacker, M., Welschinger, F. (2010). A phase-field model for rate-independent crack propagation: Robust algorithmic implementation based on operator split. *Comput. Meth. Appl. Mech. Engrg.*, 199, 2765–2778.

Moës, N., Stolz, C., Bernard, P.-E., Chevaugeon, N. (2011). A level set based model for damage growth: The thick level set approach. *Int. J. Numer. Meth. Engng.*, 86, 358–380.

Muhlhaus, H.B. and Aifantis, E.C. (1991). The influence of microstructure-induced gradients on the localization of deformation in viscoplastic materials. *Acta Mech.*, 89, 217–231.

Mumford, D. and Shah, J. (1989). Optimal approximations by piecewise smooth functions and associated variational problems. *Commun. Pur. Appl. Math.*, 42(5), 577–685.

Nagaraja, S., Elhaddad, M., Ambati, M., Kollmannsberger, S., De Lorenzis, L., Rank, E. (2018). Phase-field modeling of brittle fracture with multi-level hp-FEM and the finite cell method. *Computational Mechanics*, 63, 1283–1300.

Needleman, A. (1988). Material rate dependence and mesh sensitivity in localization problems. *Comp. Meth. Appl. Mech. Engrg.*, 67, 69–86.

Negri, M. (2005). A discontinuous finite element approximation of free discontinuity problems. *Adv. Math. Sci. Appl.*, 15, 283–306.

Nguyen, Q.S. (1993). Bifurcation and stability of time-independent standard dissipative systems. In *Bifurcation and Stability of Dissipative Systems*, Nguyen, Q.S. (ed.). Springer, Vienna.

Oliver, J. and Huespe, A.E. (2004). Theoretical and computational issues in modelling material failure in strong discontinuity scenarios. *Comp. Meth. Appl. Mech. Engrg.*, 193, 2987–3014.

Oliver, J., Huespe, A.E., Cante, J.C. (2008). An implicit/explicit integration scheme to increase computability of non-linear material and contact/friction problems. *Comp. Meth. Appl. Mech. Engrg.*, 197, 1865–1889.

Oliver, J., Dias, I.F., Huespe, A.E. (2014). Crack-path field and strain-injection techniques in computational modeling of propagating material failure. *Comp. Meth. Appl. Mech. Engrg.*, 274, 289–348.

Patzak, B.R. and Jirasek, M. (2004). Adaptive resolution of localized damage in quasi-brittle materials. *Journal of Engineering Mechanics*, 130, 720–732.

Peerlings, R.H.J., de Borst, R., Brekelmans, W.A.M., de Vree, J.H.P. (1996). Gradient-enhanced damage for quasi-brittle materials. *Int. J. Numer. Meth. Engng.*, 39, 3391–3403.

Petersson, P.E. (1981). Crack growth and development of fracture zones in plain concrete and similar materials. Report, TVBM-1006, Division of Building Materials, Lund Institute of Technology.

Pham, K., Amor, H., Marigo, J.-J., Maurini, C. (2010). Gradient damage models and their use to approximate brittle fracture. *Int. J. Damage Mech.*, 20, 618–652.

Pijaudier-Cabot, G. and Burlion, N. (1996). Damage and localisation in elastic materials with voids. *Mech. Cohes. -Frict. Mater.*, 1, 129–144.

Radovitzky, R., Seagraves, A., Tupek, M., Noels, L. (2011). A scalable 3D fracture and fragmentation algorithm based on a hybrid, discontinuous Galerkin, cohesive element method. *Computer Methods in Applied Mechanics and Engineering*, 200, 326–344.

Rice, J.R. (1968). A path independent integral and the approximate analysis of strain concentration by notches and cracks. *Journal of Applied Mechanics*, 35, 379–386.

Rinehart, J.S. (1966). Fracture of rocks. *Int. J. Fract. Mech.*, 2, 534–551.

Rodríguez-Ferran, A. and Huerta, A. (2000). Error estimation and adaptivity for nonlocal damage models. *International Journal of Solids and Structures*, 37, 7501–7528.

Rots, J.G. (1988). Computational modeling of concrete fracture. PhD Thesis, Delft University of Technology, Delft.

Scherer, J.-M. (2020). Localisation de la déformation et rupture ductile dans les mono-cristaux : application aux aciers austénitiques inoxydables irradiés. PhD Thesis, Université PSL, Paris.

Seles, K., Lesicar, T., Tonkovic, Z., Soric, J. (2019). A residual control staggered solution scheme for the phase-field modeling of brittle fracture. *Engineering Fracture Mechanics*, 205, 370–386.

Seupel, A., Hütter, G., Kuna, M. (2018). An efficient FE-implementation of implicit gradient-enhanced damage models to simulate ductile failure. *Eng. Fract. Mech.*, 199, 41–60.

Sicsic, P. and Marigo, J.-J. (2013). From gradient damage laws to Griffith's theory of crack propagation. *J. Elasticity*, 113, 55–74.

Sicsic, P., Marigo, J.-J., Maurini, C. (2014). Initiation of a periodic array of cracks in the thermal shock problem: A gradient damage modeling. *Journal of the Mechanics and Physics of Solids*, 63, 256–284.

Simo, J.C. (1989). Strain softening and dissipation: A unification of approaches. In *Cracking and Damage, Strain Localization and Size Effects*, Mazars, J. and Bazant, Z.P. (eds). Taylor & Francis, London.

Steinmann, P. and Willam, K. (1991). Localization within the framework of micropolar elastoplasticity. In *Advances in Continuum Mechanics*, Brüller, O., Mannl, V., Najar, J. (eds). Springer-Verlag, Berlin.

Svedberg, T. and Runesson, K. (1997). A thermodynamically consistent theory of gradient regularized plasticity coupled to damage. *Int. J. Plast.*, 13, 669–696.

Wu, J.-Y. (2017). A unified phase-field theory for the mechanics of damage and quasi-brittle failure. *J. Mech. Phys. Solids*, 103, 72–99.

Zhou, F., Molinari, J.-F., Shioya, T. (2005). A rate-dependent cohesive model for simulating dynamic crack propagation in brittle materials. *Engineering Fracture Mechanics*, 72, 1383–1410.

6

Extended Finite Element (XFEM) and Thick Level Set (TLS) Methods

Nicolas MOËS

École centrale de Nantes, France

6.1. Introduction

The extended finite element method (XFEM) is an extension to the finite element method, which can be used for monitoring crack propagation through a mesh without having to modify it. In linear fracture mechanics, crack propagation is driven by the stress intensity factors calculated along the front. These intensity factors are used to define both the propagation direction and the amount of progress (by loading cycle, for example). The XFEM method also allows us to take into account a cohesive zone at the tip of the crack. In this case, the progress of the crack is directly driven by the cohesive relation to which a complementary equation must be added to determine the direction of crack propagation.

In a number of situations, such as ductile fracture, crack propagation cannot be simply be reduced to evaluating stress intensity factors or the opening of a cohesive zone. It is necessary to take into account the bulk behavior of the matter in the vicinity of the tip and use the bulk softening to deduce crack propagation.

The thick level sets (TLS) method places around the crack a band in which the material softens and deformation are localized. The thickness of this band is set by the TLS model. Outside of this band, the behavior is non-softening.

Numerical Methods for Strong Nonlinearities in Mechanics,
coordinated by Jacques BESSON, Frédéric LEBON and Éric LORENTZ. © ISTE Ltd 2024.

The combined XFEM-TLS method can be used to manage softening bulk behaviors and automatically allow the presence of displacement jumps. The XFEM-TLS method also makes it possible to automatically manage crack branching and coalescence.

This chapter is organized into five sections. In section 6.2, a categorization of approaches for cracking is proposed, as well as a discussion of the advantages and disadvantages of continuous and discontinuous fracture approaches. In section 6.3, the XFEM approach is revisited and its limitations are discussed. Section 6.4 introduces the TLS method. Examples of XFEM-TLS simulation are given in section 6.5. Finally, a conclusion with final remarks completes the chapter.

6.2. Categorization of approaches to cracking

Before presenting the contributions of the extended elements method (XFEM) used on its own (for non-softening media) or in combination with the *TLS* method for softening media (XFEM-TLS), it would be appropriate to discuss different possibilities for modeling cracking. Approaches can be classified according to two criteria. The first criterion relates to kinematics and the second to the material model employed:

– Discontinuous or continuous approach to fracture: in the first case, the displacement field may be discontinuous and in the second, it will be continuous (eventually with strong gradients). The continuous approach is also sometimes referred to as diffuse.

– Softening or non-softening bulk behavior around the crack: softening behavior refers to a material that may be found in a stress–strain state such that a positive strain increment creates a negative stress increment.

Four possibilities can therefore emerge by combining these two criteria or, specifically, only three. As such, the continuous approach for a non-softening medium presents no degradation mechanism that can lead to cracking. We now examine the other three combinations.

The discontinuous approach for non-softening media is the most widespread. It concerns monitoring cracks modeled by displacement jumps. If the pre-cracking area upstream of the crack tip (*Process Zone*) is very small, the crack tip will be guided by stress intensity factors (linear fracture mechanics approach considering the elastic bulk behavior). If the *Process Zone* has a non-negligible size compared to the structure size, a cohesive area will be used, which will result in introducing a characteristic length into the cracking and therefore a different scale effect than that of linear fracture mechanics. The XFEM approach for linear fracture mechanics or for the cohesive model clearly falls into this first category of discontinuous approach in non-softening medium. The XFEM approach is also adapted to plastic (non-softening) (Elguedj et al.

2006; Prabel et al. 2007) or hyperelastic behaviors in the tip (Legrain et al. 2005). Contact and friction conditions may also be imposed on lips (Dolbow et al. 2001; Liu and Borja 2010).

Regarding the second category of continuous approaches for softening medium, the softening mechanism may be a loss of elastic stiffness due to damage and/or a reduction of the possible stress by a reduction of the elastic domain related to the plasticity criterion. In these continuous approaches, the introduction of a displacement jump is not considered. All of these approaches, on the other hand, introduce one way or another a length in order to avoid nonphysical strain localization. Work in this category concerns:

– The integral nonlocal model: damage evolution at a point is governed for a nonlocal force, calculated as the average of the local force over a certain length around the point (Bazant et al. 1984; Pijaudier-Cabot and Bazant 1987). Another look into the estimation of this distance was proposed in Rastiello et al. (2018). An integral nonlocal version on the displacement field also exists (Rodríguez-Ferran et al. 2005).

– The higher order kinematic gradient model (Aifantis 1984; Schreyer and Chen 1986; Triantafyllidis and Aifantis 1986) or including additional degrees of rotational freedom (Mühlhaus and Vardoulakis 1987).

– The higher gradient damage model: energy depends on the damage variable and its gradient (Frémond and Nedjar 1996; Pijaudier-Cabot and Burlion 1996; Nguyen and Andrieux 2005).

– The *phase-field* model derived from the physics community (Karma et al. 2001; Hakim and Karma 2005, 2009).

– The variational approach to fracture (Francfort and Marigo 1998; Bourdin et al. 2000, 2008).

The damage gradient model was compared to the integral nonlocal model by Peerlings et al. (2001) and to the *phase-field* model by de Borst and Verhoosel (2016). An energy analysis with a comparison between several forms of regularization can also be found in Lorentz and Andrieux (2003).

The third category concerns discontinuous approaches for softening media. Of the three categories, it is in the latter that there is less research, probably due the complexity of the issue.

In this category, it is actually softening-strain localization which will control the crack growth on which displacement jumps will be allowed. In this category, it is therefore necessary to manage both softening and the insertion at the right pace (and in the right place) of the displacement jump.

This is the challenge that the XFEM-TLS approach is willing to address. TLS manages softening and tells XFEM where and when to place discontinuities. In this category, other works can also be found that focus on introducing a displacement jump inside a damaged area smoothed by the gradient model. This introduction can be carried out by the XFEM approach (without remeshing) (Tamayo-Mas and Rodriguez-Ferran 2015) or with remeshing (Feld-Payet et al. 2015).

We have just discussed three very different categories for modeling cracking. Which category should be chosen? It all naturally depends on the type of material under consideration and the type of path of the observed cracks (a simple crack or very complex crack paths).

It is clear that taking a softening medium into account is much more complex than a non-softening medium. A softening behavior model cannot be used in its local version (i.e. the stress–strain relation). It is necessary to regularize the material model by introducing a length that will spread the dissipation over a certain thickness and avoid a nonphysical localization because dissipating far too little. Therefore, if in modeling, non-softening bulk behavior can be avoided, it is a good thing. There are, however, situations where this is difficult. For example, ductile fracture for which it is difficult to replace bulk softening by a cohesive zone. This is also the case for the Poisson effect cracking (Brazilian test) in quasi-brittle materials for which a cohesive area alone cannot take into account the necessary triaxiality.

Regarding the choice between continuous and discontinuous approach, the first is simpler to implement than the second but at the cost of a need to maintain a fine mesh over the entire crack path to address high gradients. The discontinuous approach, on the other hand, allows for mesh coarsening over the trajectory of the crack and gives direct access to the crack opening (which is important for taking into account conditions on the lips such as contact, friction or hydraulic loads). Mesh separation according to the crack is also necessary in the presence of fragmentation, slicing or machining.

6.3. The XFEM method for cracking in non-softening media

The XFEM was initiated 20 years ago to reduce the difficulties associated with compatible mesh generation. First applied to cracks, the method was then utilized with other types of moving or fixed surfaces. The first paper on fully modeling an arbitrarily-positioned crack on a mesh was by Moës et al. (1999). This presentation illustrates crack propagation without any remeshing required.

The XFEM method is based on using unity partition (Melenk and Babuska 1996) (to introduce the necessary discontinuous enrichments within elements). It should be noted that the XFEM method is not the only one to make use of the concept of unity partition. Methods such as the GFEM (*generalized finite element method*) (Duarte et al. 2000) also use it. For 20 years, contributions and motivations initially specific to these two methods have largely blended.

There are a number of state-of-the-art works on the XFEM method. Examples include Karihaloo (2003); Belytschko et al. (2009); Fries and Belytschko (2010). A more recent retrospective can be found in Sukumar et al. (2015). Books have also been published such as Mohammadi (2008); Pommier et al. (2011); Zhuang et al. (2014); Khoei (2015). This chapter does not intend to present the XFEM method in detail but rather to recall the interest for managing mesh-independent discontinuities. This interest will then be used in combination with the TLS method.

The basic idea of the XFEM approach is to enrich a standard displacement finite element approximation by discontinuous functions capable of modeling crack opening. Consider, in Figure 6.1, a crack placed on a mesh.

The XFEM approximation of the displacement field is written (Moës et al. 1999) as:

$$\begin{aligned} \boldsymbol{u}^h(\boldsymbol{x}) &= \textstyle\sum_{i\in I} \boldsymbol{u}_i N_i(\boldsymbol{x}) + \sum_{i\in L} \boldsymbol{a}_i N_i(\boldsymbol{x}) H(\boldsymbol{x}) \\ &+ \sum_{i\in K_1} N_i(\boldsymbol{x}) (\sum_{l=1}^{4} \boldsymbol{b}^l_{i,1} F^l_1(\boldsymbol{x})) + \sum_{i\in K_2} N_i(\boldsymbol{x}) (\sum_{l=1}^{4} \boldsymbol{b}^l_{i,2} F^l_2(\boldsymbol{x})) \end{aligned} \quad [6.1]$$

The first term of the right-hand member is the standard (continuous) finite element approximation, while the next three terms are enrichments built on the basis of the principle of unity partition (Melenk and Babuska 1996). The second term allows the crack to open at its core while the following two allow the crack to open near its tip. These two terms also make it possible to obtain a good accuracy for the solution at the crack tip.

More specifically, here follow the details of the notations used:

– I is the set of the mesh nodes.

– $\boldsymbol{u}_i$ is the classical (vector) degree of freedom at the node i.

– N_i is the (scalar) shape function associated with node i.

– $L \subset I$ is the set of nodes enriched by the discontinuity and the coefficients $\boldsymbol{a}_i$ are the corresponding (vector) degrees of freedom. A node belongs to L if its support is sliced by the crack but contains none of its tips (these nodes are circled in Figure 6.1 for a uniform and nonuniform mesh). The function $H(\boldsymbol{x})$ is discontinuous on the crack and of constant value on either side thereof: +1.0 on one side and –1.0 on the other.

– $K_1 \subset I$ and $K_2 \subset I$ are the sets of nodes to be enriched to model crack bottoms 1 and 2, respectively. The corresponding degrees of freedom are $\boldsymbol{b}^l_{i,1}$ and $\boldsymbol{b}^l_{i,2}$, $l = 1, \ldots, 4$. A node belongs to K_1 (K_2) if its media contains the first (second) crack tip. These nodes are surrounded by a square as shown in Figure 6.1.

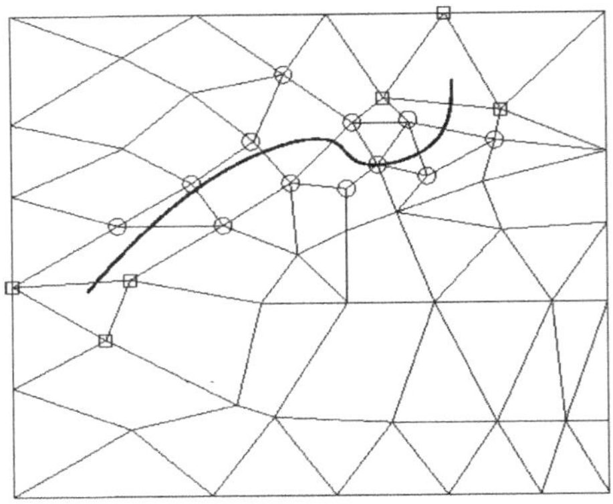

Figure 6.1. *A crack placed on a 2D mesh. Circled nodes denote the nodes enriched by the function $H(\boldsymbol{x})$ and the nodes surrounded by a square are enriched by asymptotic cracking background modes*

The functions $F_1^l(\boldsymbol{x}), l = 1, \ldots, 4$ modeling the crack bottom are given in elasticity by:

$$\{F_1^l(\boldsymbol{x})\} \equiv \left\{\sqrt{r}\sin(\tfrac{\theta}{2}), \sqrt{r}\cos(\tfrac{\theta}{2}), \sqrt{r}\sin(\tfrac{\theta}{2})\sin(\theta), \sqrt{r}\cos(\tfrac{\theta}{2})\sin(\theta)\right\} \quad [6.2]$$

where (r, θ) are the polar coordinates in local axes at the crack bottom ("tip" 1 in Figure 6.2). It should be noted that the first function is discontinuous on the crack. Similarly, the functions $F_2^l(\boldsymbol{x})$ are also given by [6.2]; the polar coordinate system is now defined for the second crack tip.

Figure 6.2. *Local axes for polar coordinates at crack tips*

For 3D cracks, the evaluation of function H and the polar coordinates close to the front appear to be more complex. It may then be interesting to store the position of the crack using two level functions (*level set field*). At each node near the crack, the signed distance is stored at the crack (or at its tangent extension). The sign indicates which side of the crack is considered. A second signed distance is also stored. It represents the distance to the crack front (following a direction tangent to the crack). Normal and tangential signed distances are denoted by ls_{n} and ls_{t}, respectively. The isocontours of these distance fields are represented in Figure 6.3a for a 2D crack. As shown in Figure 6.3(b), it gives an idea of the isocontours 0 in 3D for a surfacing crack.

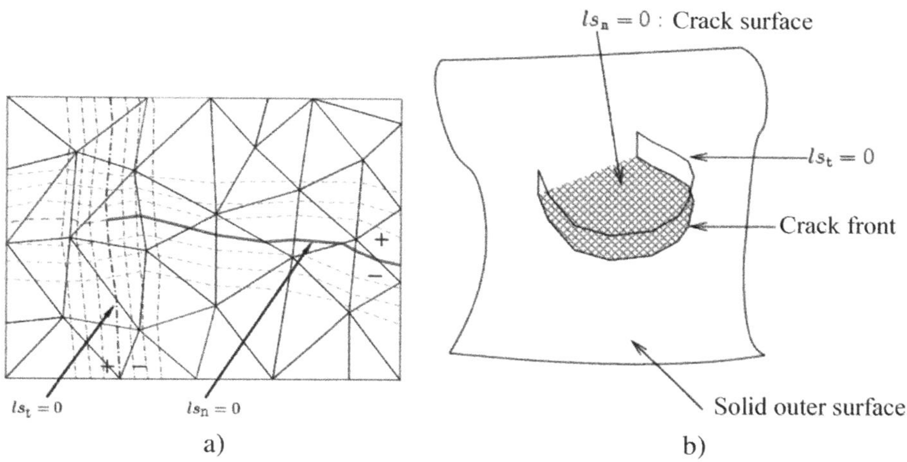

Figure 6.3. *Crack localization by two level functions (level sets). A series of isocontours are plotted in 2D (a) and the only zero contours are given for a 3D problem (b). For a color version of this figure, see www.iste.co.uk/lebon/mechanics.zip*

Historically, the first use of level function for cracking within the XFEM framework has been described in Stolarska et al. (2001) and the 3D extension was performed in Moës et al. (2002).

Knowing the levels ls_{n} and ls_{t} at a point $\boldsymbol{x}$, the values of enrichment functions can then be directly obtained by the formulas:

$$H(\boldsymbol{x}) = \mathrm{sign}(ls_{\mathrm{n}}(\boldsymbol{x})),\ r = \sqrt{ls_{\mathrm{n}}^2 + ls_{\mathrm{t}}^2},\ \theta = \tan^{-1}\left(\frac{ls_{\mathrm{n}}}{ls_{\mathrm{t}}}\right) \qquad [6.3]$$

Since the first work on crack modeling by the XFEM approach, many improvements/variants have been proposed. We expose two of them in detail. The first concerns vector enrichment.

At each node near the crack tip, 8 degrees of freedom are added and appear in equation [6.1] (because four tip enrichment scalar functions F and two directions x and y). These eight degrees of freedom can be replaced by just two additional degrees $k^l, l = 1.2$ with the use of enrichment vector functions representing displacement modes 1 and 2 at the crack tips for unit intensity factors. Clearly, expression [6.1] is assigned as follows:

$$\sum_{l=1}^{4} \boldsymbol{b}_{i,1}^{l} F_1^l(\boldsymbol{x}) \rightarrow \sum_{l=1}^{2} k_{i,1}^{l} \boldsymbol{G}_1^l(\boldsymbol{x}) \qquad [6.4]$$

The vector character is now carried by the enrichment function and more by the degree of freedom. The enrichment modes $\boldsymbol{G}$ can be visualized in Figure 6.4. In addition to reducing the number of degrees of freedom, vector enrichment provides a better stiffness matrix conditioning and direct access to stress intensity factors (admittedly less accurate than those obtained by a contour integral).

A second important improvement to the X-GEF approach concerns the concept of geometric enrichment. The enrichment of crack point nodes shown in Figure 6.1 (nodes surrounded by a square) concerns only a layer of nodes around the tip (several layers could be considered). This enrichment is said to be topological. For a number of fixed layers, the size of the enrichment area will tend toward zero if the element size decreases, making the enrichment asymptotically inoperative. On the other hand, an enrichment of all of the nodes at a given distance from the tip, known as topological enrichment, makes it possible to maintain an influence of the enrichment even asymptotically. Geometric enrichment is therefore to be preferred (Béchet et al. 2005; Laborde et al. 2005).

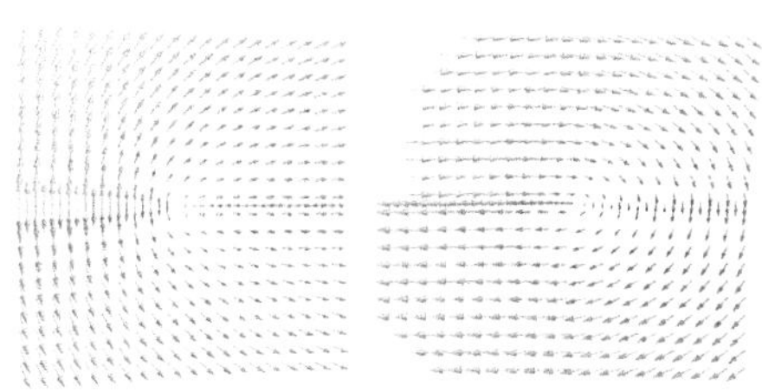

Figure 6.4. *Enrichment vector functions $\boldsymbol{G}^1$ and $\boldsymbol{G}^2$. For a color version of this figure, see www.iste.co.uk/lebon/mechanics.zip*

At this point, we considered that the crack lips were free of traction. Taking contact or friction on lip crack into account requires combining the XFEM approach with specific conditions imposed on the lips. Imposing these conditions require the use of a Lagrange multiplier (Dolbow et al. 2001; Mourad et al. 2007; Béchet et al. 2009) or a purely kinematic Nitsche-type approach (Hansbo and Hansbo 2002).

So far, we have described the kinematics of the crack opening in a non-propagation situation. Regarding its propagation, it can be based on the use of stress intensity factors along the crack front or a cohesive model on the lips (allowing a material length to be introduced into the model).

Figure 6.5 describes the geometry of the Brokenshire torsion test (Brokenshire and Barr 1996). This trial was simulated by XFEM and a cohesive lip model by Ferté et al. (2016). Figure 6.6 gives the deformation and the stress level. The numerical crack path can be compared to the experimental path given in Figure 6.7.

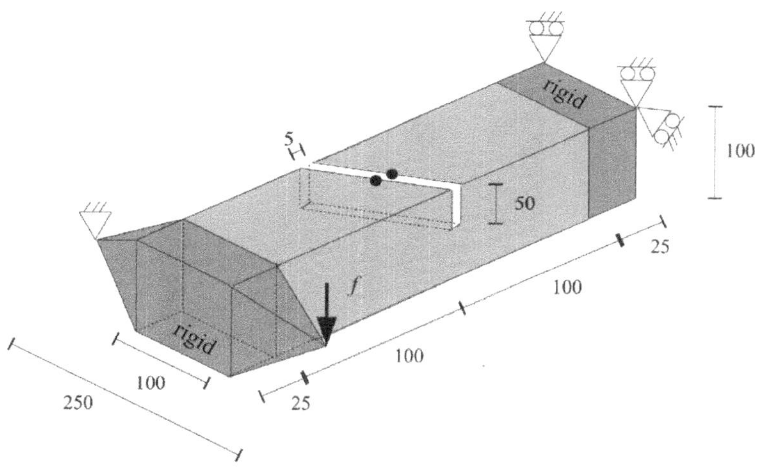

Figure 6.5. *Geometry of the Brokenshire torsion test*

The XFEM approach is a kinematic approach. Discontinuous modes are added to the continuous finite element approximation that allow the crack to open over its entire length and regardless of its position on the mesh. The bulk behavior around the tip is generally elastic but some authors have considered other (non-softening) behaviors such as plasticity or hyperelasticity. It is not the XFEM approach that indicates how the crack propagates. This must be given by the material model. For example, stress intensity factors in linear elastic fracture can be used to control the path of the crack. A cohesive model is more general and also makes it possible to introduce a characteristic length.

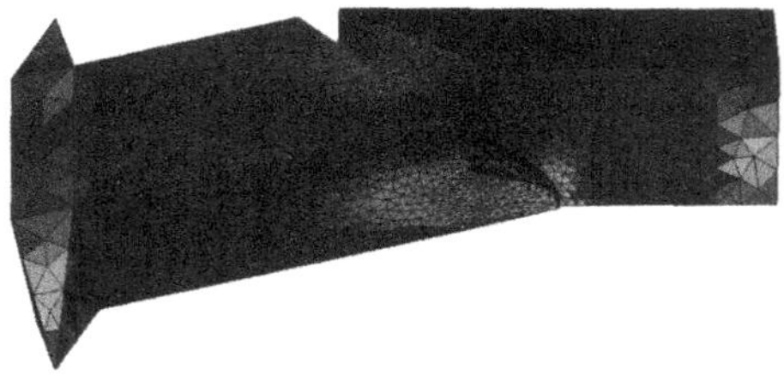

Figure 6.6. *Deformation and stress level. For a color version of this figure, see www.iste.co.uk/lebon/mechanics.zip*

Figure 6.7. *Experimental crack path*

Modeling branching for a crack tip that turns into two crack tips creates a problem. The material model must be able to predict the phenomenon. Linear fracture mechanics and cohesive models in our opinion are not suitable for this purpose. Material softening around the crack tip must be taken into account. Knowledge of the branching geometry can then be easily integrated in the XFEM approach as shown in Daux et al. (2000); Simone et al. (2006).

Section 6.4 will focus on coupling the possibilities of XFEM with an original smoothing for the surrounding softening model known as TLS. The XFEM-TLS combination will manage the (complex or not) crack propagation in a softening medium.

6.4. XFEM-TLS for cracking in softening media

In this section, a discontinuous approach to fracture adapted to softening media is presented. It combines the possibilities of XFEM with a smoothing of the softening medium known as TLS. The TLS method, first presented in Moës et al. (2011); Bernard et al. (2012); Stolz and Moës (2012a), leads to smoothing a softening material model and allows specifying when and where to place cracks. The XFEM method then manages displacement jumps along the crack.

When the medium is softening, crack propagation is the consequence of strain localization. Crack growth is no longer simply dictated by stress intensity factors or a cohesive zone. Strain localization has necessarily to be managed to ultimately know the crack paths. The underlying idea of the TLS model is not to directly follow the crack path but rather to follow the localization front.

This localization front may branch and coalesce and have a very complex geometry. To easily manage these topological changes, the level set method is employed. The localization front is the zero level (2D) curve or (3D) surfaces denoted by Γ_0, of a signed distance field, denoted ϕ. This signed distance field verifies the *eikonal* equation:

$$||\nabla\phi|| = 1 \qquad [6.5]$$

The domain of study Ω is split into a subdomain Ω^+ where ϕ is positive and a subdomain Ω^- where ϕ is negative. By convention, Ω^- corresponds to the healthy material and Ω^+ to the material being localized. The localization area Ω^+ forms a thick band in which it will be seen that the crack grows at the same rate as the localization front.

To go deeper with the description of the TLS model, the material behavior must be specified. To simplify the discussion, the theory of small displacements and small strains will be considered. Only quasi-brittle materials will be addressed. Consider an elastic-damaging material whose only internal variable is a scalar denoted by D and called damage. Free energy is given by a potential $W(\epsilon, D)$ assumed to be separately convex in strain ϵ and damage D. The state equations are obtained by differentiation:

$$\boldsymbol{\sigma} = \frac{\partial W(\epsilon, D)}{\partial \epsilon}, \; Y = -\frac{\partial W(\epsilon, D)}{\partial D} \qquad [6.6]$$

where $\boldsymbol{\sigma}$ is the stress and Y the local rate of energy restitution. Damage evolution is assumed to be independent of the strain rate and of the shape:

$$\dot{D} \geq 0, \; Y \leq Y_c H(D), \; (Y - Y_c H(D))\dot{D} = 0 \qquad [6.7]$$

An evolution depending on the strain rate is just as conceivable as well as an evolution, said to be nonassociated, where the quantity Y used in [6.7] would not be given by [6.6].

The material model considered therefore depends only on the choice of functions W, H and the parameter Y_c. In addition, it is further assumed that the model is integrable, that is, any strain path must lead to one and only one stress and damage response. Finally, to simplify the presentation, the function $H(D)$ is considered to be such that as soon as damage appears, the behavior is softening (a generalization is possible) (Moës et al. 2014; Moreau et al. 2017).

The localization front Γ_0 will separate the part of the domain Ω^- on which $D = 0$ of the area Ω^+ on which $D > 0$ (if the behavior was not directly softening for $D > 0$, the damage would not necessarily be zero on Ω^-). The minimum thickness of the localization area is obtained by binding the variable D to the variable ϕ in the localization area. Damage D is an increasing function of the signed distance ϕ:

$$D = D(\phi) \qquad [6.8]$$

An illustration of this relation is given in Figure 6.9. This function is the ingredient TLS, which allows for the smoothing of the local model. Relation [6.8] has significant consequences. The evolution of D is now completely defined by the evolution of the boundary Γ_0. The damage evolution is now linked from one point with its neighbor. More precisely, if we take the time derivative of [6.8] and [6.5], it follows that:

$$\begin{aligned} &\dot{D} = D'\dot{\phi} \\ &\nabla\dot{\phi} \cdot \nabla\phi = 0 \end{aligned} \qquad [6.9]$$

where D' indicates the derivative of D with respect to ϕ. Equation [6.9] indicates that $\dot{\phi}$ is uniform when following $\nabla\phi$, that is, according to the normal to Γ_0. If the velocity $\dot{\phi}$ of the front Γ_0 is known, the damage rate $\dot{D}$ is therefore fully defined. We now just have to express the model that gives the evolution of ϕ.

Let us look for the dual variable to the front velocity. To this end, we write the dissipation W_d which corresponds to the difference between power supplied and variation of energy:

$$W_{\text{d}} = \int_{\partial\Omega} \boldsymbol{\sigma n} \cdot \dot{\boldsymbol{u}} \, \text{d}\Gamma - \frac{\text{d}}{\text{d}t} \int_{\Omega} W(\epsilon, D) \, \text{d}\Omega = \int_{\Omega^+} Y\dot{D} \, \text{d}\Omega = \int_{\Omega^+} YD'\dot{\phi} \, \text{d}\Omega \qquad [6.10]$$

where $\boldsymbol{n}$ is the normal outside the outer boundary of the domain $\partial\Omega$. since $\dot{\phi}$ is uniform on segments orthogonal to Γ_0, we can write that:

$$W_{\rm d} = \int_{\Gamma_0} g\dot{\phi}\ {\rm d}\Gamma \qquad [6.11]$$

where:

$$g(\boldsymbol{x}) = \int_0^{\phi_{\max}(\boldsymbol{x})} YD'\left(1 - \phi/\rho_0(\boldsymbol{x})\right)\ {\rm d}\phi,\ \boldsymbol{x} \in \Gamma_0 \qquad [6.12]$$

where ρ_0 is the curvature radius of the boundary Γ_0 (it is assumed that the boundary Γ_0 is smooth) and ϕ_s is the distance between the point $\boldsymbol{x}$ of Γ_0 and the skeleton $\Gamma_{\rm s}$ associated with Γ_0 (or the edge of the domain if the skeleton is not met). The skeleton is the set of inner points to Ω^+ for which the gradient of ϕ is not defined. For a simpler interpretation of the skeleton, consider a table whose tray has been cut into the shape of Ω^+. As much sand as possible is then put onto this table. Sharp edges will appear. The normal projection of these edges onto the table indicates the skeleton position. If the table is a disk, the skeleton is reduced to the center of the disk and if the table is square, the skeleton corresponds to a cross. A skeleton is shown in Figure 6.8 (extracted from Zghal et al. (2018)). The different segments on which the integral of 6.12 will operate are indicated. Some touch the skeleton and others touch the edge of the domain.

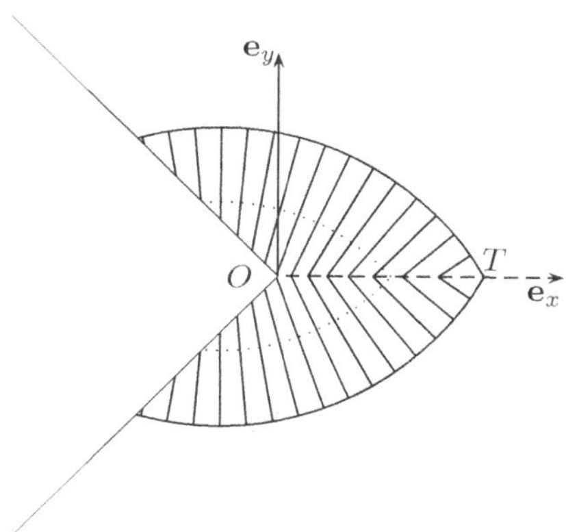

Figure 6.8. *Damaged area emanating from a reentering corner. The skeleton is indicated by segment OT*

Equation [6.11] indicates that the dual quantity to $\dot{\phi}$ according to the dissipation is g. This is a configurational force. The amount g is to $\dot{\phi}$ what the global energy restitution rate G is to the crack speed in linear fracture mechanics. The quantity g is the natural quantity to build the evolution law. For example, consider a time-independent relation:

$$\dot{\phi} \geq 0, \; g \leq g_c, \; (g - g_c)\dot{\phi} = 0, \; \text{on } \Gamma_0 \qquad [6.13]$$

where:

$$g_c(\boldsymbol{x}) = \int_0^{\phi_{\max}(\boldsymbol{x})} Y_c H(D) D' \left(1 - \phi/\rho_0(\boldsymbol{x})\right) \, \mathrm{d}\phi \qquad [6.14]$$

The mechanical behavior of the material is therefore tantamount to the following equations: the state equations in [6.6], the relation between D and ϕ, [6.8] and, finally, the evolution equations of the localization front [6.13].

6.4.1. *TLS V1 and V2 models*

Two versions of TLS are now going to be discussed which are distinguished by the choice of the function $D(\phi)$. The first version consists of letting the damage go up to 1 when ϕ reaches the distance l_c (Figure 6.9). In this version, called V1, the lips of the crack are defined as the iso $\phi = l_c$, denoted Γ_c and Ω_c denotes the domain for which $\phi > l_c$ (domain that is involved in specific at connection points).

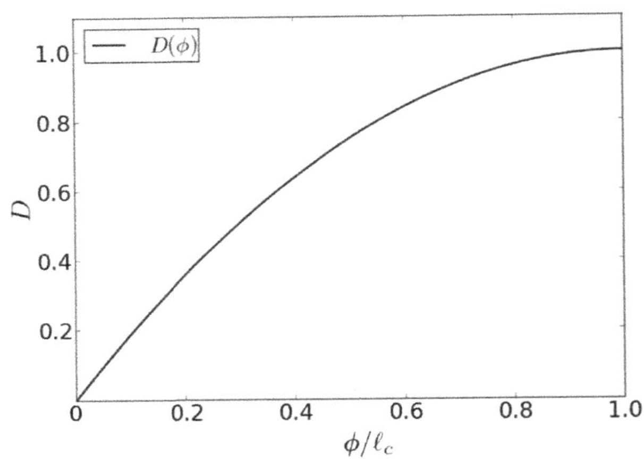

Figure 6.9. *A parabolic damage profile* $D(\phi) = 2\left(\frac{\phi}{\ell_c}\right) - \left(\frac{\phi}{\ell_c}\right)^2$. *For a color version of this figure, see www.iste.co.uk/lebon/mechanics.zip*

We now define the kinematic and equilibrium conditions. A force load applied to the portion $\partial\Omega_T$ of the boundary and zero displacement (that can be generalized without problems) is imposed on the complementary portion $\partial\Omega_u$. Loading is decomposed into a scalar parameter μ and a reference distribution $\boldsymbol{T}^0$. The quasi-static response of quasi-brittle structure that can be subject to severe *snap-backs*, if necessary, the solving algorithm will be able to take control of the loading parameter μ. The crack lips are supposed to be loading free. The kinematic condition is written as:

$$\begin{aligned}&\boldsymbol{u} \in \mathcal{U}_1 = \{\boldsymbol{u} \text{ smooth on } \Omega \setminus \Omega_{\rm c}; \boldsymbol{u} = \boldsymbol{0} \text{ on } \partial\Omega_u \\ &\text{and possibly discontinuous on } \Gamma_{\rm c}\}\end{aligned} \qquad [6.15]$$

and the equilibrium equation is:

$$\int_{\Omega\setminus\Omega_c} \boldsymbol{\sigma} : \boldsymbol{\epsilon}(\boldsymbol{u}^*)\ \mathrm{d}\Omega = \mu \int_{\partial\Omega_T} \boldsymbol{T}^0 \cdot \boldsymbol{u}^*\ \mathrm{d}\Gamma,\ \forall \boldsymbol{u}^* \in \mathcal{U}_1 \qquad [6.16]$$

where $\boldsymbol{\epsilon}(\boldsymbol{u}^*)$ is the symmetric gradient of $\boldsymbol{u}^*$.

The model just presented is suitable for different types of softening $H(D) = 1$, which corresponds to a vertical drop after elastic loading or a slower drop for increasing $H(D)$. A more gradual drop of the local model will result in a TLS damaged nose that will be increasingly visible at the crack tip and which will be long. The softening function $H(D)$ can be selected so as to ensure an identical behavior of the nose (on a 1D bar) for a given cohesive area (Parrilla Gómez et al. 2015). This cohesive capacity of TLS then makes it possible to model size and shape effects in flexural tests on three concrete beam points (Parrilla Gómez et al. 2017). An interesting discussion on taking progressive softening into account can also be found in van der Meer and Sluys (2015).

Extensive analyses of the TLS V1 (Lé et al. 2018) showed that for a more gradual softening, the numbers of elements per $l_{\rm c}$ should be larger to ensure a given accuracy. To solve this issue and maintain a softening-independent accuracy, a new version of the TLS called V2 was proposed. It consists of no longer letting damage reach 1 (and therefore the bulk behavior always maintains a certain elastic stiffness). However, a displacement jump is allowed on the skeleton when the skeleton points are at a sufficient distance from the localization front ($\phi \geq \phi^*$). The part of the skeleton that is allowed to open is denoted by $\Gamma_{\rm s}^*$:

$$\Gamma_{\rm s}^* = \{\boldsymbol{x} \in \Gamma_{\rm s} : \phi(\boldsymbol{x}) \geq \phi^*\} \qquad [6.17]$$

As in a cohesive interface model, the level of opening on the skeleton will depend on the damage at the interface on the skeleton denoted d. The dependency of D and d on the variable ϕ is shown in Figure 6.10. It is observed that unlike V1 (Figure 6.9), the variable can no longer reach 1. On the other hand, can the interface variable d ensure a completely destroyed interface by reaching 1?

In V2, the skeleton stores an energy $w([\boldsymbol{u}], d)$ depending on the displacement jump $[\boldsymbol{u}]$ and the interface damage d. Through differentiation, the state equations specifying the cohesive tension at the interface and the local rate of energy restitution are obtained:

$$\boldsymbol{t} = \frac{\partial w([\boldsymbol{u}], d)}{\partial [\boldsymbol{u}]}, \; y = -\frac{\partial w([\boldsymbol{u}], d)}{\partial d} \qquad [6.18]$$

Damage evolution is assumed to be given by:

$$\dot{d} \geq 0, \; y \leq y_c, \; (y - y_c h(d))\dot{d} = 0 \qquad [6.19]$$

where $h(d)$ describes the softening of the interface. Similarly to the volume damage D, interface damage is a function of the distance ϕ on the skeleton denoted $\phi_{\max}$ (see Figure 6.10):

$$d = d(\phi_S) \qquad [6.20]$$

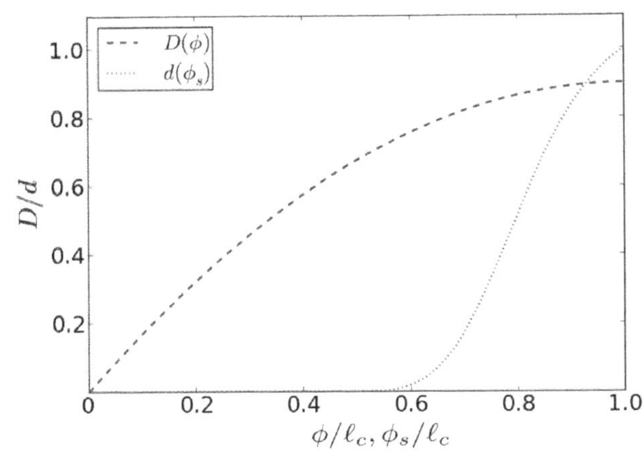

Figure 6.10. *Dependency on volume D and surface damage d according to the distance to the front in the volume ϕ and on the skeleton $\phi_{\max} = \phi(x_s)$. For a color version of this figure, see www.iste.co.uk/lebon/mechanics.zip*

COMMENT ON FIGURE 6.10.– *Volume damage limited below the value 1 prevents the bulk behavior from completely deteriorating. Furthermore, surface damage can reach 1 ensuring the final absence of force on the crack lips.*

For this V2 of the TLS, the kinematic and equilibrium equations take a slightly different shape from V1:

$$\begin{aligned} &\boldsymbol{u} \in \mathcal{U}_2 = \{\boldsymbol{u} \text{ smooth on } \Omega; \boldsymbol{u} = \mathbf{0} \text{ on } \partial\Omega_u \\ &\text{and possibly discontinuous on } \Gamma_{\mathrm{s}}^*\} \end{aligned} \qquad [6.21]$$

and the equilibrium equation:

$$\int_{\Omega} \boldsymbol{\sigma} : \boldsymbol{\epsilon}(\boldsymbol{u}^*) \, \mathrm{d}\Omega + \int_{\Gamma_{\mathrm{s}}^*} \boldsymbol{t} \cdot [\boldsymbol{u}^*] \, \mathrm{d}\Gamma = \mu \int_{\partial\Omega_T} \boldsymbol{T}^0 \cdot \boldsymbol{u}^* \, \mathrm{d}\Gamma, \; \forall \boldsymbol{u}^* \in \mathcal{U}_2 \qquad [6.22]$$

It is no longer necessary to define the completely destroyed domain Ω_{c} in V2 because the crack lips are now completely fastened. A TLS V2 crack in deformation mode is shown in Figure 6.11. The cohesive forces $\boldsymbol{t}$ can be observed on the crack lips.

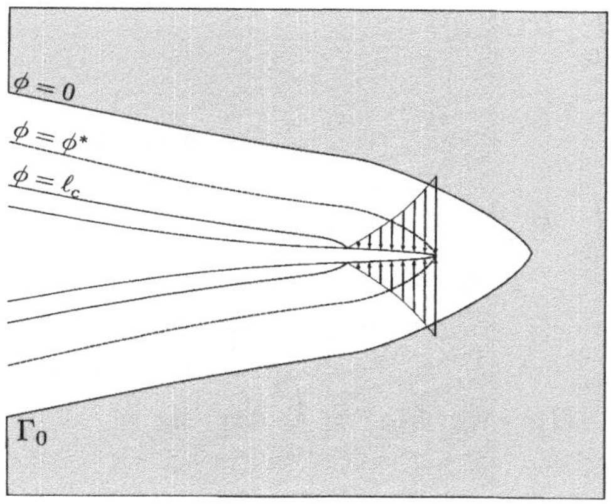

Figure 6.11. *View of a crack in deformed configuration in the TLS V2 approach*

The calculation of the dissipation in V2 gives a different expression for configurational force (since the interface now dissipates just as the volume that surrounds it). The configurational force is now:

$$g(\boldsymbol{x}) = \int_0^{\phi_{\max}(\boldsymbol{x})} Y D'\left(1 - \phi/\rho_0(\boldsymbol{x})\right) \, \mathrm{d}\phi + \frac{1}{2} y d' \,|_{\boldsymbol{x}_{\mathrm{s}}(\boldsymbol{x})} \quad \boldsymbol{x} \in \Gamma_0 \qquad [6.23]$$

where $\boldsymbol{x}_{\mathrm{s}}$ is the skeleton point associated with $\boldsymbol{x}$ (if no point in the skeleton is associated with $\boldsymbol{x}$, the second term of the right-hand side member of [6.23] no longer appears).

As for the critical value of the configurational force to allow evolution, we take:

$$g_c(\boldsymbol{x}) = \int_0^{\phi_{\max}(\boldsymbol{x})} Y_c H(D) D' \left(1 - \phi/\rho_0(\boldsymbol{x})\right) \mathrm{d}\phi + \frac{1}{2} y_c h(d) d' \,|_{\boldsymbol{x}_s(\boldsymbol{x})} \quad \boldsymbol{x} \in \Gamma_0 \quad [6.24]$$

Whether in version V1 or V2, the localization band must have a minimum width of $2l_c$ for a force-free crack to appear. This is ensured by the a priori definition of volume D and surface damage d with respect to the distance to the localization front Γ_0.

6.4.2. *Relation to the Griffith model and the cohesive model*

Consider a crack advancing rectilinearly and self-similar in a 2D medium. The energy restitution rate (energy required related to crack advance) can be calculated for V1 and V2 to give:

$$G_{c1} = 2Y_c \int_0^{l_c} \tilde{H}(D(\phi)) \, \mathrm{d}\phi \quad [6.25]$$

and:

$$G_{c2} = 2Y_c \int_0^{l_c} \tilde{H}(D(\phi)) \, \mathrm{d}\phi + y_c \tilde{h}(1) \quad [6.26]$$

where:

$$\tilde{H}(D) = \int_0^{D} H(x) \, \mathrm{d}x, \; \tilde{h}(d) = \int_0^{d} h(x) \, \mathrm{d}x \quad [6.27]$$

By appropriately choosing the parameters l_c, Y_c and the function H (and y_c, h for V2), a given level of critical energy for crack advance can thus be obtained. In the same way, the TLS model can also be combined with a given cohesive model (Parrilla Gómez et al. 2015; Lé et al. 2018), regardless of the choice of the parameter l_c.

6.4.3. *TLS: implementation aspects*

We now describe the important ingredients for a TLS implementation within the framework of the finite element method. The front velocity $\dot{\phi}$ is discretized using a given number of positive local functions M_I and forming a unity partition. These functions are called modes:

$$\dot{\phi}(\boldsymbol{x}) = \sum_I \dot{\phi}_I M_I(\boldsymbol{x}), \; \boldsymbol{x} \in \Gamma_0 \quad [6.28]$$

These individual modes are then extended within Ω^+ following the normals at the edge Γ_0. This corresponds to the standard approach to extending velocities in the *level set* method and can be efficiently achieved (without solving a matrix) by the *fast marching* method (Sethian 1999). Using a loose notation, $M_I(\boldsymbol{x})$ will also denote the modes obtained by extension across the domain Ω^+:

$$\dot{\phi}(\boldsymbol{x}) = \sum_I \dot{\phi}_I M_I(\boldsymbol{x}), \ \boldsymbol{x} \in \Omega^+ \tag{6.29}$$

The modes will best satisfy the condition of extension:

$$\nabla M_I \cdot \nabla \phi = 0 \tag{6.30}$$

on Ω^+ for a known ϕ.

The modes are then used to define the discrete configuration forces associated:

$$g_{\mathrm{I}} = \int_{\Omega^+} Y D' M_I \ \mathrm{d}\Omega + \frac{1}{2} \int_{\Gamma_s} y d' M_I \ \mathrm{d}\Gamma \tag{6.31}$$

$$g_{\mathrm{cI}} = \int_{\Omega^+} Y_{\mathrm{c}} H(D) D' M_I \ \mathrm{d}\Omega + \frac{1}{2} \int_{\Gamma_s} y_{\mathrm{c}} h(d) d' M_I \ \mathrm{d}\Gamma \tag{6.32}$$

Only the expression of quantities for V2 has been considered (more general than V1). The discrete quantities $\dot{\phi}_I$ and g_{I} are dual for the dissipation in the same way as their continuous analogue. We have:

$$W_d = \sum_I g_{\mathrm{I}} \dot{\phi}_I \tag{6.33}$$

The discrete model for front evolution is quite naturally chosen as its continuous analogue. For each mode I, it is imposed that:

$$\dot{\phi}_I \geq 0, \ g_{\mathrm{I}} \leq g_{\mathrm{cI}}, \ (g_{\mathrm{I}} - g_{\mathrm{cI}}) \dot{\phi}_I = 0 \tag{6.34}$$

The implementation requires finite element sliced at two levels. The first concerns the localization front Γ_0. It goes through elements that will be sliced to accurately integrate their contribution into the calculation of the g_{I} and the elementary stiffness matrix. These cuts are fairly standard and reuse the methods developed within the XFEM framework.

Less standard cuts are required along the crack lips. In V1, the two crack lips are extracted. They can be separated by a small distance (in particular at the branching points of the cracks). An algorithm known as double slicing has been specifically set up for 2D and 3D elements (Salzman et al. 2016). The different examples of slicing a tetrahedron are given in Figure 6.12.

Regarding V2, the challenges are a slightly different. The crack is not positioned on an isocontour but on a skeleton. The skeleton is obtained by *fast marching*.

Whether for V1 or V2, the mesh nodes for which the support is sliced by the crack into two or more pieces must be enhanced by discontinuous modes (as much modes as pieces minus one per spatial direction). These enrichments are very similar to the enrichments already in place for crack branchings in linear fracture mechanics (Daux et al. 2000) or for taking into account discontinuities at the edge of the grains in polycrystals (Simone et al. 2006).

Before showing some XFEM-TLS cracking simulations, the solving algorithm should be specified (Moreau et al. 2017; Lé et al. 2018). The calculation starts as a standard elastic computation. As long as the variable Y does not exceed the value Y_c, there is no damage and the field ϕ is not defined. As soon as, at any point in the structure, the variable Y does exceed the value Y_c, a small circle is centered thereon, with a radius in general of one or two elements. This circle gives a distance field ϕ.

In knowing the displacement, *level sets* field as well as the loading parameter at time t_n: $(\boldsymbol{u}^n, \phi^n, \mu^n)$, an attempt is made to find the same quantities at time t_{n+1}. The load parameter is left as unknown because in the occurrence of *snap-back* this parameter must decrease to follow the quasi static curve. However, a given maximum damage increment is imposed (thereby of ϕ) between two time steps which will be denoted $\Delta\phi_{\max}$.

The algorithm consists of two steps. The first leads to calculating ϕ^{n+1}. The quantities g_{I} and g_{cI} are linearized around their value in t_n:

$$g_{\mathrm{I}}^{n+1} \sim g_{\mathrm{I}}^{n} + \alpha_I \Delta\mu \qquad [6.35]$$

$$g_{\mathrm{cI}}^{n+1} \sim g_{\mathrm{cI}}^{n} + \beta_I \Delta\phi_I \qquad [6.36]$$

In the first relation, the dependency of the displacement field was neglected relatively to a change in damage and in the second relation only the influences of $\Delta\phi_I$ on g_{cI} are retained (and the influence of the other modes are ignored). Injecting expressions [6.35] and [6.36] into the evolution conditions of [6.34] yields (with imposition of the maximal increment of ϕ):

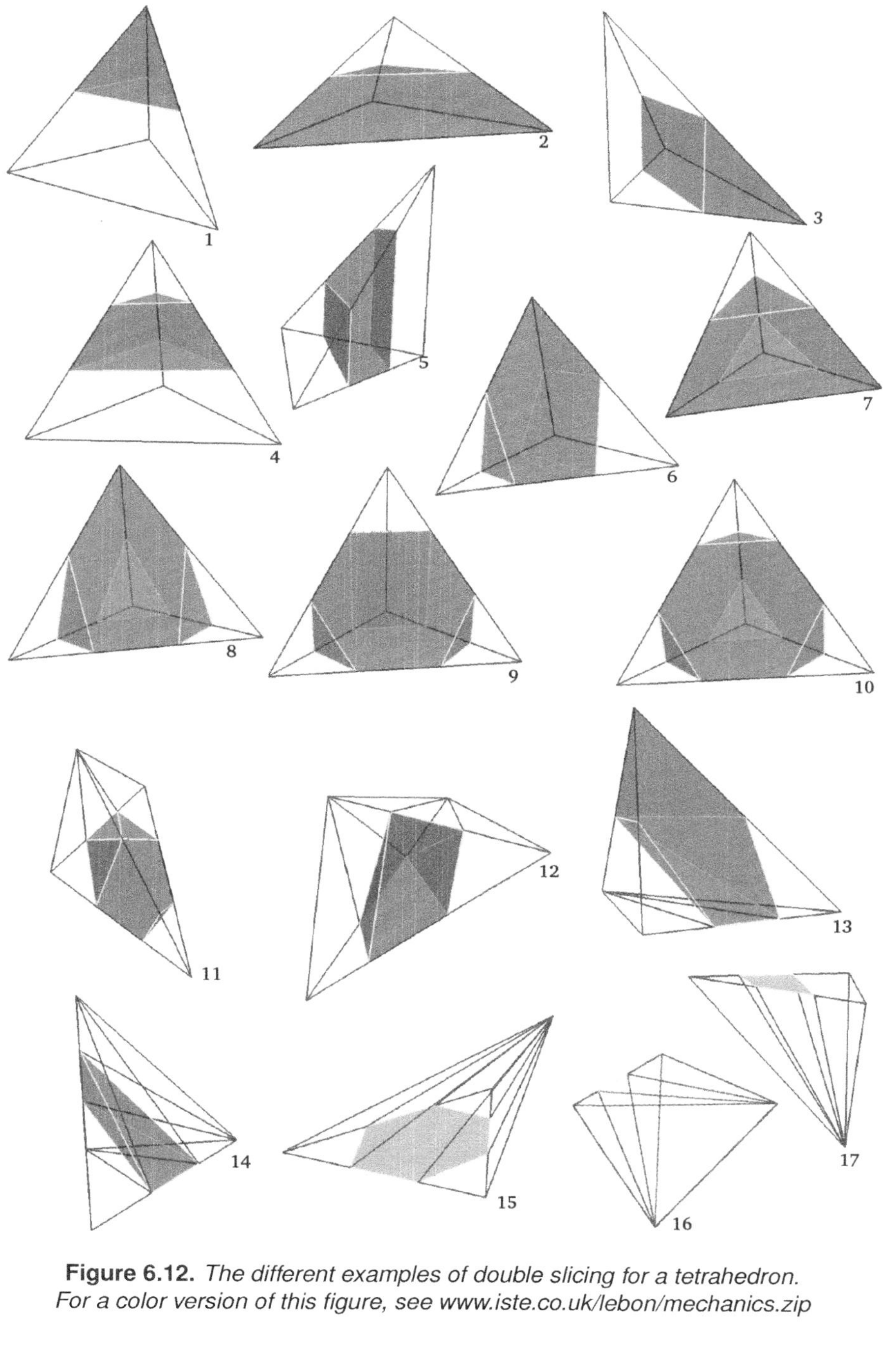

Figure 6.12. *The different examples of double slicing for a tetrahedron. For a color version of this figure, see www.iste.co.uk/lebon/mechanics.zip*

Find $\Delta\phi_I$ and $\Delta\mu$ such that:

$$\Delta\phi_I \geq 0, z_I = g_{\rm I}^n - g_{\rm cI}^n + \alpha_I \Delta\mu - \beta_I \Delta\phi_I \leq 0, z_I \Delta\phi_I = 0$$
$$\max_I \Delta\phi_I = \Delta\phi_{\max} \quad [6.37]$$

The above system is a *linear complementarity problem* (LCP) parameterized by $\Delta\mu$ whose matrix is diagonal (therefore very simple to solve). Given the convexity of W in D and the imposition of the convexity of $\tilde{H}(D)$, it can be shown that the coefficients β_I are all positive ensuring uniqueness and existence of the LCP solution. This solution is:

$$\Delta\mu = \min_I\left(\frac{\beta_I \Delta\phi_{\max} - g_{\rm I} + g_{\rm cI}}{\alpha_I}\right), \ \Delta\phi_I = \max\left(\frac{g_{\rm I}^n - g_{\rm cI}^n + \alpha_I \Delta\mu}{\beta_I}, 0\right) \quad [6.38]$$

Since the front advance is known, the front can be advanced by *fast marching* to figure out ϕ^{n+1}. The second phase of the algorithm consists of calculating $\boldsymbol{u}^{n+1}$ and μ^{n+1} by solving the following problem. Knowing $\phi^{n+1}, \Gamma_{\rm s}^{n+1}$, find $(\boldsymbol{u}^{n+1}, \mu^{n+1})$, such that:

$$\int_\Omega \boldsymbol{\sigma}(\boldsymbol{\epsilon}(\boldsymbol{u}^{n+1}), D(\phi^{n+1})) : \boldsymbol{\epsilon}(\boldsymbol{u}^*) \mathrm{d}\Omega + \int_{\Gamma_{\rm s}^{n+1}} \boldsymbol{t}([\boldsymbol{u}^{n+1}], d(\phi^{n+1})) \cdot [\boldsymbol{u}^*] \mathrm{d}\Gamma =$$
$$\mu^{n+1} \int_{\partial\Omega_T} \boldsymbol{T}^0 \cdot \boldsymbol{u}^* \ \mathrm{d}\Gamma, \ \forall \boldsymbol{u}^* \in \mathcal{U}_2 \quad [6.39]$$

$$\max_I(g_{\rm I}^{n+1} - g_{\rm cI}^{n+1}) = 0 \quad [6.40]$$

where the quantities $g_{\rm I}^{n+1}$ and $g_{\rm cI}^{n+1}$ are calculated by using expressions [6.31]–[6.32] (not the approximated expressions of [6.35]–[6.36]).

6.5. XFEM-TLS simulation examples

We present three simulations. The first concerns fracture in chalk. There are no cracks initially. Damage and then crack initiation leading to total fracture of chalk are shown. The second simulation addresses multiple cracking in a cube with cavities.

This example is used to illustrate the automatic coalescing capability of TLS. Cracks bypass cavities and automatically reconnect on the opposite side. These two simulations are taken from Salzman et al. (2016) and illustrate the use of TLS V1. Cohesive behavior is not sought for at the tip. The softening function is simply $H(D) = 1$. The third simulation features TLS V2 with a three-point bending example with a cohesive-type crack lip opening (presence of recall tension).

6.5.1. *Torsional chalk fracture*

Geometry and loading are given in Figure 6.13. This problem had already been simulated by Bordas et al. (2008) with a meshfree method. A small initial damage surface Γ_0 is placed halfway up on the surface of the chalk to force symmetrical cracking. The final state of the chalk after rupture is given in Figure 6.14.

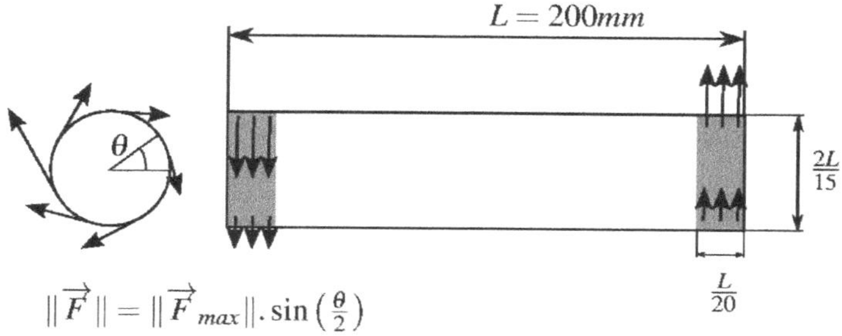

Figure 6.13. *Geometry and loading for the chalk problem under torsion (source: Salzman et al. 2015)*

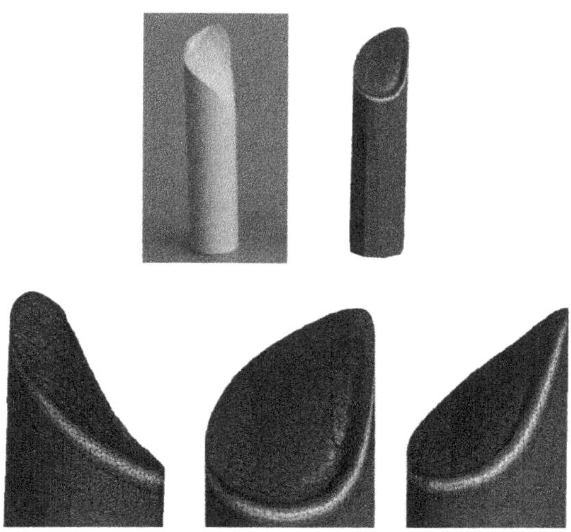

Figure 6.14. *Final state of the chalk after fracture. The color indicates the damage level (blue D = 0, red D = 1) (source: Salzman et al. 2015). For a color version of this figure, see www.iste.co.uk/lebon/mechanics.zip*

6.5.2. *Multiple cracking in a block with holes*

This example is also taken from Salzman et al. (2016). The geometry is given in Figure 6.15. Loading is applied according to direction y on both sides normal to the y axis. The calculation starts without boundary Γ_0. It is discovered by searching for points such that $Y \geq Y_c$. During the computation, the existing fronts propagate but may appear again. The use of XFEM is clearly shown in Figure 6.16 where the crack opening can be seen. Figure 6.17(a) gives the damage condition of the cube after complete cracking. Figure 6.17(b) shows the capacity of the TLS method for coalescing two localization fronts and then for coalescing the cracks therein, without any special treatment.

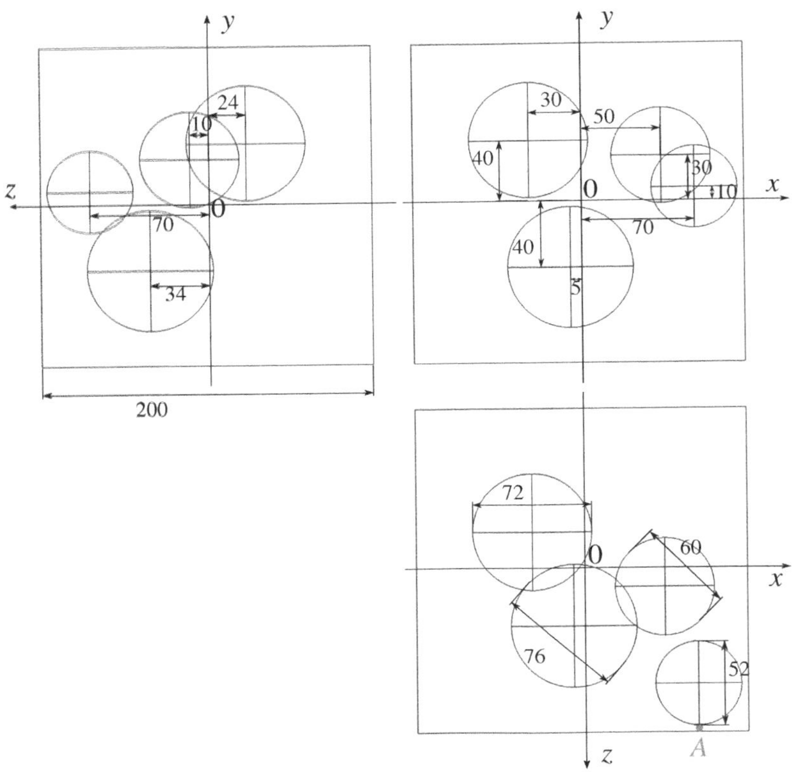

Figure 6.15. *Localization of hollows in the cube, sizes in mm (source: Salzman et al. 2015). For a color version of this figure, see www.iste.co.uk/lebon/mechanics.zip*

Figure 6.16. *Crack opening during cube cracking (source: Salzman et al. 2015). For a color version of this figure, see www.iste.co.uk/lebon/mechanics.zip*

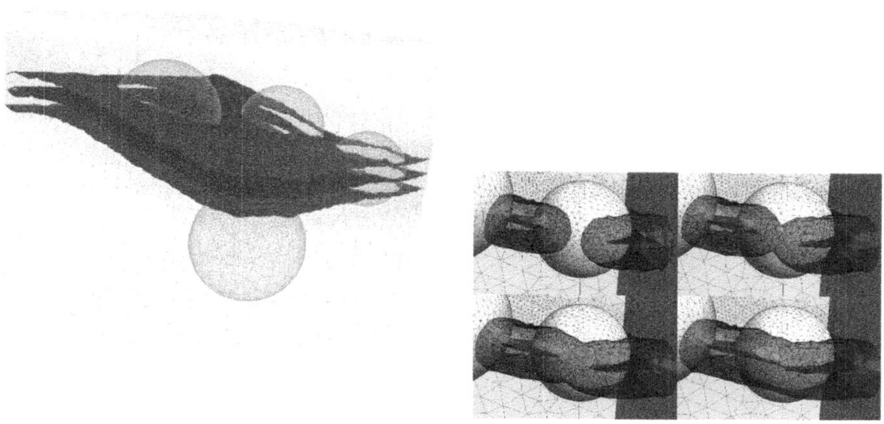

a) b)

Figure 6.17. *Final damage state at yield. The blue color indicates the localization front* Γ_0 *and the red the crack lips* Γ_c *(a). Enlargement of the coalescence of two localization fronts followed by crack coalescence, the figures are chronologically placed, from top to bottom and left to right (b) (source: Salzman et al. 2015). For a color version of this figure, see www.iste.co.uk/lebon/mechanics.zip*

6.5.3. *Three-point bending for a beam and cohesive crack*

The two previous examples were carried out within the framework of TLS V1 for a very brittle behavior model $H(D) = 1$. The focus us now on another type of behavior with very gradual softening. In this case, the *Process Zone* at the crack tip is going to be quite long and the interest of V2 can be seen because it makes it possible to considerably improve the computation time and accuracy.

Consider a three-point (symmetric) bending problem as shown in Figure 6.18.

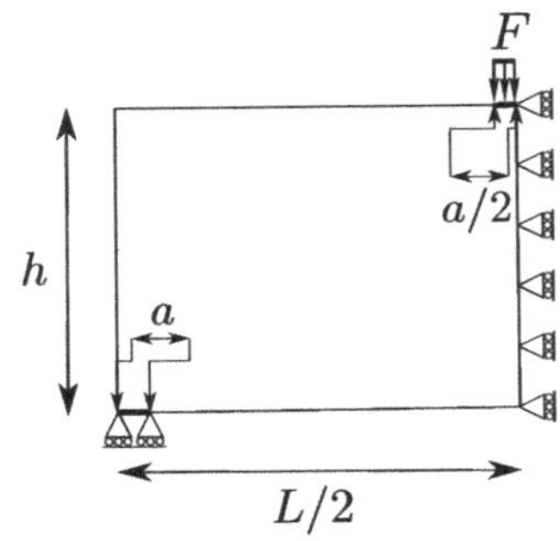

Figure 6.18. *Symmetric geometry of the three-point bending problem (source: Le et al. 2018)*

Figure 6.19(a) gives the deformation of the beam near the crack. The crack opening is clearly visible as well as the cohesive tensile field on the lips. This field is also drawn inside the domain where it merely ensures the non-opening of the crack. It is observed that on the top of the beam, this field becomes a compression field.

Figure 6.19(b) gives the interface damage D. A value of 1 corresponds to the absence of traction. The presence of cohesive tractions corresponds to the zone where d has not yet reached 1.

6.6. Conclusion

In this chapter, we first presented the use of the XFEM method on its own for non-softening media with or without cohesive area. In a second step, XFEM capabilities were combined with TLS smoothing to create the cracking in softening media. The XFEM-TLS method is original because it combines the advantages of a discontinuous approach to failure with the benefits of a diffuse approach for managing initiation, coalescence and branching.

Regarding the computation cost, it is limited because the smoothing work is limited to a thin band around the crack and, in addition, updating the damage variable in this band does not require solving a system. A comparison between a *phase-field* model and a TLS model can be found in Cazes and Moës (2015).

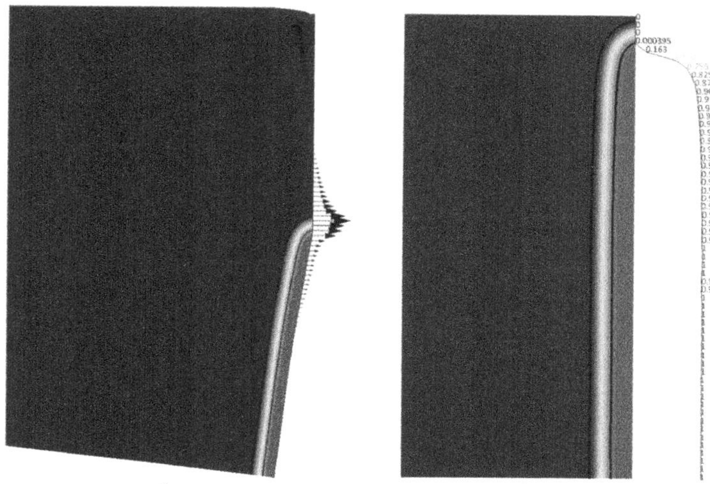

a) b)

Figure 6.19. *Beam deformation with tensile forces (a) and enlargement of the distribution of the interface damage* d *(b). The background color indicates the level of volume damage (source: Le et al. 2018). For a color version of this figure, see www.iste.co.uk/lebon/mechanics.zip*

We have chosen to present the TLS method through its fundamental tool which is the *level set* technology (Sethian 1999). It is also possible to understand the method as a particular velocity formulation on a moving front (Stolz and Moes 2012b) or as a variational approach where the condition on the damage is taken into account by a Lagrangian multiplier field (Frémond and Stolz 2017).

The chapter focused on quasi-static loading. XFEM-TLS results in fast dynamics and fragmentation can be found in Moreau et al. (2015) and Stershic et al. (2017). The important advantage of the XFEM-TLS approach in explicit dynamics is the fact that updating damage does not require solving a global system. The explicit nature of the approach is not called into question by damage smoothing.

Finally, an application of the TLS approach at the interface level was proposed by Latifi et al. (2017); Voormeeren et al. (2017); Liu et al. (2018), with applications to fatigue and dynamics.

6.7. References

Aifantis, E. (1984). On the structural origin of certain inelastic models. *Journal of Engineering Materials and Technology*, 106, 326–330.

Bazant, Z., Belytschko, T., Chang, T. (1984). Continuum theory fo strain-softening. *Journal of Engineering Mechanics*, 110, 1666–1692.

Béchet, E., Minnebo, H., Moës, N., Burgardt, B. (2005). Improved implementation and robustness study of the X-FEM for stress analysis around cracks. *International Journal for Numerical Methods in Engineering*, 64(8), 1033–1056.

Béchet, E., Moës, N., Wohlmuth, B. (2009). A stable Lagrange multiplier space for stiff interface conditions within the extended finite element method. *International Journal for Numerical Methods in Engineering*, 78(8), 931–954.

Belytschko, T., Gracie, R., Ventura, G. (2009). A review of extended/generalized finite element methods for material modeling. *Modelling and Simulation in Materials Science and Engineering*, 17(4), 043001.

Bernard, P.-E., Moës, N., Chevaugeon, N. (2012). Damage growth modeling using the Thick Level Set (TLS) approach: Efficient discretization for quasi-static loadings. *Computer Methods in Applied Mechanics and Engineering*, 233/236, 11–27.

Bordas, S., Rabczuk, T., Zi, G. (2008). Three-dimensional crack initiation, propagation, branching and junction in non-linear materials by an extended meshfree method without asymptotic enrichment. *Engineering Fracture Mechanics*, 75, 943–960.

de Borst, R. and Verhoosel, C.V. (2016). Gradient damage vs phase-field approaches for fracture: Similarities and differences. *Computer Methods in Applied Mechanics and Engineering*, 312, 78–94.

Bourdin, B., Francfort, G.A., Marigo, J.-J. (2000). Numerical experiments in revisited brittle fracture. *Journal of the Mechanics and Physics of Solids*, 48(4), 797–826.

Bourdin, B., Francfort, G.A., Marigo, J.-J. (2008). The variational approach to fracture. *Journal of Elasticity*, 91, 5–148.

Brokenshire, D. and Barr, B. (1996). A study of torsion fracture tests. PhD Thesis, University of Cardiff, Cardiff.

Cazes, F. and Moës, N. (2015). Comparison of a phase-field model and of a thick level set model for brittle and quasi-brittle fracture. *International Journal for Numerical Methods in Engineering*, 103, 114–143.

Daux, C., Moës, N., Dolbow, J., Sukumar, N., Belytschko, T. (2000). Arbitrary branched and intersecting cracks with the extended finite element method. *International Journal for Numerical Methods in Engineering*, 48(12), 1741–1760.

Dolbow, J., Moës, N., Belytschko, T. (2001). An extended finite element method for modeling crack growth with frictional contact. *Finite Element in Analysis and Design*, 36(3), 235–260.

Duarte, C., Babuska, I., Oden, J. (2000). Generalized finite element methods for three-dimensional structural mechanics problems. *Computers & Structures*, 77(2), 215–232.

Elguedj, T., Gravouil, A., Combescure, A. (2006). Appropriate extended functions for X-FEM simulation of plastic fracture mechanics. *Computer Methods in Applied Mechanics and Engineering*, 195(7/8), 501–515.

Feld-Payet, S., Chiaruttini, V., Besson, J., Feyel, F. (2015). A new marching ridges algorithm for crack path tracking in regularized media. *International Journal of Solids and Structures*, 71, 57–69.

Ferté, G., Massin, P., Moës, N. (2016). 3D crack propagation with cohesive elements in the extended nite element method. *Computer Methods in Applied Mechanics and Engineering*, 300, 347–374.

Francfort, G.A. and Marigo, J.-J. (1998). Revisiting brittle fracture as an energy minimization problem. *Journal of the Mechanics and Physics of Solids*, 46, 1319–1412.

Frémond, M. and Nedjar, B. (1996). Damage, gradient of damage and principle of virtual power. *International Journal of Solids and Structures*, 33(8), 1083–1103.

Frémond, M. and Stolz, C. (2017). On alternative approaches for graded damage modelling. *Models, Simulation, and Experimental Issues in Structural Mechanics*, 2, 87–104.

Fries, T.P. and Belytschko, T. (2010). The extended/generalized finite element method: An overview of the method and its applications. *International Journal for Numerical Methods in Engineering*, 84(3), 253–304.

Hakim, V. and Karma, A. (2005). Crack path prediction in anisotropic brittle materials. *Physical Review Letters*, 95(23), 235501.

Hakim, V. and Karma, A. (2009). Laws of crack motion and phase-field models of fracture. *Journal of the Mechanics and Physics of Solids*, 57(2), 342–368.

Hansbo, A. and Hansbo, P. (2002). An unfitted finite element method, based on Nitsche's method, for elliptic interface problems. *Computer Methods in Applied Mechanics and Engineering*, 191, 5537–5552.

Karihaloo, B.L. (2003). Modelling of stationary and growing cracks in FE framework without remeshing: A state-of-the-art review. *Computers & Structures*, 81(3), 119–129.

Karma, A., Kessler, D., Levine, H. (2001). Phase-field model of mode III dynamic fracture. *Physical Review Letters*, 87(4), 045501.

Khoei, A.R. (2015). *Extended Finite Element Method: Theory and Applications*. John Wiley & Sons, New York.

Laborde, P., Pommier, J., Renard, Y., Salaün, M. (2005). High order extended finite element method for cracked domains. *International Journal for Numerical Methods in Engineering*, 64(3), 354–381.

Latifi, M., van der Meer, F., Sluys, L. (2017). Fatigue modeling in composites with the thick level set interface method. *Composites Part A: Applied Science and Manufacturing*, 101, 72–80.

Lé, B., Moës, N., Legrain, G. (2018). Coupling damage and cohesive zone models with the thick level set approach to fracture. *Engineering Fracture Mechanics*, 193, 214–247.

Legrain, G., Moës, N., Verron, E. (2005). Stress analysis around crack tips in finite strain problems using the eXtended finite element method. *International Journal for Numerical Methods in Engineering*, 63(2), 290–314.

Liu, F. and Borja, R.I. (2010). Stabilized low-order finite elements for frictional contact with the extended finite element method. *Computer Methods in Applied Mechanics and Engineering*, 199(37/40), 2456–2471.

Liu, Y., van der Meer, F., Sluys, L. (2018). Cohesive zone and interfacial thick level set modeling of the dynamic double cantilever beam test of composite laminate. *Theoretical and Applied Fracture Mechanics*, 96, 617–630.

Lorentz, E. and Andrieux, S. (2003). Analysis of non-local models through energetic formulations. *International Journal of Solids and Structures*, 40, 2905–2936.

van der Meer, F.P. and Sluys, L.J. (2015). The thick level set method: Sliding deformations and damage initiation. *Computer Methods in Applied Mechanics and Engineering*, 285, 64–82.

Melenk, J. and Babuska, I. (1996). The partition of unity finite element method: Basic theory and applications. *Computer Methods in Applied Mechanics and Engineering*, 139, 289–314.

Moës, N., Dolbow, J., Belytschko, T. (1999). A finite element method for crack growth without remeshing. *International Journal for Numerical Methods in Engineering*, 46(1), 131–150.

Moës, N., Gravouil, A., Belytschko, T. (2002). Non-planar 3D crack growth by the extended finite element and level sets – Part I: Mechanical model. *International Journal for Numerical Methods in Engineering*, 53(11), 2549–2568.

Moës, N., Stolz, C., Bernard, P.-E., Chevaugeon, N. (2011). A level set based model for damage growth: The thick level set approach. *International Journal For Numerical Methods in Engineering*, 86, 358–380.

Moës, N., Stolz, C., Chevaugeon, N. (2014). Coupling local and non-local damage evolution with the thick level set model. *Advanced Modelling and Simulation in Engineering Sciences*, 2(16), 21.

Mohammadi, S. (2008). *Extended Finite Element Method for Fracture Analysis of Structures*. Blackwell Publishing, Oxford.

Moreau, K., Moës, N., Picart, D., Stainier, L. (2015). Explicit dynamics with a non-local damage model using the thick level set approach. *International Journal for Numerical Methods in Engineering*, 102(3/4), 808–838.

Moreau, K., Moës, N., Chevaugeon, N., Salzman, A. (2017). Concurrent development of local and non-local damage with the thick level set approach: Implementation aspects and application to quasi-brittle failure. *Computer Methods in Applied Mechanics and Engineering*.

Mourad, H., Dolbow, J., Harari, I. (2007). A bubble-stabilized finite element method for Dirichlet constraints on embedded interfaces. *International Journal For Numerical Methods in Engineering*, 772–793.

Mühlhaus, H.B. and Vardoulakis, L. (1987). The thickness of shear bands in granular materials. *Geotechnique*, 37(3), 271–283.

Nguyen, Q.-S. and Andrieux, S. (2005). The non-local generalized standard approach: A consistent gradient theory. *Comptes rendus de l'Académie des sciences : mécanique, physique, chimie, astronomie*, 333, 139–145.

Parrilla Gómez, A., Moës, N., Stolz, C. (2015). Comparison between thick level set (TLS) and cohesive zone models. *Advanced Modeling and Simulation in Engineering Sciences*, 2(1), 18.

Parrilla Gómez, A., Stolz, C., Moës, N., Grégoire, D., Pijaudier-Cabot, G. (2017). On the capability of the thick level set (TLS) damage model to fit experimental data of size and shape effects. *Engineering Fracture Mechanics*, 184, 75–87.

Peerlings, R., Geers, M., De Borst, R., Brekelmans, W. (2001). A critical comparison of nonlocal and gradient-enhanced softening continua. *International Journal of Solids and Structures*, 38(44/45), 7723–7746.

Pijaudier-Cabot, G. and Bazant, Z. (1987). Nonlocal damage theory. *Journal of Engineering Mechanics*, 113, 1512–1533.

Pijaudier-Cabot, G. and Burlion, N. (1996). Damage and localisation in elastic materials with voids. *Mechanics of Cohesive-frictional Materials*, 1, 129–144.

Pommier, S., Gravouil, A., Combescure, A., Moës, N. (2011). *Extended Finite Element Method for Crack Propagation*. ISTE Ltd, London, and John Wiley & Sons, New York.

Prabel, B., Combescure, A., Gravouil, A., Marie, S. (2007). Level set XFEM non-matching meshes: Application to dynamic crack propagation in elastic-plastic media. *International Journal For Numerical Methods in Engineering*, 69(8), 1553.

Rastiello, G., Giry, C., Gatuingt, F., Desmorat, R. (2018). From diffuse damage to strain localization from an Eikonal Non-Local (ENL) Continuum Damage model with evolving internal length. *Computer Methods in Applied Mechanics and Engineering*, 331, 650–674.

Rodríguez-Ferran, A., Morata, I., Huerta, A. (2005). A new damage model based on non-local displacements. *International Journal for Numerical and Analytical Methods in Geomechanics*, 29(5), 473–493.

Salzman, A., Moës, N., Chevaugeon, N. (2016). On use of the thick level set method in 3D quasi-static crack simulation of quasi-brittle material. *International Journal of Fracture*, 202, 1–29.

Schreyer, H. and Chen, Z. (1986). One-dimensional softening with localization. *Journal of Applied Mechanics*, 53(4), 791–797.

Sethian, J. (1999). *Level Set Methods and Fast Marching Methods: Evolving Interfaces in Computational Geometry, Fluid Mechanics, Computer Vision and Material Science*. Cambridge University Press, Cambridge.

Simone, A., Duarte, C.A., Van der Giessen, E. (2006). A generalized finite element method for polycrystals with discontinuous grain boundaries. *International Journal for Numerical Methods in Engineering*, 67(8), 1122–1145.

Stershic, A.J., Dolbow, J.E., Moës, N. (2017). The thick level-set model for dynamic fragmentation. *Engineering Fracture Mechanics*, 172, 39–60.

Stolarska, M., Chopp, D.L., Moës, N., Belytschko, T. (2001). Modelling crack growth by level sets in the extended finite element method. *International Journal for Numerical Methods in Engineering*, 51(8), 943–960.

Stolz, C. and Moës, N. (2012a). A new model of damage: A moving thick layer approach. *International Journal of Fracture*, 174(1), 49–60.

Stolz, C. and Moes, N. (2012b). On the rate boundary value problem for damage modelization by thick level-set. In *ACOME 2012*. Ho Chi Minh, 205–220.

Sukumar, N., Dolbow, J.E., Moës, N. (2015). Extended finite element method in computational fracture mechanics: A retrospective examination. *International Journal of Fracture*, 196(1/2), 189–206.

Tamayo-Mas, E. and Rodriguez-Ferran, A. (2015). A medial-axis-based model for propagating cracks in a regularised bulk. *International Journal For Numerical Methods in Engineering*, 101, 489–520.

Triantafyllidis, N. and Aifantis, E. (1986). A gradient approach to localization of deformation. I. Hyperelastic materials. *Journal of Elasticity*, 16, 225–237.

Voormeeren, L., van der Meer, F., Maljaars, J., Sluys, L. (2017). A new method for fatigue life prediction based on the thick level set approach. *Engineering Fracture Mechanics*, 182, 449–466.

Zghal, J., Moreau, K., Moës, N., Leguillon, D., Stolz, C. (2018). Analysis of the failure at notches and cavities in quasi-brittle media using the thick level set damage model and comparison with the coupled criterion. *International Journal of Fracture*, 211(1/2), 253–280.

Zhuang, Z., Liu, Z., Cheng, B., Liao, J. (2014). *Extended Finite Element Method*. Tsinghua University Press, Beijing.

7

Damage-to-Crack Transition

Sylvia FELD-PAYET
DMAS-ONERA, Université Paris-Saclay, Châtillon, France

7.1. Introduction

7.1.1. *Continuous damage models and their limitations*

In the local approach to fracture based on continuous damage models, material degradation is characterized by a quantity, here referred to by the umbrella term as the *damage variable*, which can be represented as a spatial field. The continuous damage models considered here are of various natures: they can be phenomenological or micromechanical, adapted to ductile or quasi-brittle materials, regularized or not. The damage variable can have a direct influence on the material behavior (in the case of *coupled damage model*), or serve only as a *damage indicator* (for an *uncoupled damage* model).

The main interest of these continuous damage models is to allow for crack initiation without the need to introduce pre-cracking, unlike models used in linear fracture mechanics. This is now possible thanks to the study of the damage variable. A simple indicator of damage (or uncoupled model) may suffice to characterize crack initiation if failure occurs abruptly after reaching the ultimate load. On the other hand, coupled damage models may become necessary when material degradation strongly influences the structural response, for example, with a noticeable decrease of the pre-initiation force response due to material softening.

Numerical Methods for Strong Nonlinearities in Mechanics,
coordinated by Jacques BESSON, Frédéric LEBON and Éric LORENTZ. © ISTE Ltd 2024.

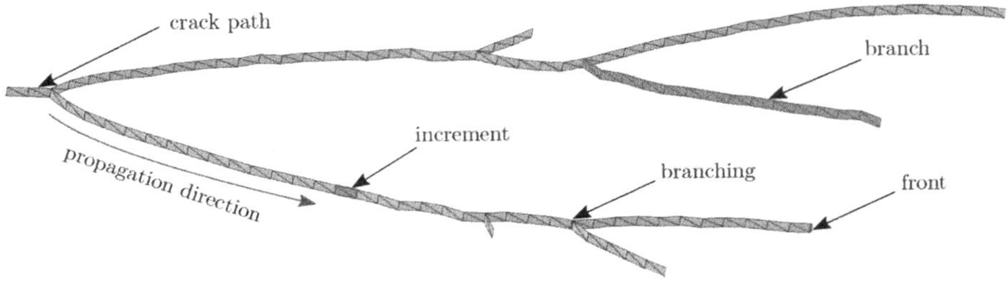

Figure 7.1. *Diagram illustrating the terminology associated with the representation of a discrete crack. For a color version of this figure, see www.iste.co.uk/lebon/mechanics.zip*

In both cases, after initiation, continuous damage models reach their limits:

1) They do not make it possible to correctly represent the propagation of a discrete front (a three-dimensional line or a two-dimensional point, see Figure 7.1) and even less to measure a crack front advance rate.

2) They do not enable us to visualize crack surfaces (in three dimensions, or two-dimensional lines), nor to measure crack lip opening (or a flow through a crack surface).

3) They do not make it possible to *easily* model crack closure and even less contact between the crack lips.

4) From a numerical point of view, it can be difficult to run simulations up to a truly zero effort level.

5) For some types of regularized models (such as the implicit gradient models originally proposed by Peerlings et al. (1996)), the damage localization band tends to widen non-physically over time, due to the continuity of the material through the surface that should correspond to the crack. This problem, mentioned in particular by Geers et al. (1998) and Comi et al. (2007), is shown in Figure 7.2.

To overcome these limitations, it is possible to consider so-called *continuous-discontinuous* approaches or *damage-to-crack transition* approaches for the special case of damage models. The principle of these approaches consists in defining, based on a variable of the continuous model, a *discrete continuous* path which can be used as a support for discontinuity, and in inserting at the appropriate time the corresponding crack increment in the model.

7.1.2. Modeling a discontinuity

Before explaining in more detail what a *damage-to-crack transition* strategy is, let us first consider how discontinuity modeling can be achieved and what the advantages and drawbacks of the different options are by using the example in Figure 7.2.

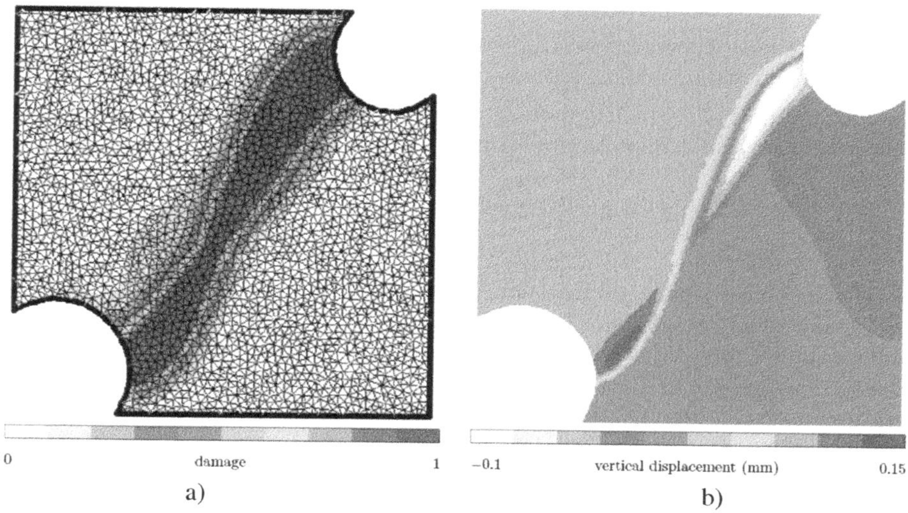

Figure 7.2. *Damage (left) and vertical displacement field (right) distribution in a two-notch specimen submitted to shear stress with the continuous regularized coupled damage model described in Feld-Payet et al. (2015). For a color version of this figure, see www.iste.co.uk/lebon/mechanics.zip*

COMMENT ON FIGURE 7.2.– *The width of the localization band is related to the use of an implicit gradient smoothing method (Engelen et al. 2003). The vertical displacement exhibits a strong gradient in the area where the damage is most significant: this corresponds to the area where a crack should appear. However, this crack is not represented in any form and has not been introduced into the model.*

Different methods are commonly used to represent a discontinuity within the framework of the finite element method[1]. The choice between these methods ultimately results from a trade-off between the possibility to accurately represent the crack lip kinematics, the difficulties in ensuring a gradual decrease of the overall

1 It should be noted that other discontinuity modeling choices are also possible in different frameworks such as the *cracking particle* meshless method of Rabczuk and Belytschko (2007) or the discontinuous Galerkin method used by Wu et al. (2014).

system energy for nonlinear models, as well as the cost and challenges associated with the numerical implementation.

Three types of strategies can be namely distinguished:

– removal of a material volume on an area of uncontrolled thickness when the latter no longer transmits energy;

– localization of degradation in a controlled thickness around the crack to dissipate energy while modeling gradual crack opening;

– localization of degradation at an interface, which can be combined with a cohesive area model to dissipate energy if needed.

7.1.2.1. *Removal of a material volume*

7.1.2.1.1. Principle

The most basic way to represent a discontinuity is to eliminate, from the computational model, a volume (in three dimensions, or a surface in two dimensions) deemed sufficiently degraded by an insertion criterion (see section 7.3 for more details). The position of the discontinuity is directly determined by the model and the insertion criterion, without any control from the user. The width of this area is in particular not a priori controlled: the completely cracked area can therefore extend over several elements, thus causing an uncontrolled loss of structural mass. Moreover, it should be noted that this technique still presents some of the disadvantages of continuous models, especially regarding of the nonoptimal representation of the crack surface and the corresponding lips.

7.1.2.1.2. Erosion

Removal of material can be directly achieved by eliminating from the computation any element that has satisfied the insertion criterion: this is the erosion technique. This solution has been widely applied in the literature with coupled damage models, in particular by Tvergaard (1982), Gao et al. (1998), Hambli (2001), Tvergaard and Needleman (2006), Saanouni (2008), Teng (2008), Xue and Wierzbicki (2009) and Yun et al. (2019). One of the main disadvantages of this method is that the shape of the crack path is highly dependent on the mesh, as shown in Figure 7.3. Indeed, the crack path presents a *sawtooth* effect when the crack does not follow the orientation of the mesh and this effect is all the more marked when the removed element is large. This explains why a very small element size is often preferred in the damaged area (as proposed by Cao (2014)). However, in order to limit the computational cost associated with very small elements, it is preferable to use erosion in combination with an adaptive remeshing strategy, as was done by El Khaoulani and Bouchard (2012) or Yang et al. (2018). A post-erosion remeshing procedure was also proposed for smoothing the *sawtooth*-shaped edges by Labergère et al. (2014). It should be noted, however, that this procedure cannot easily be extended to the three-dimensional case. Another important problem associated with erosion, and particularly visible in three dimensions, is that the continuity of the damage field under consideration does not suffice to guarantee the continuity of the resulting crack path.

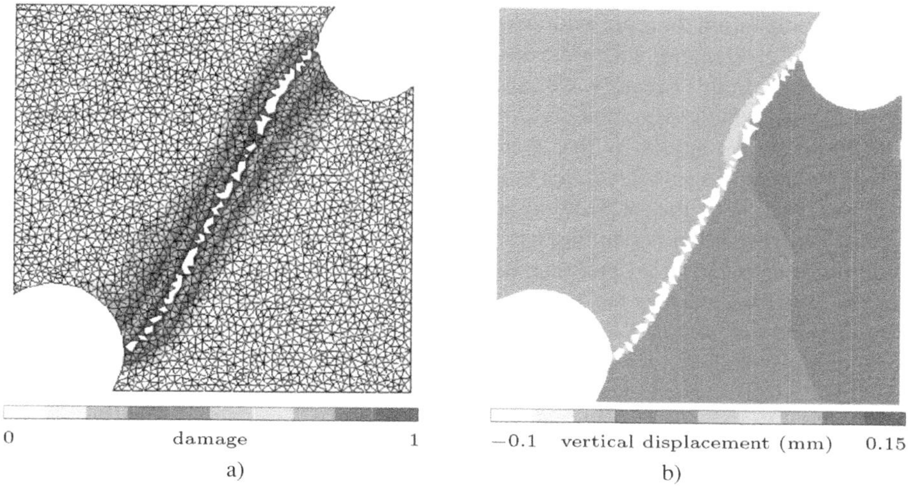

Figure 7.3. *Damage (left) and vertical displacement field (right) distributions with the same continuous model as for Figure 7.2 combined with erosion. For a color version of this figure, see www.iste.co.uk/lebon/mechanics.zip*

COMMENT ON FIGURE 7.3.– *The width of the localization band has been decreased. The elements have been eroded where the vertical displacement field showed a strong gradient, which corresponds to the area where damage is most significant. The crack is represented as a set of eroded elements and its shape is therefore highly dependent on the mesh. The adjacency of the elements observed in this 2D case was not a priori ensured. The* extrema *reached by the displacement field are lower than the ones obtained without transition.*

7.1.2.1.3. Extended finite element method for *thick level set*

Another way to eliminate an area from the computation is to use enrichment functions of the extended finite element method (X-FEM) (Moës et al. 1999), as proposed by Moës et al. (2011) and Bernard et al. (2012) with the *thick level set* model. The idea consists in relying on a model of damage distribution that imposes a damage profile (Cazes and Moës 2015). The iso-value corresponding to the maximum damage value then defines the cracked area to be removed from the calculation. The size of this area is therefore a result of the computation and cannot be a priori controlled.

7.1.2.2. *Degradation localization in a restricted volume around the crack*

7.1.2.2.1. Principle

Another solution for representing a discontinuity is to define a restricted area, often the width of an element, where the dissipative phenomena are concentrated *around a discrete crack*. In this category, we can also consider the strategies for modeling crack by an already totally degraded area of controlled position this time. The difference with the previous models is that there is a determination of a *discrete* crack path before associating a *restricted* area to this crack. There is therefore additional control over the position of the localization area, in addition to the continuous model. This type of model results in obtaining a more realistic crack representation, at the cost of additional computation to determine its path (see section 7.2). Some examples of this type of modeling strategy are proposed here, but the literature on the subject is still evolving and the proposed list is far from exhaustive. Since this category is intended to be quite general, the various corresponding methods can give very different results with regard to crack lip kinematics and dissipative phenomena.

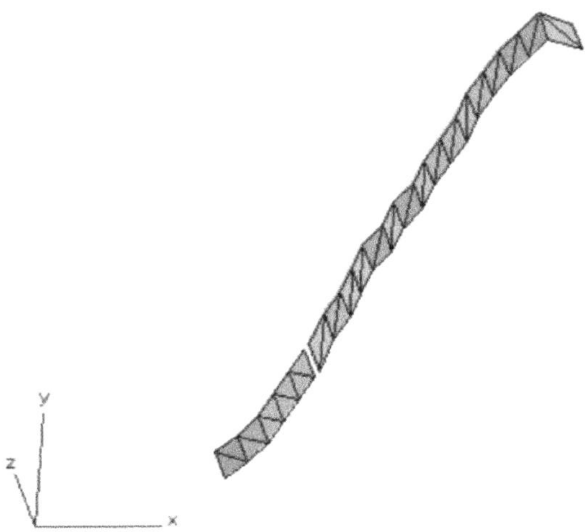

Figure 7.4. *Crack path obtained with the Marching Ridges technique presented in section 7.2.4 for the same model as in Figure 7.2, visualized in 3D. For a color version of this figure, see www.iste.co.uk/lebon/mechanics.zip*

7.1.2.2.2. Controlled erosion

It is possible to control the erosion of elements by ensuring that only elements that satisfy a chosen insertion criterion (see section 7.3) *and located on the path determined otherwise* (see Figure 7.4) are eliminated from the computation. This results in making sure that only a band of contiguous elements is removed from the computation and thus that *continuity* of the crack path is ensured. This is the approach employed by Yun et al. (2019b). This technique has the advantage of not requiring enrichment of the finite element formulation but leads to a crack whose shape is still mesh-dependent (see Figure 7.5).

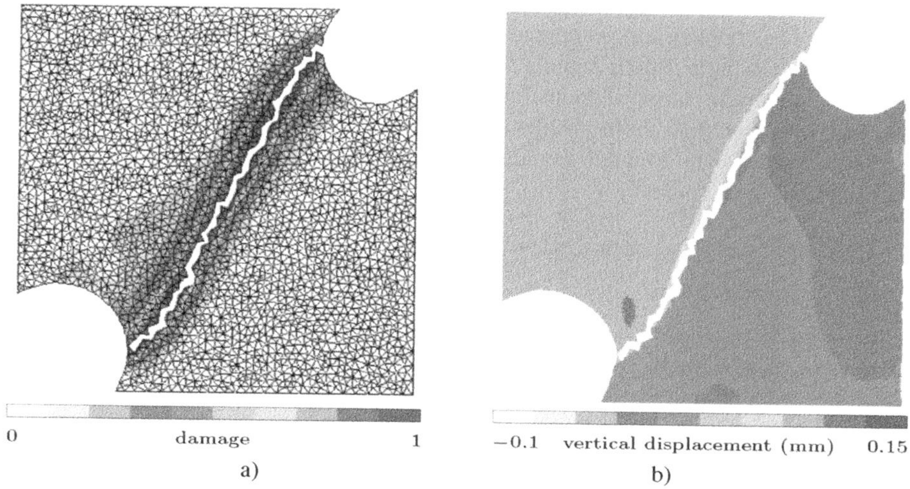

Figure 7.5. *Damage (left) and vertical displacement field (right) distributions with the same continuous model as for Figure 7.2 combined with controlled erosion. For a color version of this figure, see www.iste.co.uk/lebon/mechanics.zip*

COMMENT ON FIGURE 7.5.– *The width of the localization band is similar to that obtained for uncontrolled erosion. The crack is represented as a set of adjacent eroded elements (those that the crack crosses, Figure 7.4); its shape remains mesh dependent. The* extrema *reached by the displacement field are lower than the ones obtained without transition.*

7.1.2.2.3. REXFEM

A regularized XFEM formulation was proposed by Benvenuti et al. (2008) and used with continuous damage models in Benvenuti (2011) and Benvenuti and Tralli (2012) by localizing damage in a band of elements (one element wide, or even more depending on the chosen parameters). One of the advantages of this method compared to the conventional X-FEM formulation is that it does not require special techniques for integration: since the fields are continuous, a classic integration procedure is sufficient.

7.1.2.2.4. Embedded elements

Another enrichment technique consists in modifying the finite element formulation of the elements crossed by the crack. The enriched area here is therefore limited to a band of elements around the crack. This procedure is common to a wide family of techniques known as *E-FEM (embedded finite element method)*. Different versions were compared by Jirásek (2000). More recently, this type of technique was employed in the context of damage models by Jirásek and Zimmermann (2001), Oliver et al. (2004, 2014), Huespe et al. (2009, 2012), Lloberas-Valls et al. (2016) and Dias et al. (2018) and compared to other techniques notably by Oliver et al. (2006) and Essongue-Boussougou (2017). Although this family includes a wide variety of methods, most of them have limited kinematics. Techniques are being developed to overcome these problems, but they are still too recent and remain to date rarely applied to coupled damage models to allow for the evaluation of their contribution in this context.

7.1.2.3. *Interface level degradation localization*

7.1.2.3.1. Principle

It is finally possible to localize degradation phenomena at the interface level, in other words a surface in a three-dimensional space or a line in a two-dimensional space. From the point of view of the crack lip kinematics, this is probably the most realistic representation. Regarding dissipation, this means that a cohesive zone model may have to be used, with a law that may be independent of the material behavior model. It is nevertheless possible to propose cohesive zone models that are built in coherence with the coupled continuous damage model (see in particular the work of Cuvilliez (2012) and Cuvilliez et al. (2012)). It should be noted that the construction and insertion of these equivalent models still presently represent a challenge, especially for mixed mode loads. This is why, in many works, the authors have preferred to insert an interface without any interaction between the lips when material degradation is almost complete.

As with methods that localize degradation in a restricted volume around the crack (presented in section 7.1.2.2), the crack path must be known (see section 7.2) to use these techniques.

Here again, new propositions are still being made on the subject and the list of examples hereafter is merely for illustrative purposes and is by no means exhaustive.

7.1.2.3.2. Inter-element cracking

The simplest way to introduce discontinuity into a model is to allow cracks to appear along the interfaces between the elements. However, with this strategy, the geometry of the crack is highly dependent on mesh topology. Two options are then possible: to use a mesh pre-adapted to the crack or to automatically adapt the mesh. The solution of a mesh initially adapted to the crack geometry involves knowing the geometry of the crack in advance, which is not obvious when the crack does not propagate in a straight line.

A more expensive, but also more robust, solution consists of automatically adapting the mesh during computation. This is called adaptive remeshing or mesh adaptation. This process can be described as *global* if it concerns the entire structure, or *local* if it concerns a limited region only. This solution experienced a resurgence of interest in the 2000s with the development of more efficient mesh generators. It should be noted that mesh adaptation techniques require the result to be mesh independent. It is therefore more natural to use it for *regularized* coupled damage models, as did, in 2D, Mediavilla (2005) and Mediavilla et al. (2006) and in 3D, Peerlings et al. (2008), Feld-Payet (2010), Areias et al. (2011), Feld-Payet et al. (2015) and Javani et al. (2016). Figure 7.6 illustrates the application of mesh adaption to insert cracks determined by the *Marching Ridges* technique presented in section 7.2.4 for the same model as Figure 7.2.

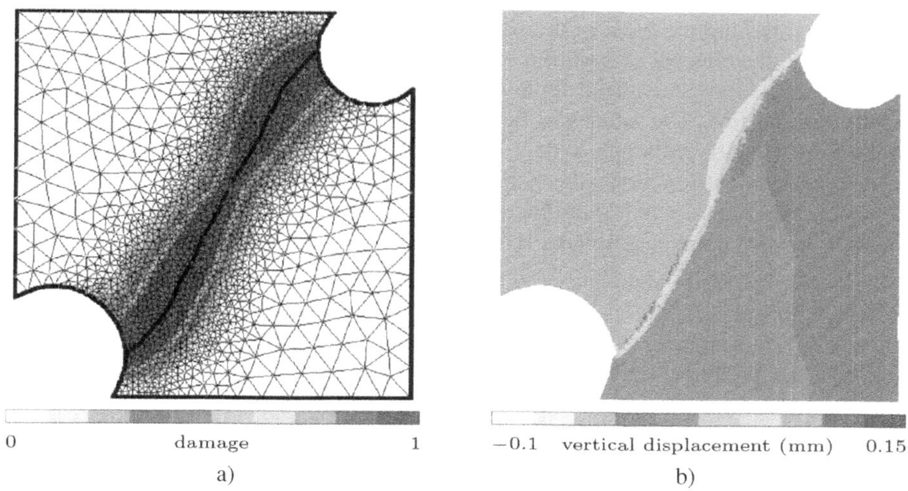

Figure 7.6. *Damage (left) and vertical field displacement (right) distributions with the same continuous model as for Figure 7.2 combined with adaptive remeshing. For a color version of this figure, see www.iste.co.uk/lebon/mechanics.zip*

COMMENT ON FIGURE 7.6.– *The elements are concentrated in the damaged area thanks to a discretization error estimation which results in having fewer elements than the mesh of Figure 7.2. The width of the localization band is similar to that obtained without transition. It could be smaller for other parameters. The structure boundaries are highlighted in order to visualize the cracks in the reference configuration. The crack path is smooth. The* extrema *reached by displacements are smaller than without transition. The crack lip opening could be easily measured and visualized in the deformed configuration.*

The main disadvantage usually associated with mesh adaptation is the need to transfer fields after a change of mesh topology. As explained in the following (see section 7.4), this operation is anyway necessary for the fields stored at integration points, at least locally, as soon as any discretization change is considered. For mesh adaptation, this problem can be limited by resorting to local adaptation only in the area of interest. Adaptive remeshing can even prove to be advantageous if used in combination with an error estimator (see, in particular, Askes and Sluys 2000; Rodríguez-Ferran and Huerta 2000; Feld-Payet 2010; Feld-Payet et al. 2015) to optimize the accuracy/cost ratio. Finally, mesh adaptation is essential when subjected to large strains in order to maintain a good element shape factor[2].

7.1.2.3.3. Enrichment methods

The most common enrichment methods used to represent strong discontinuities consist in enriching the basis of the shape functions associated with the nodes. They are based on the concept of partition of unity (Melenk and Babuška 1996). The most widespread are the generalized finite element method (Strouboulis et al. 2000) and mostly the X-FEM. The latter was developed by Belytschko and Black (1999) and Moës et al. (1999). Although widely used in linear elastic fracture mechanics with specific enrichment functions, the X-FEM technique relies mainly on the addition of Heaviside functions (or unit step functions), so that their application remains possible even for coupled damage problems. This method was used as follows:

– in 2D, by Wells and Sluys (2002), Patzák and Jirásek (2003), Simone et al. (2003), Comi et al. (2007), Pourmodheji and Mashayekhi (2012), Broumand and Khoei (2013), Seabra et al. (2011, 2013), Crété et al. (2014), Roth et al. (2015), Bobiński and Tejchman (2016a, 2016b), Wang et Waisman (2016), Beese et al. (2018) and Wolf et al. (2018);

– in 3D by Areias and Belytschko (2005), Simatos (2010), Tamayo-Mas and Rodríguez-Ferran (2015) and Javanmardi and Maheri (2019).

Although the position of the node remains unaffected, the enrichment of the interpolation functions requires the overall stiffness matrix as well as the distribution of the integration points around the crack tip to be adapted (Elguedj et al. 2007; Broumand and Khoei 2013). Using this technique therefore requires a transfer of the state variables from the old integration points to the new ones.

There are other methods that also consist of progressively enriching the finite element formulation, such as the approaches based on the phantom-node method (Hansbo and Hansbo 2004; Song et al. 2006) or the initial method of augmented finite elements (Ling et al. 2009), as well as the method of the floating

---

2 According to the author, adaptive remeshing is then more effective when focusing on the current configuration mesh.

node (Chen et al. 2014). These methods are mentioned in a recent article by Ma et al. (2019), which presents a review of existing methods in early 2019, before proposing a new formulation for augmented finite elements that ensure conformity of crack displacements. The crossed elements are then divided into two sub-elements that are connected by a cohesive element.

7.1.3. *Definition of a damage-to-crack transition strategy*

7.1.3.1. *Elements of a damage-to-crack transition strategy*

Once the continuous model and the type of discontinuity to be introduced have been selected, it is possible to define a strategy for a continuous-discontinuous transition, or in this case, a damage-to-crack transition, to combine these two models. To this end, it is necessary to answer several questions:

1) Where should the discontinuity be placed? (a) Which orientation criterion should be chosen? (b) How can a continuous discrete crack representation be obtained?

2) When should the discontinuity be inserted? And how far?

3) What are the necessary steps to resume computation after insertion?

These points will be discussed, respectively, in sections 7.2–7.4.

7.1.3.2. *The constraints guiding the choice of strategy*

The choice between the different possible options is made according to the constraints experienced by the user, such as:

– the type of finite element formulation that can be considered (standard, regularized, mixed, enriched, etc.);

– the material behavior (with or without nonlocal variables, with or without viscosity, whether plasticity is confined or not, etc.);

– the modeling framework: in two or three dimensions, with small or finite strains, in quasi-statics or dynamics, etc.;

– whether or not it is possible to intervene intrusively in the computation code;

– the need for the methodology to be suitable for a nonexpert user;

– the implementation and/or computation cost;

– etc.

7.1.3.3. *Damage-to-crack transition limitations*

To fully understand what a damage-to-crack transition strategy is, it is useful to also define its limits, while specifying what it does not cover.

In this regard, the most important point is to clearly dissociate the ability of a continuous model to localize strains in a relatively thin band of elements and the contribution of a damage-to-crack transition strategy. Actually, standard damage models, as well as some regularized models (and in particular *phase-field* type approaches), lead to the localization of maximal damage in a one-element-wide band. This localization depends entirely on the continuous damage model without any intervention of any additional criteria. It might then be tempting to consider the localization band as representative of a crack. This would only be a choice of interpretation based on a visual (and therefore qualitative) exploitation of the fields. This is does not however constitute a damage-to-crack transition strategy.

Indeed, the representation in a discrete, numerical form of the crack would require the definition of a criterion to select the elements that belong to the crack. Based on this information, it would be possible to change the model and introduce some sort of discontinuity in the area under consideration using the techniques mentioned in section 7.1.2. This approach could then be considered as a damage-to-crack transition strategy.

It is therefore important to distinguish between the continuous model (which can be more or less satisfactory) and the operations to shift to a discontinuous model. Notably, it is not the use of a *thick level set* model that can be associated with the damage-to-crack transition, but the choice to then use the X-FEM technique to remove a volume of material. As for *phase-field* approaches, they can be associated with continuous models and share has many similarities with the *thick level set* model (see in particular the comparison of the two models in Cazes and Moës (2015)).

Phase-field type models, as well as nonlocal models, can, in a second step, be combined with damage-to-crack transition strategies.

7.1.4. *Study objective and framework*

The objective of this chapter is to propose a guide that allows readers to analyze and compare the different strategies proposed in the literature. The chapter is meant to help them choose the option that best suits their constraints for each of the already mentioned questions:

1) Where should the discontinuity be placed? (a) Which orientation criterion should be chosen? (b) How can a continuous discrete crack representation be obtained?

2) When should the discontinuity be inserted? And how far?

3) What are the necessary steps to resume computation after insertion?

These three questions are addressed successively in the following three sections.

Some of the strategies presented here can be used within the broader framework of continuous–discontinuous transition. However, the bibliographic references in this chapter mainly regard the damage-to-crack transition, in particular from a *scalar* damage field[3], obtained from coupled damage models. For a damage *tensor*, the same strategies could be applied by considering one of the principal components. The range of considered models is still very wide, encompassing models with different physical meanings for the scalar damage field (e.g. Kachanov damage and porosity).

Here, only quasi-static modeling is taken into consideration, which already provides a sufficiently broad framework to explore the problem of damage-to-crack transition (although applications for computations in dynamics are also mentioned, in particular in relation with branching).

For the sake of brevity, the options presented here are only summarized and the interested reader is invited to refer to the associated reference for further details.

7.2. Localizing discontinuity

When it is desirable to control the extent of the discontinuity, it is necessary to define precisely where it should be inserted. This section therefore focuses on the tools used to define a discontinuity orientation and to ensure its continuity. If, on the other hand, only the removal of an uncontrolled volume of material is of interest, these tools do not need to be used and the concerned reader can move on directly to the insertion of cracks discussed in sections 7.3.

7.2.1. *Formulation of an orientation criterion*

7.2.1.1. *Introduction*

To determine the orientation of a crack increment, it is already necessary to be able to determine propagation direction(s) for the points of the crack front. This direction can be defined equivalently by a vector or by two points (namely the initial front and a point further on the crack path).

The determination of this propagation direction must rely on a criterion, referred to here as an *orientation criterion*. In the context of linear fracture mechanics, they are usually referred to as *bifurcation criteria*: see, for example, Bouchard et al. (2000) for an overview of the criteria available within this framework. They can rely on variables defined at the global level or at the local level (Lemaitre and Chaboche 2004). The criteria resulting from a global approach are restricted to problems where it is not

3 Please note in particular that this chapter does not deal with models using *smeared cracks* for which modeling diffuse degradation at the level of a few elements already requires a continuous–discontinuous transition approach.

necessary to precisely know the stress states or the degradation level in the vicinity of the crack front[4]. However, this type of approach can be found to be inadequate for short cracks, when the plastic area is larger, when there are some history effects related to loading or temperature or because of the nature of the deterioration mechanisms (multiple cracking in concrete, delamination in composites). A local approach based on the state of the variables in the vicinity of the crack front must then be considered.

In the present case, only local approaches to fracture are studied. The orientation criteria considered here rely on a result of the local mechanical problem, in other words, a quantity estimated near the crack front. The use of an orientation criterion may require the application of an operation on this result in order to obtain an indication of the direction of the propagation that occurs:

– either as a vector field (tangent or normal to the propagation direction of the crack) (in blue);

– or as a scalar field with a maximum at the point where the discontinuity is to be inserted (in red).

See Table 7.1 for a summary.

7.2.1.2. *Criteria based on a vector field*

A first category of criteria associates the desired direction with a field of discrete vectors. These vectors are therefore provided at several points, usually the integration points of the elements. This vector information is, in general, derived from a tensor analysis.

7.2.1.2.1. Stability and bifurcation analysis

With models for which the damage variable is *coupled* to the behavior *without regularization*, there is a loss of ellipticity in quasi-statics (or hyperbolicity in dynamics) of the equations of the problem. The crack orientation can then be determined by a stability and bifurcation analysis, according to the work of Hadamard (1902), Thomas (1957), Hill (1958) and Mandel (1966). This approach was adopted by Crété et al. (2014) and Wolf et al. (2018) for small strains and by Huespe et al. (2012) for large transformations. In dynamics, it was used by Belytschko et al. (2003) and Song et al. (2006, 2008) for materials with rate-independent behavior and has even been used to capture crack branching.

The crack orientation is given in quasi-statics by the vector $\vec{n}$ for which the determinant of the acoustic tensor $\vec{n} \cdot \mathbf{D} \cdot \vec{n}$, built from the tangent stiffness tensor $\mathbf{D}$[5], becomes null (or numerically negative) for the first time; similar conditions are

4 This is for example the case for quasi-elastic two-dimensional media in which the size of the plastic zone is negligible compared to the size of the crack for three-dimensional media with proportional loading or periodic fatigue loading.

5 The tangent operator being defined by the relation $\dot{\underline{\sigma}} = \mathbf{D} : \dot{\underline{\varepsilon}}$.

exploited in dynamics. It should be noted that solving this equation requires us to understand the behavior of the material. Moreover, it is important to know whether the latter is strain rate dependent (as can often be the case in dynamics), as this approach is only suitable for materials with rate-independent behavior.

Input	Operation	Obtained field
Acoustic tensor (Belytschko et al. 2003; Song et al. 2006, 2008; Huespe et al. 2012; Crété et al. 2014; Wolf et al. 2018).	Stability and bifurcation analysis	*Vector field*
- Strain tensor (Areias and Belytschko 2005; Dufour et al. 2012). - Stress tensor (Belytschko et al. 2003; Song et al. 2006; Bobiński and Tejchman 2016a). - Rate gradient field (Beese et al. 2018). - Anisotropic damage tensor (Javanmardi and Maheri 2019).	Main direction	*Vector field*
- Displacement field (Lloberas-Valls et al. 2016).	Displacement scalar function gradient	*Vector field*
- Strain tensor (Simone et al. 2003; Bobiński and Tejchman 2016b). - Stress tensor (Wells and Sluys 2002).	Eigenvalue scalar function or norm	Scalar function with a ridge
- Damage field (Feld-Payet 2010; Broumand and Khoei 2013; Seabra et al. 2013; Feld-Payet et al. 2015; Bottoni et al. 2015).	–	Scalar field with a ridge

Table 7.1. *Operations required to obtain the fields on which the orientation criteria are based*

Indeed, viscosity prevents the determinant of the acoustic tensor used for bifurcation analysis from taking negative or zero values. In this case, an additional step where the material state is frozen may then be necessary: see the works of Crété et al. (2014).

This type of analysis, although motivated by excellent arguments, turns out to be quite intrusive and expensive.

7.2.1.2.2. Main direction

A less expensive technique than bifurcation analysis and applicable to any kind of coupled damage model (regularized or not, with or without rate effect) is spectral decomposition analysis. This spectral decomposition is carried out in order to exploit use the main directions and can be applied to different tensors:

– to the *strain tensor* as was done in two dimensions by Dufour et al. (2012) and in three dimensions by Areias and Belytschko (2005);

– to the *stress tensor*, as was done for example by Bobiński and Tejchman (2016a) in quasi-statics and by Belytschko et al. (2003) and Song et al. (2006) in dynamics;

– to the *gradient tensor of the damage rate field*, as proposed by Beese et al. (2018);

– to an *anisotropic damage tensor* as proposed by Javanmardi and Maheri (2019).

7.2.1.2.3. Gradient of a displacement-based scalar function

Lloberas-Valls et al. (2016) considered that an *acceptable extension* of the normal to the crack could be given by the gradient of a scalar field that is equal to the sum or product of the Cartesian components of the displacements (in two dimensions). This approximation is not justified and its use is very limited.

7.2.1.2.4. Analysis

Vector information is generally obtained using tensor analysis, which provides, for each evaluation point, not *a single* vector, but *a set* of vectors. To the author's knowledge, there is no general methodology for selecting a vector from those provided. The selection methods used are heuristic and not always specified. However, several authors indicated that they followed the recommendation of Rabczuk and Belytschko (2007) and selected the vector corresponding to the direction that maximizes the displacement gradient or an equivalent strain, such as Crété et al. (2014).

The next step is to exploit this vector information. The different possible options will be presented further in section 7.2.2. Nonetheless, it can already be specified that the most obvious procedure is to evaluate the advance of a point of the crack front using the direction associated with the element containing this point, thus advancing from one element to the next.

7.2.1.3. *Criteria considering a scalar field maximum*

A second category of criteria is constituted by criteria that consider scalar fields.

7.2.1.3.1. Construction of the field of interest

As an example, a list of fields considered in the literature is proposed here. Depending on the nature of the studied field, there may or may not be a first tensor analysis step:

– *With prior spectral decomposition*: some authors propose to build the analyzed scalar field from the eigenvalues of a strain tensor. For example, Simone et al. (2003) have chosen to rely on the equivalent nonlocal strain obtained by regularization of the local equivalent strain $\varepsilon_{eq} = \sqrt{\sum_{i=1,2,3} \langle \varepsilon_i \rangle^2}$, where the ε_i denote the eigenvalues of the strain tensor. Bobiński and Tejchman (2016b) chose to rely on the maximal value reached in time by the equivalent strain $\varepsilon_{eq} = \frac{\max \sigma_i^{eff}}{E}$ where the σ_i^{eff} are the eigenvalues of the effective stress tensor. It should be noted that, unlike the spectral decomposition mentioned above, the result of the analysis used subsequently is not the eigen*vector* but the eigen*value*.

– *Without prior spectral decomposition*: it is also possible to choose scalar fields that do not require prior spectral decomposition. For example, Wells and Sluys (2002) proposed to study the equivalent stress $\sigma^{eq} = \sqrt{3/2}||\underset{\sim}{\sigma}^{dev}||$ where $\underset{\sim}{\sigma}^{dev}$ denotes the deviatoric part of the stress tensor. The simplest choice for coupled damage models, however, is to consider the scalar field corresponding to the damage variable: see in particular the works of Broumand and Khoei (2013); Seabra et al. (2013); Feld-Payet (2010); Feld-Payet et al. (2015); and Bottoni et al. (2015). Another option, even more advantageous, suggested by the latter authors, is to study a field involved in the damage calculation, but with steepest slopes on either side of the ridge.

7.2.1.3.2. Analysis

The principle here is slightly different from that of vector approaches. It is no longer a matter of calculating a vector field for all of the elements near the front to use only a few of them on the crack path. The idea here is to determine the location of the points corresponding to a local maximum of a scalar field.

In the ideal case, we could imagine determining the crack by assembling the sides (in two dimensions or faces in three dimensions) of the elements where the nodes have reached the maximal damage value. Obviously, this case is very rare because it requires having:

– a piecewise linear scalar field so that the maximal values are at the mesh nodes;

– a scalar field sufficiently smooth in the direction of the advance and having a sufficiently sharp ridge, that is, with sufficiently steep sides (without, however, having problems with strain evaluation or localization);

– a mesh perfectly adapted to the geometry of the crack.

In general, there is no guarantee that these conditions are all met. The use of a scalar field therefore requires advanced methods based on modeling the advance of a point of the crack front by a mesh-independent segment that goes through several elements. The methods used to determine the position of this segment will be detailed later in section 7.2.2.

7.2.1.4. *Relations between vector and scalar fields*

The range of possibilities for formulating an orientation criterion is therefore very broad and remains largely to be explored. From a local quantity, we may as well choose to build a vector field or a scalar field with a ridge to use as a support for the orientation criterion. The nature of the local quantity does not matter since various operations can be employed to reach the desired (vector or scalar) nature of the output field including:

– spatial derivation operations to shift from a scalar field to a vector field or from a vector field to a tensor field;

– tensor analyses: for example, eigenvalue analyses to obtain, from a tensor, an eigenvector (often corresponding to a normal vector in the literature since mode I has been the most widely studied) or an eigenvalue, or stability analyses to obtain vectors, or why not the study of invariants to obtain scalar fields;

– projection operations to shift from vectors to a scalar field;

– or even norm calculations to shift from vectors or tensors to a scalar field.

In the end, either a vector field or a scalar field with a ridge should be obtained. These fields are neither arbitrary nor independent: ideally, the output vector field should coincide with the normals to the iso-values of the scalar field with a ridge corresponding to the crack, assuming that these iso-values are parallel to the considered portion of the crack. This ideal equivalence is illustrated in Figure 7.7. The crack then corresponds to a particular iso-value where the normal is zero and changes direction depending on the side.

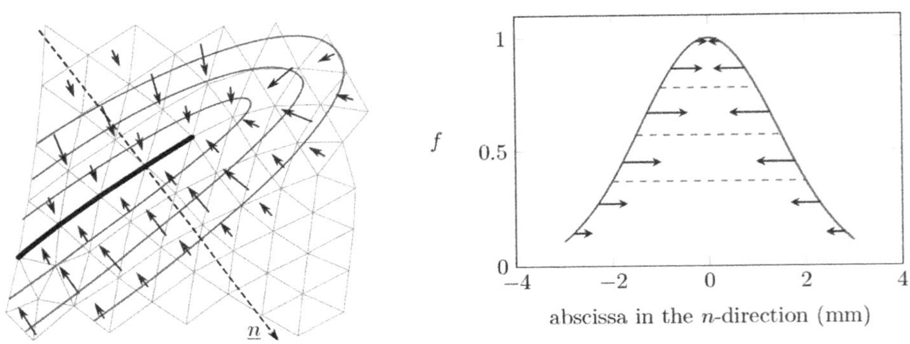

Figure 7.7. *The orientation criteria use either a vector field (in blue) or a scalar field with a ridge f (in red) to find the orientation of the crack (shown in thick black line). Ideally, there would be a correspondence between the vectors studied and the normals to the iso-values of the scalar field. For a color version of this figure, see www.iste.co.uk/lebon/mechanics.zip*

In practice, the vectors should be as close as possible to the normals to the crack, in the vicinity of the crack. It is actually important to differentiate the gradient component in the normal direction to the ridge and the gradient component in the direction of the crack front advance. Any information about the evolution in the tangent direction (in other words, along the crack) is not important when looking for a ridge and should even be eliminated as much as possible to avoid polluting the information (especially near the crack front where the tangential component of the gradient of the scalar field can be significant).

7.2.2. *From orientation criterion to crack surface*

7.2.2.1. *Development of the* Crack Path Tracking *algorithms*

The orientation criterion is used to define the direction to which a front point should advance based on a vector field or a scalar field. However, it is still necessary to clarify how to go from this theoretical definition to the practical construction of a continuous discretized representation of the crack surface, which can be either explicit (in the form of a mesh) or implicit (in the form of a function). This is done using *Crack Path Tracking* algorithms.

It should be noted that, when these algorithms were first proposed, the main objective was already to ensure crack continuity from a field of vectors resulting from a tensor analysis and defined element by element. It was only later that the crack path regularity requirement was added.

Indeed, whatever orientation criterion is chosen, the result is a crack path that generally lacks regularity (compared to the regularity of the continuous fields used for the orientation criterion) if the considered information is elementwise. This problem is related to discretization, but it does not seem to diminish when the mesh is finer, as illustrated by the work of Cervera and Chiumenti (2006).

7.2.2.2. *The different approaches*

To ensure continuity and regularity of the discontinuity surface, two families of approaches were then considered:

– The first family mainly focuses on continuity: it aims at building the crack surface from the information on the current mesh, element by element, making corrections as the elements are cut to ensure continuity.

– The second family focuses on regularity, which necessarily implies continuity, before considering the construction of the surface.

7.2.2.2.1. Continuity-based approaches

For orientation criteria based on tensor analysis and providing a vector per element, the idea is to start from an element (either an element where a criterion has been met for crack initiation, or an element adjacent to the front for propagation) and build the crack element by element.

The output may be:

– either a mesh of the crack (which in two dimensions corresponds to a series of tangent segments or vectors deriving from intersections with the current mesh): this approach is known in the literature as the *local method*;

– or a locally monotonic scalar function defined at the nodes of the elements intersected by the crack, which is implicitly defined as an iso-value of this function: this is the *partial domain crack tracking algorithm* of Feist and Hofstetter (2006).

The procedure consists in using the normal and a point of the first element to find the first crack increment. It is then necessary to consider neighboring elements that share one side with the previous element and use the normal direction *and* the previously determined intersection to find the position of the crack increments in these elements. Continuity is ensured by repeating this procedure. This approach may seem inexpensive at first, since it processes the raw data directly without any pre-processing. However, the management of intersections on a case-by-case basis may actually prove to be more difficult than it seems, especially in three dimensions: for example, there may be cases where the normal direction of an element is not taken into account at all if enough edges are already intersected by crack increments from neighboring elements, which may depend on the order in which the elements are considered.

7.2.2.2.2. Methods for more regularity

There are three types of methods that can result in a more regular crack path:

– First of all, there are methods that are considered here as *basic*: relatively inexpensive, they provide minimal additional regularity. These methods rely on averaging or smoothing techniques to give an output of the same nature as the input field. They do not provide an output with a format corresponding to a crack representation and are therefore rather used as a pre-processing step. They are described in section 7.2.3.

– There are the more advanced techniques that also take into account not only the neighborhood around the point under consideration but this time change the nature of the input field. In two dimensions, they lead to an output format adapted to the discrete representation of the crack. These techniques are relatively more expensive or involve some form of modeling, but they guarantee continuity while improving regularity. They are described in section 7.2.4.

– Finally, the last type of method takes as input some information on the advance of a few front points (some tangent vectors extending over several elements at once) to reconstruct a three-dimensional surface. These methods are detailed in section 7.2.5.

In any case, the main difficulty is to improve regularity while remaining able to capture relevant local variations, and thus be able to detect branching if necessary. When choosing a method, in addition to its ability to capture branching, the cost of the method (for both implementation and use) and its robustness are also to be taken into account.

7.2.2.3. *How to use these methods*

All of the methods mentioned above could be referred to in the literature as *Crack Path Tracking* algorithms. This can lead to confusion because these methods do not all have the same function: some can serve as alternatives to others (e.g. one advanced technique can be replaced by another), while some can be combined with each another (e.g. an average based pre-processing followed by a local method to cut the elements as done by Areias and Belytschko (2005)). The classification in the literature (see in particular the review by Saloustros et al. (2018a)) into *local*, *nonlocal* or *global* algorithms and damage field-based algorithms is therefore not entirely satisfactory. Indeed, these methods may share common points at various levels: the type of input, the type of output, their principle or their use. The author of this chapter proposes instead to describe and compare these methods according to the use, leading to distinguishing four categories:

1) methods for ensuring continuity which simply cut the elements;

2) basic pre-processing methods for more regularity;

3) advanced methods for more regularity;

4) methods for constructing a three-dimensional crack representation from a few tangent vectors.

Figure 7.8 shows a diagram illustrating the links between the first three types of method and the fourth is illustrated in Figure 7.13. The objective here is not to propose a new classification, but to help the reader to better understand the similarities and differences between the methods. Within each of these categories, it is possible to consider subdivisions based on the way the information is processed, the latter being often related to the nature of the field considered as input or the nature of the desired output field.

7.2.2.4. *Converting the nature of output fields*

Ultimately, there are two types of output field that can be used to describe the crack surface: meshes (explicit description) or locally monotonic scalar fields (implicit description). The choice between these two descriptions depends on the way chosen to model discontinuity (see section 7.1.2) or on its exploitation:

– A surface that is described implicitly on from the current mesh enables a direct use of controlled erosion or enrichment techniques (at the node or element level).

– An explicit mesh makes it possible to better visualize the crack surface and to perform flow or surface calculations more directly. This mesh can be obtained: either by cutting the current mesh, or by constructing of an independent mesh, or even by intersecting this independent mesh with the current mesh of the structure (with or without subsequent meshing).

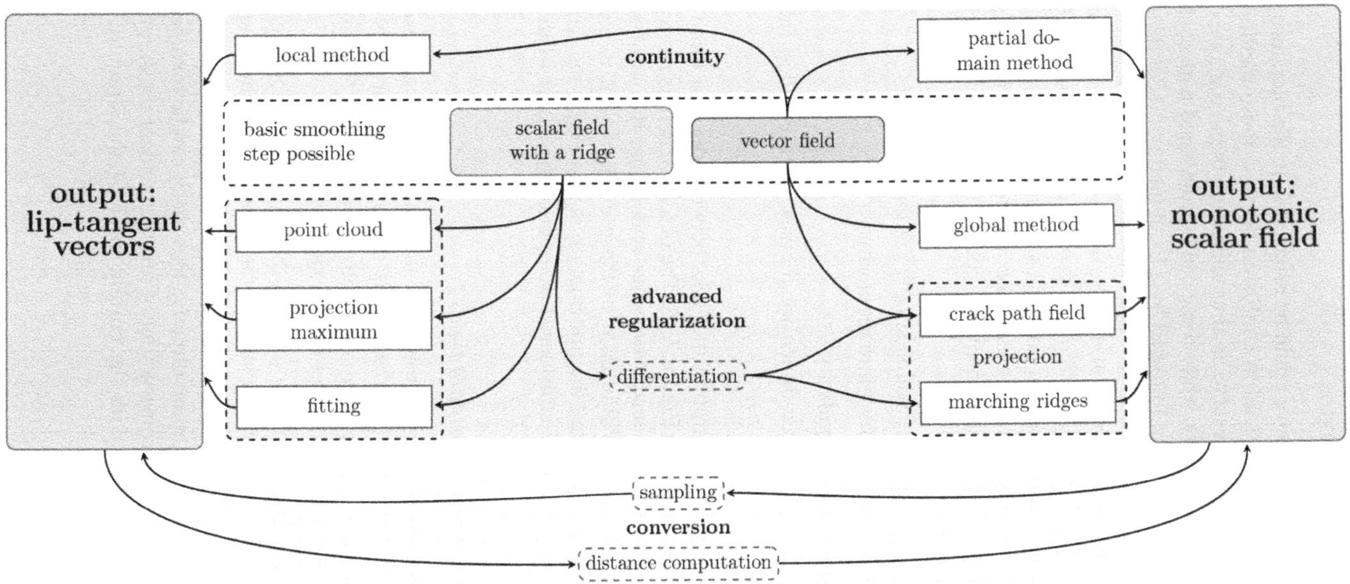

Figure 7.8. *Diagram summarizing the different methods for obtaining a continuous discrete crack path in two dimensions. For a color version of this figure, see www.iste.co.uk/lebon/mechanics.zip*

It is possible to switch from one description to another without changing the regularity of the discretized surface and this is relatively easily done in two dimensions. Therefore, from a locally monotonic function, the study of the values at the nodes of the current mesh enables us to find the intersection points of the crack (corresponding to an iso-value) with the edges of the elements and thus to build the mesh of the crack element by element. This is referred to as the sampling operation in Figure 7.8. Conversely, from a mesh of the crack, it is possible to define a function, such as the *distance to the surface*, and to calculate its values at the mesh nodes of the structure: the result is a locally monotonic function. It is therefore possible to derive a monotonic function from an independent mesh, thus enabling subsequent enrichment by the X-FEM (Fries and Baydoun 2012).

In conclusion, the nature of the output fields should not be an obstacle to the choice of a method since it is possible, for a given field, to build another field better suited to a given method or an intended use (in the same way as for input fields).

7.2.3. *Basic evaluation methods for more regularity*

7.2.3.1. *Notion of neighborhood*

The most basic techniques consist of taking into account, instead of a purely local value, the mechanical state in the neighborhood. This notion is essential in a local approach. Two levels of *neighborhood* can be distinguished:

– The largest neighborhood corresponds to an area around the crack in which no other cracks can appear. It is therefore a concept used when considering crack initiation. In practice, the definition of this area requires the choice of a threshold (on the level of damage, for example) or a distance from the crack front or crack lips (*exclusion radius criteria*). This is what makes the simultaneous management of several crack fronts possible.

– The second neighborhood level is included in this area but restricted to a part ahead of the crack front. This notion is used for the calculation of the orientation criterion in the propagation stage. The neighborhood considered is often a portion of disk that is symmetrical with respect to the crack tangent direction with an opening varying between 90° (for Simone et al. 2003) and 180° (for Wells and Sluys 2002).

It should be noted that this notion can also cause issues when the neighborhood is physically reduced: for example, when the crack is close to an edge. The amount of information is then limited, and this impacts the efficiency of the measures taken to introduce more regularity.

7.2.3.2. *The different basic methods*

7.2.3.2.1. Tensor field: taking the neighborhood into account before tensor analysis

Some authors chose to perform a tensor analysis on *nonlocal* quantities, which therefore already takes into account the influence of the neighborhood. Some works proposed applications in two dimensions:

– Jirásek and Zimmermann (2001) and Patzák and Jirásek (2003) considered nonlocal strains.

– Saloustros et al. (2018b) used a stress tensor obtained from the mean values at the nodes of each element and Roth et al. (2015) considered one of the principal direction of a nonlocal stress tensor;

– In dynamics, Song et al. (2006) smoothed the tensor field of interest (the stresses) in the vicinity of the crack tip by a method of moving least squares.

7.2.3.2.2. Vector field: tensor analysis then average over a domain

The simplest method to improve continuity of the crack path in the propagation direction is to average the directions found by tensor analysis in a certain area (with a radius corresponding to a few elements) around the crack tip. It is a technique that can be described in the literature as *nonlocal*. It was used in two dimensions by Crété et al. (2014) and Bobiński and Tejchman (2016a) and in three dimensions by Areias and Belytschko (2005) and Javanmardi and Maheri (2019). In dynamics, the crack path can be smoothed by taking into account in the average the previously found directions, except when branching occurs (Belytschko et al. 2003).

7.2.3.2.3. Scalar field: smoothing and/or averaging

For scalar fields, the equivalent of the averaging operation would be the smoothing operation. For example, Feld-Payet (2010) and Feld-Payet et al. (2015) used a second-order polynomial approximation on patches of elements following the work of Zienkiewicz and Zhu (1992a, 1992b) in a pre-processing step and then used the smoothed scalar field to determine the direction of the crack by an advanced method.

7.2.3.3. *Analysis*

The most basic methods presented here are based on notions of averaging or smoothing. Only spatial averaging or smoothing have been mentioned here, but we could add in this category time averaging or smoothing as well as any form of correction restricted to the neighborhood without additional modeling. Regardless of the nature of the field under consideration, these methods provide no guarantee of sufficient continuity or regularity of the crack surface. The output fields remain of the same nature as the input fields, so that they could be considered as prior and post correction versions. In particular, when manipulating vector fields from tensor analysis to cut the current mesh element by element, this correction can be either systematic or performed only under certain conditions: for example when the curvature of the crack path exceeds a fixed tolerance as proposed by Cervera et al. (2010). Although inexpensive, these methods are not very effective and can be thought of as a simple pre-processing step before applying more advanced methods.

7.2.4. *Advanced evaluation methods ensuring more regularity*

The final objective is to obtain a continuous discretized representation of the crack either in the explicit form of a mesh (independent or not of the current mesh) or in the implicit form of an iso-value of a locally monotonic field in the vicinity of the current front. For the author, these are the two types of outputs that enable us to qualify the advanced methods of *Crack Path Tracking* algorithms, the others being rather pre-/post-processing algorithms. This section first considers the methods for obtaining a continuous discretized representation of the crack for two-dimensional propagation problems.

7.2.4.1. *Building a monotonic scalar field from a normal field*

7.2.4.1.1. Principle

Let us consider the possible equivalence between a normal vector field and a scalar field with a ridge as presented in Figure 7.7. The main difficulty with the scalar field with a local maximum is that it is not possible to study the associated nodal values to determine the position of the iso-value corresponding to the maximal value, in other words, the ridge. The solution proposed by Oliver et al. (2004) then involves building another field, of the same order as the considered scalar field, but with a monotonic evolution, as illustrated in Figure 7.9. The idea here is to keep locally the shape of the iso-values of the scalar field (therefore the direction of normals), but to choose the same orientation for all normal vectors, which would otherwise change direction when crossing the desired ridge. The monotonic field is constructed from the normal vectors while imposing zero variation in the direction tangent to the iso-value. This is the principle of the global method, which is a priori the only one that proceeds this way.

7.2.4.1.2. Global method

Name	Global Method
Input	Normal vector field and at least two nodal values of the solution field
Output	Locally monotonic scalar field (2D/3D) with iso-zero value of interest to be determined

Table 7.2. *Summary of input and output fields for the global method*

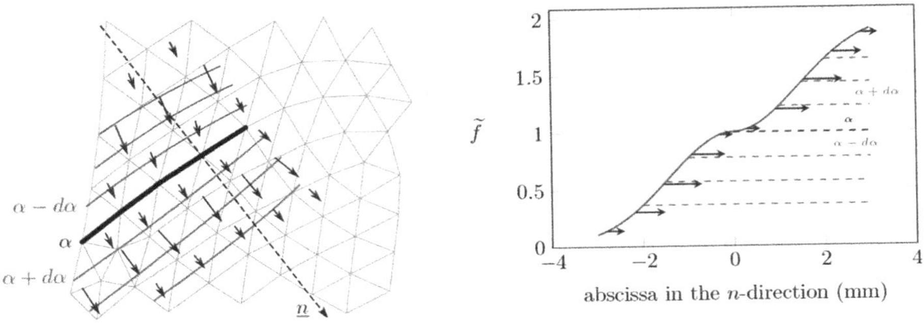

Figure 7.9. *Schematic representation of the idea behind the global method to make a comparison with the idealized situation presented in Figure 7.7 (the norm of normal vectors and therefore the slope variation are here only illustrative). The idea here is to build a monotonic function $\tilde{f}$ to more easily find the iso-value associated with the crack. For a color version of this figure, see www.iste.co.uk/lebon/mechanics.zip*

To better understand this method, let us consider the two-dimensional case. The zero variation of the monotonic scalar field $\tilde{f}$ along the iso-values is then written as: $\frac{\partial \tilde{f}}{\partial T} = 0$ with $\underline{\mathbf{T}}$ a unit vector orthogonal to the normal vector (for the element under consideration); the normal vectors are represented in blue in Figure 7.9. To build the monotonic scalar field $\tilde{f}$, Oliver et al. (2004) proposed a method for solving a thermal conduction problem on the current mesh. The authors actually demonstrated (Oliver et al. 2002) that, at least for two-dimensional problems, the solution of a thermal conduction type problem with adiabatic flow boundary conditions and no internal heat source has a zero gradient in the tangent direction. Solving this auxiliary problem requires the construction of a tensor (similar to a conductivity tensor), which is derived from the unit vectors in the tangent direction(s). The problem associated with the construction of the global matrix is that its rank is one degree lower than the number of nodes: it is then necessary to impose also as input the value of the solution field (which can be assimilated to temperature) for at least two nodes. The output is a monotonic field that makes it possible to localize the points of intersection of the edges with an iso-value of interest (which depends on the values of the field imposed as input to the auxiliary problem and is defined from a point assumed to belong to the current front).

Since this problem is solved on the entire mesh, this method is referred to as the *global*[6] method. It should be noted, however, that the validity of the solution is still

6 It is also referred to as the *scalar field method* by Yun et al. (2019b). However, this name can be quite confusing since there are other methods that involve scalar fields and it would be preferable to specify how the field was constructed and its characteristics to clearly distinguish the different methods.

local[7]. Indeed, the basic assumption is that the field does not vary in the direction of the iso-value corresponding to the crack. Yet, this hypothesis has its limitations within the framework of the local approach. Schematically, this assumption no longer corresponds to the description of the scalar field with a ridge as presented in Figure 7.7 as soon as the scalar values diminish ahead of the crack front.

This type of approach was followed in two dimensions by Cervera and Chiumenti (2006) and Dufour et al. (2012) for coupled damage models, but without discontinuity insertion, and by Huespe et al. (2009) within the framework of small strains and by Huespe et al. (2012) within the framework of finite transformations. In three dimensions, it was more recently used by Beese et al. (2018).

This method was particularly interesting at the time it was proposed. In the early 2000s, the most studied problems were still those of linear fracture mechanics and the propagation direction was often given by a vector (rather than by a criterion based on the maximal value of a scalar associated with material damage): this would explain the nature of the input. For the output, the implicit representation provided by the study of a scalar field with an iso-value of interest on the current mesh avoided the need for a mesh generator, less robust at the time. However, the issues associated with this method now seem to outweigh its advantages and some of its authors have subsequently proposed other techniques described below. Indeed, solving the auxiliary thermal problem presents two main difficulties. The first is the imposition of Dirichlet boundary conditions. This point was addressed in particular by Jäger et al. (2008a, 2008b, 2009) and Annavarapu et al. (2016). The second difficulty lies in the fact that the matrix resembling the stiffness matrix is singular: a perturbation parameter can then be introduced during the resolution. This problem prompted Riccardi et al. (2017) to propose a modified version of the method.

It should be noted that this method indeed leads to more regularity of the crack surface by modifying the direction of the normal vectors given as input. Actually, at the end of the problem, the crack is represented by an iso-value with normals per element that are not necessarily those initially given. It could even be considered that the new normal vectors are another output of the method, in addition to the scalar field.

7.2.4.2. *Advanced methods using projection of the scalar field gradient*

7.2.4.2.1. Principle

Obtaining a monotonic scalar field with the global method proposed by Oliver et al. (2004) introduces a number of difficulties associated with solving the auxiliary problem. Subsequently, other more direct approaches were proposed to obtain a

7 This is the case for all of the methods considered in this framework.

monotonic scalar field as output: their principle is to use a projection of the gradient of a scalar field along an appropriate direction and to focus on the iso-zero of the resulting field. The common point shared by the methods mentioned here is that smoothing techniques are necessary to limit the noise resulting from the differentiation of discretized data. The difference between the methods resides in particular in the choice of the vector used for the projection. This vector is chosen so as to be as close as possible to the crack normal *in the vicinity of the crack*. It is indeed the normal direction that appears in the definition of a ridge according to Furst and Pizer (1998): a one-dimensional height ridge of a function $f : \mathcal{X} \rightarrow \mathcal{R}$ defined in a space $\mathcal{X}$ of dimension d is the locus of dimension $(d - 1)$ corresponding to a relative maximum of the function f in a direction $\underline{\mathbf{v}}$ transverse to the putative ridge, that is:

$$\begin{cases} \underline{\mathbf{v}} \cdot \underline{\nabla} f(\underline{\mathbf{x}}) = 0 \\ \underline{\mathbf{v}}^t \cdot \underline{\underline{\mathcal{H}}}\left(f(\underline{\mathbf{x}})\right) \cdot \underline{\mathbf{v}} < 0 \end{cases} \qquad [7.1]$$

where $\underline{\underline{\mathcal{H}}}$ denotes the Hessian matrix and $\underline{\mathbf{x}}$ is a point of the ridge. This definition is more frequently used in the computer vision domain. The methods proposed in mechanics are rather simplified versions to avoid a second derivative calculation and therefore rely only on the gradient of the scalar field, with some precautions.

7.2.4.2.2. Gradient projection onto a vector assumed to be normal: *Crack Path Field*

This method proposed by Oliver et al. (2014) and referred to as the *Crack Path Field* method consists in starting:

– from a *smoothed scalar field* which is localized in the same way as strains (in other words that reaches a maximum at the crack level); the recommended smoothing operation is the simple projection onto the mesh nodes;

– from a vector field (acting as the normal vector field).

Name	*Crack Path Field*
Input	Normal vector and gradient vector of a scalar field with a local maximum
Output	Locally monotonic scalar field (2D/3D) with iso-zero value of interest

Table 7.3. *Summary of input and output fields for the Crack Path Field method*

After an intermediate step for calculating the scalar field gradient followed by a step for smoothing the calculated gradient field, a scalar field is obtained, after projection of the vector fields, supposed to be locally monotonic around the crack and its iso-zero corresponds to the crack.

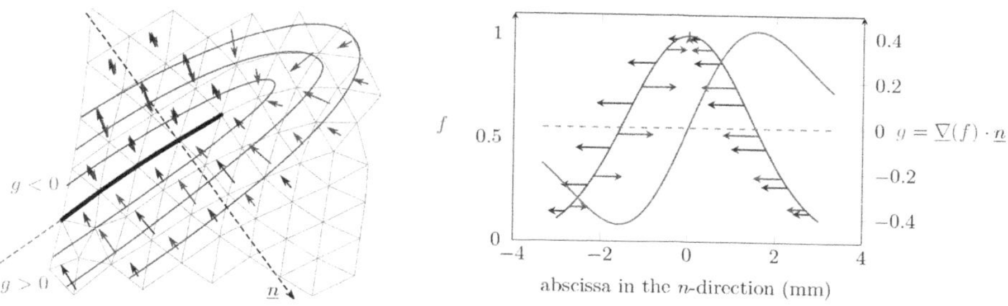

Figure 7.10. *Schematic representation of the idea on which the Crack Path Field method is based for comparison with the idealized situation presented in Figure 7.7. For a color version of this figure, see www.iste.co.uk/lebon/mechanics.zip*

COMMENT ON FIGURE 7.10.– *The main challenge is to propose a vector field (in blue) close enough to the crack normal such that the change of sign of the scalar product (denoted g) with the gradient of the scalar function $\underline{\nabla}(f)$ (represented by red arrows) is representative of the crack position.*

In principle, this method could be seen as an improved version of the global method better suited to a local approach. The common point is the use of the normal vector field as input. Again, the directions of the normals are kept, but the normals must all have the same orientation to observe a change in the sign of the projected field (see Figure 7.10). The type of output field is also the same: a field with an iso-value of interest. Due to the nature of the output, this method should be directly usable in three dimensions: see the work of Dias et al. (2018) within the quasi-static framework. On the other hand, there are important differences with the global method: this time, the iso-value of interest (iso-zero) is known a priori and it is no longer necessary to solve an auxiliary problem. It should be noted that another significant improvement is to allow the capture of branching phenomena based on the proximity of two iso-zeros: see the work of Lloberas-Valls et al. (2016) for an implementation in dynamics in two dimensions.

However, the efficiency of the method entirely depends on the choice of vector field used to act as iso-value normals. Furthermore, to date, a major drawback is that the choice of the input vector appears to be problem dependent. For example, in dynamics, Lloberas-Valls et al. (2016) relied on the gradient of a scalar field equal to the sum *or* to the product of the Cartesian components of the displacements, while the result of a bifurcation analysis has been used by Oliver et al. (2014) and Dias et al. (2018). This lack of robustness in the choice of the input vector field remains for this method, yet promising in principle, a major weakness.

7.2.4.2.3. Gradient projection onto a polar grid: *Marching Ridges*

Name	*Marching Ridges*
Input	Gradient vector of a scalar field with a ridge and polar grid
Output	Locally monotonic scalar field (2D) with iso-zero value of interest

Table 7.4. *Summary of input and output fields for the Marching Ridges method*

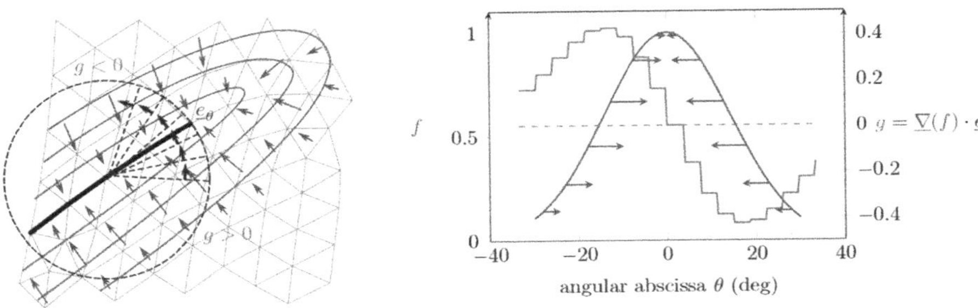

Figure 7.11. *Schematic representation of the idea on which the Marching Ridges method is based in comparison with the idealized situation presented in Figure 7.7. For a color version of this figure, see www.iste.co.uk/lebon/mechanics.zip*

COMMENT ON FIGURE 7.11.– *The use of a polar grid (shown in blue) eliminates the need to choose the vectors for the projection of the gradient of the scalar function $\underline{\nabla}(f)$ (represented by red arrows). Angular discretization leads to a piecewise constant scalar product g. Studying its change of sign enables us to identify the crack.*

Another method using the projection of the gradient of a scalar field was proposed by Feld-Payet (2010) and Feld-Payet et al. (2015) and is illustrated in Figure 7.11. This method has been referred to as *Marching Ridges* to signify its proximity to the methods of the same name originally used in computer vision. However, this name could actually be generalizable to all of the methods in this category that look for ridges based on spatial derivatives of the scalar field. Here again, we consider smoothed fields (with a ZZ2-type technique) (Zienkiewicz and Zhu 1992a, 1992b). The distinctive feature of the method proposed by Feld-Payet et al. (2015) is to consider, as an input, the scalar field as well as *a polar grid centered at the considered front point.* Indeed, this time, the vectors acting as normals for the projection are the orthoradial vectors $\vec{e}_\theta$ associated with the polar grid. This choice offers several advantages. On the one hand, this choice of vectors is more robust and much less expensive than that made with the *Crack Path Field* method. On the other hand, it makes it possible to control the discretization of the evaluation space. Therefore, an inherent part of the method is to divide the polar grid into angular sectors (with an angular step of about 5–10°) to average the projection result in each sector making the method more robust. The output is a piecewise continuous locally monotonic field in the considered plane. The position of a crack increment is therefore given by the iso-zero of the field (when the sign of the scalar product changes from positive to negative). This position corresponds to an *average* position for *the entire length* studied (as opposed to an average position of *a point*). This length (or radius) can be calculated in order to ensure consistency between the problem discretization (thus, the amount of discrete information available) and the desired angular accuracy. It should be noted that this length is greater than one element, as for all of the methods that exploit a scalar field with a ridge (presented in section 7.2.4.3). This method also provides a field on (but within the plane), quite similarly to the global method and the *Crack Path Field* method. This is a trade-off resulting from the introduction of a model (since the crack increment is approximated by a fixed length segment passing through several elements) that leaves only one degree of freedom in two dimensions: the angle.

One of the main advantages of this method is its ability to detect several directions simultaneously, and thus to manage branching. In addition, it is very robust to the possible evolution of the scalar field in the tangent direction and can therefore detect a ridge even for relatively low values (i.e. even ahead of the crack tip). Its limit is mathematically well defined (see Feld-Payet et al. 2015): this method is valid only if the local radius of curvature is smaller than the distance between the evaluation point and the considered front point (namely, the radius corresponding to the evaluation point). This method is therefore not suitable for quasi-plateau areas with small evaluation radii. Let us underline that, although this method does not produce a three-dimensional field but a field in the evaluation plane, this does not prevent its use in three dimensions: the solution proposed by Feld-Payet (2010) and Feld-Payet et al. (2015) is explained later (in the chapter see section 7.2.5).

7.2.4.3. *Advanced methods based only on a scalar field with a ridge*

7.2.4.3.1. Principle

Advanced methods based only on a scalar field with a ridge were developed after those solely relying on vectors. In this case, it is not possible to study the values at the nodes to determine where the iso-value corresponding to the ridge intersects the mesh. In order to exploit the scalar field, the crack increment is modeled by a segment passing through several elements. In the propagation phase, the origin of this segment is the current front, which guarantees crack continuity. To determine the position of the other end of the segment, two types of approaches are possible:

– either focus on the position of the other *endpoint* of the segment by requiring it to be on the ridge;

– or search for the best possible approximation of the ridge by the *segment* with the constraint that the origin of the segment has a fixed position.

These approaches can be considered equivalent when the radius of curvature of the path is small enough for the approximation by a segment to be satisfactory. In general, if the input data are widespread enough, it is more likely to approximate the position of a segment.

Although there is no obvious way of classifying these methods, it is proposed here, on the basis of the best known methods (bearing in mind that other types of propositions may appear in the future), that three categories be distinguished:

– methods that rely on a point cloud to place *a* point, *a* segment or *a* mesh increment directly on the ridge: these methods are relatively less expensive but are not capable of easily capturing crack branching;

– methods using the projection of the field onto a line to evaluate its maximum;

– methods that seek to fit a shape or profile to deduce, in a following step, the ridge position thanks to an assumption of symmetry: these methods can capture branching but have certain limitations, notably due to the assumption of symmetry.

In all cases, additional regularity is obtained due to the fact that the segment is passing through several elements, whereas in the case of a vector field, the vector directions for neighboring elements are independent.

7.2.4.3.2. Exploiting a point cloud

Name	Point cloud
Input	Scalar field with a ridge seen as scattered points of different values
Output	Tangent vector to crack lips

Table 7.5. *Summary of input and output fields for the method exploiting a point cloud*

The most commonly used method in the literature to find points of the crack surface is to calculate, for each point of the current front under consideration, a weighted average of the directions associated with damaged points ahead of the crack front. This weighting involves:

– the considered scalar field to give more weight to the points where its value is more important;

– a Gaussian function, with a radius of the order of a few elements, to give more weight to the points that are closer to the crack tip;

– optionally the area (in two dimensions or the volume in three dimensions) associated with the considered damaged points (usually integration points) to take into account the discretization effect.

The output is *a* position associated with a point of the current front, which leads to a tangent vector to the crack lips.

This method was used by Wells and Sluys (2002), Simone et al. (2003), Mediavilla (2005), Mediavilla et al. (2006), Seabra et al. (2011), Bobiński and Tejchman (2016b) and Wang and Waisman (2016) in two dimensions and by Peerlings et al. (2008), Javani (2011) and Javani et al. (2016) in three dimensions (for the propagation phase). It has the advantage of being inexpensive but does not capture branching since only one position can be predicted at a time.

Similarly, a point cloud could be used to fit, or even build, not an endpoint, but a given geometric shape representing a whole crack increment, therefore an approximation of the ridge. In two dimensions, the idea would be to fit a whole crack segment starting from the current front, and in three dimensions to fit a simple 2D shape, such as a quadrangle. In three dimensions, this technique would be particularly advantageous with a relatively simple shape (quadrangle, ellipse, etc.) during the initiation phase. This type of fitting requires calculations such as minimization or decomposition into eigenvalues that are more complex than simple weighted averages, but has the advantage of taking into account the entire increment and not just the position of the endpoint.

7.2.4.3.3. Projection maximum

Name	Projection maximum
Input	Scalar field with a ridge and choice of line for the projection
Output	Tangent vector to crack lips

Table 7.6. *Summary of input and output fields for the projection maximum method*

Another possible method consists of searching for the maximum projection of the scalar field along a line (this is therefore a study carried out in two dimensions, in the propagation plane of the considered front point, see section 7.2.5 for the adaptation to three dimensions). There are different existing options for the choice of this evaluation line. Therefore, Broumand and Khoei (2013) chose to study the projection onto a semicircle at a distance equal to the characteristic length introduced in the smoothed model. Bottoni et al. (2015) considered the projection along a segment perpendicular to the crack tangent direction at a given distance (or on a circle for the initiation step). This projection is often coupled with smoothing or averaging operations, in pre- or post-treatment in order to limit the dependence of the result on discretization. Various options for smoothing the projection result are proposed in the literature: Comi et al. (2007) proposed to use (without applying it, however) a fourth-order polynomial, Seabra et al. (2013) used third-order Bézier curves and Bottoni et al. (2015) used a convolution product by a Gaussian function. Since these methods provide additional regularity only along the direction of the considered line, it is possible to also introduce more regularity in the crack advance direction by calculating the average of the *maxima* obtained after projection onto several lines. For example, Seabra et al. (2013) considered the average of the positions corresponding to maximum damage on three semicircles close to each other, in the half-plane ahead of the crack front. The largest evaluation radius has been set to include all of the elements where a critical damage value has been reached ahead of the crack tip.

7.2.4.3.4. Fitting manifolds of the same dimension with an assumption of symmetry

Name	Registration
Input	Scalar field with a ridge with focus on a particular iso-value or the shape of the field
Output	One or more points on the crack lips

Table 7.7. *Summary of input and output fields for the fitting method*

The idea here is to proceed in two stages:

1) fit a given manifold with a manifold of the same dimension associated with the considered field;

2) by assumption of symmetry, position the points sought on the ridge at the center (or axis) of symmetry of the fitted shape.

For example, Tamayo-Mas and Rodríguez-Ferran (2015) and Tamayo-Mas et al. (2019) proposed to fit balls (two-dimensional circles or three-dimensional spheres) so that they are bi-tangent to the domain defined by a given iso-value. The ridge is then approximated by the skeleton obtained by connecting the centers: this is the principle of the median axis. The authors' proposition is in fact a version that they describe as *simplified*. Their proposition makes it possible to diminish the number of undesirable branches by introducing a limit angle parameter. One of the main advantages of this method, compared to those based on averages, is to capture crack branching. Moreover, this method has the advantage of not requiring the definition of a propagation plane for each point of the crack front in three dimensions. It should be particularly interesting if the damage surface shows a plateau and the iso-value associated with maximum damage is considered. But it also has some drawbacks. In particular, it is implicitly assumed that there is some symmetry in the distribution of damage around the median axis: this is actually not always true. Consequently, the choice of the considered iso-value can impact the final result. Finally, an important restriction is that the crack must necessarily start from an edge: this method is therefore not recommended when initiation can occur completely inside the structure, that is to say, away from any boundary. Besides, it should be noted that this method provides as an output, scattered points along the crack surface (without any direct link to points of the current front in three dimensions). An additional step is therefore required to use these points and obtain tangent vectors from the crack front. Another example of this type of strategy is the method proposed by Geelen et al. (2018). This method is based on minimizing the difference between the considered scalar field and a so-called *auxiliary* field. The value of this auxiliary field varies with the distance to the segment, which therefore implies a symmetric distribution with respect to the segment. The procedure enables us to find the optimal orientation and length of the segment which is then considered as the approximation of the desired ridge. This method has the advantage of fitting a whole crack increment (and not just a few points): it therefore exploits more information than the simplified median axis method. It enables us to capture branching if the number of possible branches is not underestimated. However, it has so far only been used in two dimensions with a phase field damage model. Its robustness to other damage profiles remains to be evaluated. It is nonetheless already clear that this method would not be the most suitable for a nonsymmetrical distribution.

7.2.4.4. *A challenging topic: branching*

The development of the latest methods is particularly motivated by the possibility of capturing branching in the most direct way possible. Indeed, with the methods only capable of predicting one direction at a time, several steps may be required to capture all branches. Capturing branching is a major challenge, often deliberately avoided in the literature (Wells and Sluys 2002; Wolf et al. 2018).

In any case, the difficulty is to distinguish relevant branches from those that are due only to a numerical artifact. This artifact can be related to discretization (especially because the corners of highly damaged elements can be mistaken for small damage branches) or to the way in which the criterion is calculated (e.g. many branches may appear with the simplified version of the median axis proposed by Tamayo-Mas and Rodríguez-Ferran (2015) and Tamayo-Mas et al. (2019) if the separation angle parameter is not large enough). The elimination of branches corresponding to artifacts requires heuristic criteria, which are not necessarily related to the insertion criteria presented later (see section 7.3.2). Distinguishing between different branches generally requires the introduction of a limit angle to be defined. This angle corresponds to the minimum angle separating two cracks and is used to ensure more robustness regarding mesh dependency. The smaller this angle, the more likely the method is to capture all existing branches. Here are some examples of limit angles found in the literature: for Rabczuk et al. (2010), the angle is 60° (for a mesh-free method); for Saloustros et al. (2018b), it is 40°, and for Feld-Payet (2010) and Feld-Payet et al. (2015), it is equal to twice the angular discretization, namely 10–20°. It is also possible to consider other heuristic criteria to verify the validity of the branch. For example, it can be verified whether, in the direction under consideration, damage is significant enough over a sufficiently long distance. This latter type of criterion is already similar to an insertion criterion, which can also help thereafter to eliminate spurious branches.

It should be noted that in three dimensions, in addition to the difficulty of eliminating irrelevant branches, this may be challenging when the branch orientation varies along the front.

7.2.4.5. *Summary*

The advanced methods providing more regularity have been subdivided here into three categories according to the nature of the studied field: monotonic scalar field, field obtained by the projection of the gradient of the scalar field or scalar field with a ridge. These methods however could be grouped in many other ways, making their classification relatively arbitrary. The main point here is to understand some similarities in the philosophy behind data processing.

In the end, the choice between these methods results mainly from a trade-off between cost (implementation and computation) and performance (especially in terms of ability to capture branching). The most expensive methods are those that rely on normal vectors, mainly because these vectors are usually derived from tensor analysis.

The numerical implementation of the global method is challenging and the *Crack Path Field* method, which uses the projection of the gradient of the scalar field onto normal vectors, lacks robustness in the choice of the normal vector. Nevertheless, these methods have the advantage of directly providing a locally monotonic field, with a specific iso-value corresponding to the crack surface, both in two and three dimensions.

It is possible to significantly reduce computation costs by introducing slightly more modeling hypotheses. Notably, it is possible to model the crack increment by a segment that passes through several elements (rather than having a normal direction per element).

In this context, the least expensive methods are those that rely on a point cloud deriving from a scalar field with a ridge or on the maximum of the projection of the scalar field onto a line. These methods may be sufficient if there can only be one crack branch to follow (provided that the weights for the exploitation of a point cloud be carefully chosen).

When it is desirable to capture multiple branches, the *Marching Ridges* technique, which is based on the projection of the scalar field gradient onto a polar grid, proves to be a good compromise. The fact that the tested increment length is fixed leaves only one degree of freedom (the angle), which is directly associated with a normal vector to be tested. Nonetheless, robustness and performance come at the price of a user-defined accuracy on the curvature of the crack path (from 5 to 10° in first approximation).

Other methods involving the shape of the scalar field with a ridge corresponding to the crack propose to introduce even more model hypotheses, such as the method consisting in fitting manifolds of the same dimension. These methods can be effective if the damage distribution is symmetric with regard to the crack, which is not always the case.

All of the methods presented here provide as an output either a locally monotonic function or a set of vectors that are tangent to the crack path (and pass through one or more elements). In two dimensions, these two types of information are enough to understand which elements are crossed by the crack and to calculate the intersection of their edges with the crack path. In the latter case, an explicit form of crack mesh can be obtained. In three dimensions, the knowledge of a locally monotonic function also enables us to understand the intersection of the crack with the edges of the elements. However, to shift from a set of a few tangent vectors to a three-dimensional representation of the crack surface, whether implicit or explicit, an additional step is required.

7.2.5. *Constructing a continuous discretized surface in 3D*

This section first reviews all of the ways of defining a continuous discretized surface in dimension three, before highlighting the challenging issues specifically associated with this dimension.

7.2.5.1. *Obtaining tangent vectors along the 3D front for 2D methods*

Among the advanced methods presented in section 7.2.4, some lead (directly or as a result of sampling) to three-dimensional tangent vectors, while others have been designed to obtain tangent vectors in a plane. The extension to three dimensions of advanced methods defined only in two dimensions is little discussed in the literature. However, this should not be an impediment to the use of these methods.

To provide a solution to this problem, the author of this chapter proposes to generalize the method employed in Feld-Payet (2010) and Feld-Payet et al. (2015) to obtain three-dimensional tangent vectors from advanced regularization methods defined in a plane.

The idea is to approximate the crack front by a sequence of segments (the increment size being independent of the mesh size). Each segment can then be associated with an orthogonal plane (positioned for example at the center of the segment), which acts as the research plane for the chosen advanced method. The resulting tangent vectors are then approximations of the intersections of the crack path with these planes that are orthogonal to the front line.

For more robustness, the author of this chapter suggests avenging the results rather than making a single approximation for each front segment. For example, in Feld-Payet (2010) and Feld-Payet et al. (2015), this approach (illustrated in Figure 7.12) was used, with the Marching Ridges technique to:

– compute, for each segment, a tangent vector by averaging the results for different planes distributed along the segment (in this case 3, but this may vary depending on the relative size of the segment and of the mesh elements);

– compute the average of the tangent vectors at the points between the segments.

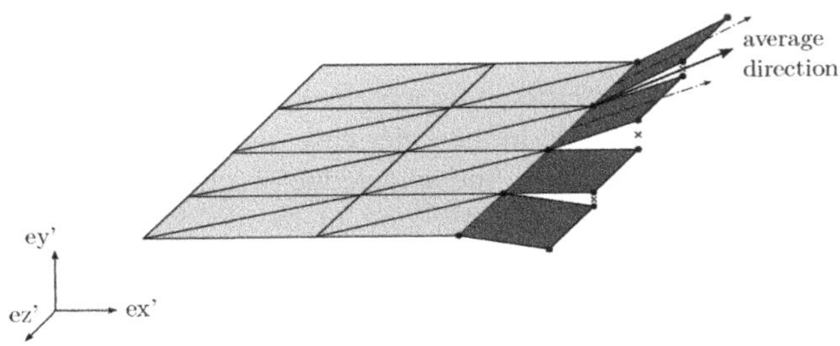

Figure 7.12. *Schematic diagram representing the discretization of a crack front into segments, the orientation that each crack increment would have if an advanced method developed for two dimensions was applied in a plane orthogonal to the segment and the resulting points (marked with red crosses) of the new crack front if the average directions between segments are considered*

7.2.5.2. *Crack surface construction from tangent vectors*

7.2.5.2.1. Principle

The idea here is to start from a selection of a few tangent vectors distributed along the front which is, in three dimensions, a line. Regardless of the method used to obtain this set of tangent vectors, its norm[8] is assumed to exceed one element. The objective is then to interpolate, from these vectors, a continuous surface. This surface can be explicitly defined by a mesh or implicitly by one or more locally monotonic functions, as summarized in Figure 7.13. In all cases, the resulting crack representation is less sensitive to changes in the discretization of the problem than a division of the elements of the structure mesh.

8 The norm can be final or only be used to construct a potential crack path on which an insertion criterion will be evaluated later.

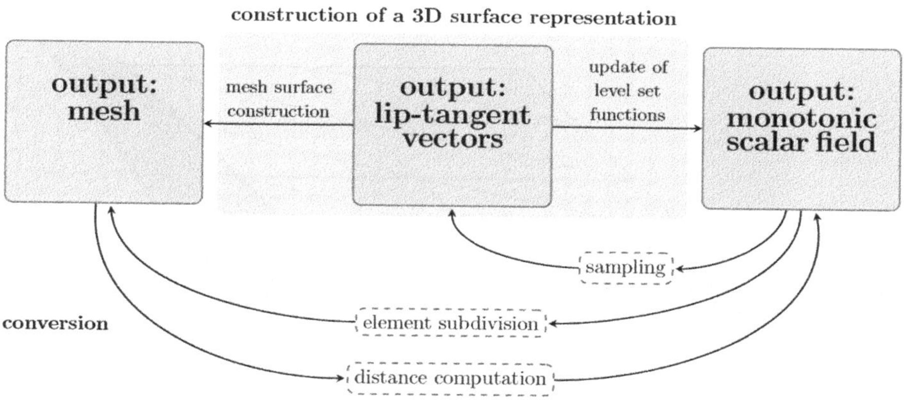

Figure 7.13. *Diagram summarizing the different methods to obtain a three-dimensional continuous discrete crack path from the outputs of the algorithms presented in section 7.2.2. For a color version of this figure, see www.iste.co.uk/lebon/mechanics.zip*

7.2.5.2.2. Construction of a crack surface mesh

It is possible to explicitly represent the crack by a mesh independent of the mesh used for mechanical computation, as done by Peerlings et al. (2008), Javani (2011), Javani et al. (2016), Feld-Payet (2010) and Feld-Payet et al. (2015). To this end, different means can be employed and it is not possible to mention them all here. The simplest solution is probably to determine the position of each point of the new front by linear interpolation between the endpoints of the tangent vectors and to pave the surface with triangular elements that present the advantage to be perfectly planar, as shown in Figure 7.14. For more regularity, it is possible to opt for building the new front using more regular functions, such as B-splines. However, trying to smooth the information has two disadvantages, namely, (1) a larger distance from the original information and (2) an increased difficulty in representing the crack surface by pieces of plane. This second point can be problematic if it is necessary to calculate the signed distance of a point from the crack surface (e.g. when building a locally monotonic function to determine the intersection of the structure mesh with the crack mesh). There is a trade-off to be found depending on the situation.

In all cases, the main difficulties lie in the management of situations where:

– two vectors corresponding to the advance of adjacent front points intersect;

– the propagation of an edge point corresponds to a vector that is not aligned with the edge but is oriented towards the interior of the domain;

– there are obstacles (holes, for example) in the crack path.

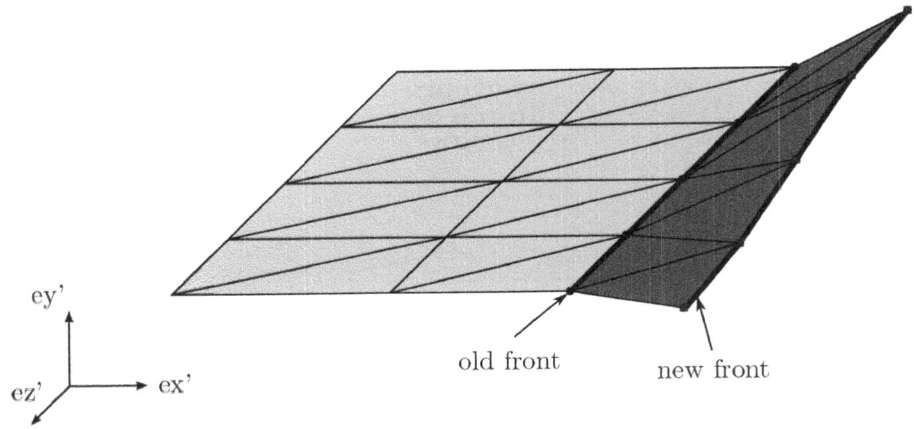

Figure 7.14. *Diagram representing the construction of an explicit crack mesh, independent of the structure mesh, from simple linear interpolations between the endpoints of the tangent vectors*

Solutions have already been proposed for some of these problems (Pereira et al. 2009; Chiaruttini et al. 2010; Garzon et al. 2014; Gupta and Duarte 2014).

7.2.5.2.3. Surface construction using level set functions

The crack can also be modeled implicitly thanks to level set functions, as introduced by Stolarska et al. (2001). The crack advance then requires updating these functions used to locate the position of the crack from their iso-zero, based on the tangent vectors provided on the front. Different sub-steps for reinitialization and orthogonalization are then necessary: see for example (Duflot 2007) for a comparison of the associated techniques. These operations are relatively more complex than a simple interpolation, but they ensure a good regularity of the crack surface.

It should be noted that this type of technique is only mentioned, in this section on the *Crack Path Tracking* algorithms, when considering three dimensions (and not two dimensions): it is only through the reconstruction of a surface from discrete tangent vectors distributed along the front that this level-set technique introduces more regularity.

7.2.5.3. *Conversion*

Output format conversion operations are also slightly different in three dimensions than in two dimensions (see Figure 7.13). It is still possible to calculate a distance to change from a mesh to a monotonic scalar field (Fries and Baydoun 2012). However, sampling in three dimensions, that is, selecting a few intersection points between the edges of the structure and the iso-value of interest of a monotonic scalar field, leads to tangent vectors to the lips, and not to an explicit crack mesh (unlike the systematic division of all of the crossed elements).

As in the case of dimension two, these conversion operations involving distance calculation and element division do not make the output more regular but only change the format. On the other hand, the sampling operation, insofar as it leads to a reconstruction from *a few* tangent vectors, can lead to more smoothness. This operation is only mentioned because it is possible, but is not or rarely used (the author's knowledge).

7.2.5.4. *Summary of 3D surface construction methods*

There are different methods, seen in section 7.2.4 and in this section, to define a continuous discretized crack surface in three dimensions. The author of this chapter propose characterizing these methods according to:

– whether or not they use an intermediate regularization step;

– whether or not they rely on a model of a tangent vector model crossing through several elements;

– whether they directly provide an explicit surface representation or an implicit one (keeping in mind that latter conversion is always possible).

These possibilities are summarized in the first column of Table 7.8 and explained in sections 7.2.5.4.1–7.2.5.4.3.

	Partial domain method	Local method	Global method/ *Crack Path Field*	Construction of an independent mesh	Update of level set functions
Regularization	No	No	Yes	Yes	Yes
Increment model	No	No	No	Yes	Yes
Explicit	No	Yes	No	Yes	No

Table 7.8. *Characterization of the 3D surface construction methods according to whether or not they involve an intermediate regularization step, whether or not they rely on a model of a tangent vector crossing several elements, and whether they directly provide an explicit representation of the crack surfaceor or an implicit one*

COMMENT ON TABLE 7.8.– *All of the methods that are not mentioned do not allow us to obtain directly a 3D crack surface but require an additional step for building an independent mesh or updating level set functions.*

7.2.5.4.1. Intermediate regularization step

As seen in section 7.2.2, following the formulation of a criterion for the crack orientation, it is possible:

– either to directly use methods ensuring mainly continuity, with local division based on a field of normal vectors using the local method or the partial domain method;

– or to go through an intermediate step, with advanced methods ensuring more regularity (see section 7.2.4).

7.2.5.4.2. Model with tangent vectors crossing several elements

As described above, with advanced methods designed to ensure more regularity, it is possible to build the crack surface from vectors that are tangent to the crack lips and that cross several elements. In two dimensions, this is equivalent to modeling a crack increment by a segment. This model necessarily leads to an additional form of regularity. The end result is a crack surface with a representation that is relatively independent of the current mesh discretization. In contrast, methods that use normal vector fields defined by elements lead to a representation of the crack surface that is relatively more dependent on the current structure mesh (since it is constructed by dividing its elements). However, these methods have the advantage of being able to better capture high curvature of the crack path compared to the methods that model crack increments with segments.

7.2.5.4.3. Explicit or implicit representation

Regardless of the choices previously made, it is possible to have:

– either an explicit representation (as a mesh that is independent or not of the current mesh);

– or an implicit crack representation (which may strongly depend on the current mesh or not).

7.2.5.4.4. Analysis

Based on Table 7.8 summarizing the possible options for reaching a three-dimensional surface, it is obvious that it would be difficult to propose a general classification for three-dimensional surface construction methods. Moreover, it can be seen that the local method is just as much the opposite of the method using the update of level set functions as the partial domain method can be the opposite of the construction of an independent mesh. On the other hand, the methods proposed by Oliver et al., namely, the global method and the *Crack Path Field* method, share some of the characteristics of the other methods[9].

9 This contradicts the statement made by Oliver et al. (2004) specifying that methods based on level set functions share characteristics of the global method and the local method.

Other methods can still be proposed, with even more exchanges with the computer vision community. Actually, in principle, the *Crack Path Tracking* methods amount to using discrete data representing relatively regular fields for which *extrema* (*minima* or *maxima*) are sought for. This is a common goal for several domains.

7.2.5.5. *Challenging issues*

7.2.5.5.1. 3D core initiation

The matter of crack initiation inside the structure (far from any boundary) has long been a challenging problem often avoided in the literature, even in two dimensions. The definition of the initiation site is a difficult topic and many authors deliberately limit themselves to crack initiation on an edge (with propagation in only one possible direction). This may be due to the difficulty of managing two crack branches propagating in opposite directions or it may be due to an artifact of the method[10]. One of the main challenge is related to the intent to be mesh-independent and not simply choose the point (node or integration point or center of an element) for which an initiation criterion would be verified. To this end, Seabra et al. (2013) proposed, in case of initiation on a boundary, to approximate the damage for the most damaged element and its two closest neighbors by a Bézier curve on the boundary and set the initiation point where the smoothed damage is maximal. Feld-Payet et al. (2015) and Javani et al. (2016) used a weighted average of the position of the points where damage exceeds a critical value. Many other solutions can still be devised for obtaining an initiation point that does not necessarily coincide with the mesh topology.

From this starting point, in three dimension, it is necessary to build at least one segment to constitute a first front, or even directly the surface of a first crack increment. Methods based on a point cloud obtained from a scalar field with a ridge are then a good solution. For example, Javani et al. (2016) proposed to build from a point cloud and its center a rectangular surface that would serve as a first crack increment. Feld-Payet (2010) and Feld-Payet et al. (2015) proposed to use the *Marching Ridges* technique in the three sagittal planes centered on the initiation point. The searches in these three planes provide at least one starting segment. When the search yields several segments, it is recommended to choose the longest, in order to faster generate the crack surface, but all of the segments are valid and can be chosen as the starting segment. It would be conceivable to use this approach for any search method in a plane that approximates the crack path by a sequence of segments, if we desire to use the same approach for both initiation and propagation. Here again, many other solutions can still be imagined.

10 This is particularly the case for the simplified version of the median axis proposed by Tamayo-Mas and Rodríguez-Ferran (2015) and Tamayo-Mas et al. (2019), which has a tendency to create several branches near the edges of the structure.

7.2.5.5.2. 3D branching

Capturing branches has already been mentioned as a difficult issue in two dimensions and this is obviously even more challenging in three dimensions. In the latter case, it is necessary to be able to group together directions or segments that correspond to different potential crack branches. Based on the author's experience, it is useful to define a local coordinate system associated with the last increment of the crack (rather than considering a global coordinate system) and to compare directions between adjacent segments. As in the case of two dimensions, it is also necessary to define a separation angle to associate directions corresponding to the same branch. For example, with the *Marching Ridges* technique the separation angle is twice the imposed angular accuracy (in other words, if the angular accuracy is 10°, two adjacent segment that are less than 20° apart will be associated with the same branch; beyond 20°, they are considered to belong to two different branches).

7.3. Inserting a discontinuity

7.3.1. *Objectives and connection with the orientation criterion*

The first objective of an insertion criterion is to indicate that the time has come to insert the crack increment under consideration[11], whether the user has chosen to directly open the crack lips or to include a cohesive zone model. In practice, this moment is chosen first according to the behavior model, then to the numerical issues occurring when resuming the computation after insertion and finally to more physical considerations. Two cases must be distinguished: the increment length is imposed by the user or remains to be determined.

If we choose to insert fixed-length increments (see, for example, Feld-Payet 2010; Feld-Payet et al. 2015; Wang and Waisman 2016), then the insertion criterion is only used to indicate when the increment can be *entirely* inserted. This approach can be interesting to separate the orientation and insertion criteria and the associated difficulties and thus to be able to focus on the development of only one of these two criteria. In this context, Bobiński and Tejchman (2016b) even proposed to look for the orientation of the crack only when the insertion criterion is reached in an element that is adjacent to the front or in the surrounding area. This solution has the advantage of limiting the cost associated with the evaluation of the orientation criterion. However, this procedure is rarely applied and seems limited to the two-dimensional framework. Moreover, this approach, which consists of fixing the length of crack increments has limitations in terms of the representativeness of the results. Indeed, in three dimensions, this leads to inserting constant length increments

11 It is considered here that the discontinuity is introduced incrementally, depending on the progression of the crack.

along the front, which can be problematic when all of the front points do not progress at the same speed and the front should in reality therefore be curved.

If the length of the crack increment is not fixed by the user, its value (which corresponds a priori to several elements) should be determined. Unlike linear elastic fracture mechanics, this length cannot be determined from global quantities, such as the energy release rate (Lemaitre and Chaboche 2004). With the local approach, the insertion criterion makes use of the local state to determine the length of the increment. The difficulty, in three dimensions, is that it is then necessary, before the actual insertion of the crack increment, to reconstruct the crack surface from tangent vectors with a length corrected by the insertion criterion. In this case, the separation between the steps associated with the orientation criteria and those associated with the insertion criterion become blurred.

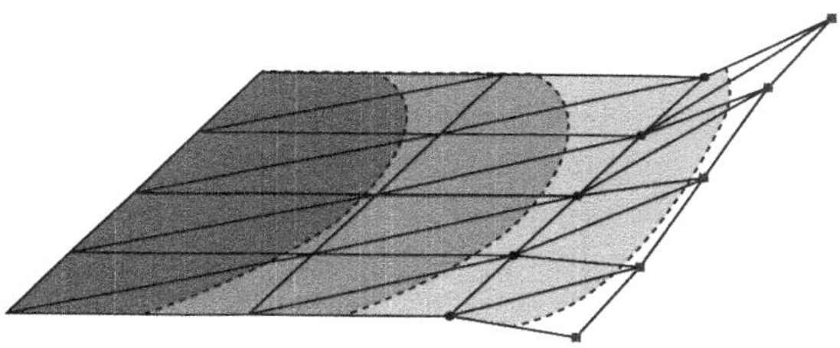

Figure 7.15. *Diagram illustrating crack advance in time based on an insertion criterion along a potential path determined by an orientation criterion*

In both cases, whether the crack has a controlled thickness (of one element in general) or no thickness (discrete crack)[12], one possible approach is to evaluate the insertion criterion on the potential cracking path, defined by an orientation criterion, to determine the advance of the current front at the end of each time step, as shown in Figure 7.15. It is even possible to freeze the potential crack path at a certain point to evaluate only the insertion criterion at the end of each time step. This is, for example, the strategy chosen by Feld-Payet (2010) and Feld-Payet et al. (2015). There are two main advantages to the latter solution:

1) the cost associated with determining the potential crack path thanks to the orientation criterion is reduced;

2) this solution avoids waiting for the formation of a ridge plateau to evaluate where the position of the crack needs to be evaluated. This situation may occur for some regularized models and affect the evaluation of the crack path position, as discussed in section 7.2.

12 For a crack with uncontrolled thickness, the criterion can be evaluated in a wider area of the damaged zone.

Nonetheless, this strategy requires defining an additional criterion to signal that it is time to freeze the potential crack path since it will no longer evolve (or negligibly).

7.3.2. *The different insertion criteria*

7.3.2.1. *Principle*

The choice between the different insertion criteria and the associated values strongly depends on the chosen behavior model:

1) For non-regularized coupled models, the insertion criterion must allow for the insertion of the discontinuity before the solution loses its uniqueness. The remainder of the energy in the area corresponding to the discontinuity is then dissipated via a cohesive zone model or a band of elements.

2) For uncoupled or regularized models that take into account, through the insertion of some form of discontinuity, a strain relocalization phase from a volume to a surface or to a zone of very small thickness (which would correspond to a coalescence phase of defects), the remaining energy in the zone corresponding to the discontinuity must also be dissipated via a cohesive zone model, or even a band of elements. The choice of the time of insertion is in theory less constrained by the model. This leads to different choices of insertion criteria.

3) For regularized coupled models without a cohesive zone model for the opening of lips, the insertion criterion must allow us to insert the discontinuity when the area is sufficiently damaged and can no longer (or hardly) enable force transmission.

7.3.2.2. *Stability and bifurcation criterion*

For standard models with coupled damage (in other words, local or nonregularized, case 1), it is essential to determine the direction before localization. Indeed, after localization:

– for orientation criteria based on the damage or strain field, the localization band will be influenced by the mesh orientation;

– for orientation criteria based on the stress field, the latter will decrease and may therefore provide less and less relevant information (see, for example, Wells and Sluys 2002).

To choose the right moment, the most logical approach is then to use a stability and uniqueness criterion (also often used for the orientation criterion in such cases). The difficulty, highlighted in particular by Areias and Belytschko (2005), is that there is still no robust general procedure existing to verify the stability and uniqueness conditions (the analysis proposed by Boussaa and Aravas (2001) is however mentioned because it is sufficiently general for the particular case of elastoplasticity).

7.3.2.3. *Critical value of a local quantity*

Methods based on a critical value of a quantity are the most widely used.

7.3.2.3.1. The critical value of equivalent strain

A possible insertion criterion is to consider a critical equivalent strain: see, in particular, the work of Jirásek and Zimmermann (2001) and Jirásek (2004). The interest of this proposition is to consider a quantity that, in principle, could be measured by image correlation analysis. Yet, in reality, the experimental measurement of an appropriate equivalent strain is not easy and, to this day, would depend on the methods used both for image correlation and for computing strain from displacement fields. In practice, Wells and Sluys (2002) and Bobiński and Tejchman (2016b) chose to set the critical value as a percentage of the material yield strain. However, this type of approximation is not valid for a ductile material, that is, a material exhibiting significant plastic strains at failure. This approach therefore remains to be developed in the future.

7.3.2.3.2. Critical degradation value

Methods based on a critical degradation value, on the other hand, do not allow for a choice of value that is based directly on a local measurement during an experimental test. Although it may be possible to try to measure local porosity in the vicinity of the crack after the test, the vast majority of the work cited here uses a damage value as Kachanov defined it. The critical value is then less material dependent since one considers normalized quantities of the damage model: regardless of the material, the fully degraded state is generally represented by a critical degradation value equal to one.

If damage still evolves after inserting a discontinuity (in a cohesive zone or a band of elements), the threshold values may be relatively lower (cases 1 and 2): 0.4286 for Pourmodheji and Mashayekhi (2012) (determined based on so-called *Compact Tension* tests and three-point bending tests on notched specimens of A53B1 steel), 0.45 for Hambli (2001), 0.5 for Javanmardi and Maheri (2019) and between 0.6 and 0.7 for Roth et al. (2015). It should be noted that, when damage is coupled and the problem is not regularized (case 1), the choice of a low critical value does not guarantee avoiding the loss of uniqueness of the solution in the same way as a stability and bifurcation study and this may affect the crack orientation. This problem may have more limited consequences for a ductile material since the development of plasticity will guide damage, but these consequences remain undesirable. With such low values for degradation, there is still a significant amount of energy to be dissipated. This can be a problem even in the presence of cohesive zones to ensure a gradual decrease of the forces transmitted through the crack. For example, Bobiński and Tejchman (2016b) and Roth et al. (2015) indicate that they had problems with the choice of the critical value. They concluded that a critical value between 0.6 and 0.7 would allow for a consistent transition. If special measures are not taken to dissipate this energy (in other words, if the discontinuity directly represents a discrete open crack or a crack with a thickness for the erosion method), then such low damage values would be accompanied by large jumps in the force–displacement curve with each new insertion, as highlighted notably by Seabra et al. (2011, 2013).

In the cases where the discontinuity is inserted when the damage is complete or almost complete (case 3), the variable representing degradation must have a value *close* to its maximal value. However, the critical value cannot be *exactly* set to the maximal value for numerical reasons. The values vary according to the authors: 0.9 for El Khaoulani and Bouchard (2012), 0.99, 0.999 or 0.99995 for Simone et al. (2003). The higher the critical value, the longer the insertion time is delayed. The problem, identified by Simone et al. (2003), is that the same critical value can lead to different values of stress at failure depending on the considered model. This choice of critical value is therefore insufficient in ensuring that the stresses in the area to be fractured are sufficiently low. Again, inserting a discontinuity into an area that always transmits forces can cause undesirable jumps in the force–displacement curve.

7.3.2.3.3. Where and how should the insertion criterion be evaluated?

The insertion criteria considered here are based on local quantities associated with critical values. However, as has already been pointed out for the orientation criterion, these local quantities generally lack regularity. The question of where and how to evaluate the insertion criterion is therefore not to be neglected here either. Wolf et al. (2018) and Bobiński and Tejchman (2016b) highlighted some of the problems in their attempt to estimate the insertion criterion too locally near the current crack tip. Broumand and Khoei (2013) and Bobiński and Tejchman (2016b) thus proposed to evaluate the insertion criterion from data on an area larger than a single element, ahead of the crack front. It should be noted that, depending on the method for crack representation, it may be more relevant to evaluate the insertion criterion with values projected onto the future crack surface rather than in a neighboring volume. For example, for the construction of an explicit independent mesh representation of the crack, Feld-Payet (2010) and Feld-Payet et al. (2015) evaluated the criterion on the crack increment to be inserted. In any case, it seems necessary to focus on the evolution of local variables near the potential crack path and its vicinity in order to be able to propose a robust insertion criterion. This is an issue that has so far received little attention in the literature, but which deserves more consideration.

7.3.2.4. *Critical value of transmitted force*

A possible solution to the problem of choosing a critical value for the damage variable is to consider a critical value for the force transmitted through the increment to be inserted. Again, this is not a local quantity that can be measured during the test. This solution is of particular interest when the discontinuity to be introduced is represented by a discrete crack. From the numerical point of view, the force transmitted through the surface of the increment to be inserted can then actually be measured (although how to integrate the stress over along the increment remains an open question). From the physical point of view, this is consistent with a final relocalization of the strains on a surface which, *in fine*, must no longer transmit force (as opposed to the degradation of an entire volume in the vicinity).

This solution:

– is not sufficient on its own in case 1, as mentioned above;

– may be a valid option in case 2 and, in principle, should not constraint the choice of the critical value of transmitted force;

– is particularly interesting in case 3 where the idea would be to wait until the forces transmitted by the increment have fallen below a critical value.

In practice, one of the main challenges is to set the critical value, especially in case 3. Models are indeed often implemented so as to avoid reaching a zero stress state and it is difficult to predict what minimal stress value the chosen model can reach and whether this value depends on the loading state. To get an idea of the order of magnitude of the critical value, it can be mentioned that this method was used by Wolf et al. (2018) to insert a crack[13] for a critical value of 100 MPa.

7.3.2.5. *Other criteria*

Other less common criteria exist, such as the energy criterion proposed by Crété et al. (2014): this review is not intended to be exhaustive, but is only meant to give an idea of the most commonly used criteria.

It is also possible to combine several criteria. For example, for a non-regularized coupled damage model (case 1), Wolf et al. (2018) combined a first criterion, based on stability and bifurcation, with a critical damage criterion (critical porosity of 0.03) before inserting a cohesive zone. Then the crack was opened before complete loss of strength for a critical value of transmitted force (100 MPa). For models with regularized coupled damage, it may be interesting to use a mixed criterion with sufficiently significant damage and sufficiently low transmitted force.

7.3.3. *Challenges associated with front determination*

The insertion criterion may play a role in determining the length of the crack increment and, in particular, the position of the new front. Yet, the resulting position of this new front can be in the end modified due to a number of constraints related to numerical issues.

7.3.3.1. *Correction of the position of the new front points*

For discrete crack representation techniques, the fact that the front points are not necessarily nodes of the mesh will require a sub-division of the elements containing

13 In this case, the crack was inserted to finish opening the cohesive band, instead of waiting for its complete loss of strength, which is quite unusual. This choice was motivated by two reasons. The first is that the bonds between the porosities that connect the crack lips will abruptly break beyond a certain level of loading. The second is that this approach allowed Wolf et al. (2018) to improve the convergence rate.

the new front. For insertion by mesh adaptation, this measurement is accompanied by a change in the mesh topology that becomes obvious when comparing the position of the nodes. And in situations where the front is on an edge, it would still be necessary to duplicate the nodes along this edge to get a compatible mesh. For a representation using the X-FEM technique, the position of the front is defined by a second specific monotonic function, in addition to the function used to position the crack lips. Based on this information, the element containing the front is also sub-divided, but only for data management at integration points. This approach has for example been applied by Seabra et al. (2013). Other authors using the X-FEM technique tried to limit the complexity of the problem by truncating the crack path to stop at an edge. Examples include Wells and Sluys (2002), Comi et al. (2007), Crété et al. (2014), Tamayo-Mas and Rodríguez-Ferran (2015), Wang and Waisman (2016), Wolf et al. (2018) and Javanmardi and Maheri (2019). In the latter case, the initially predicted position is modified for purely numerical reasons.

In three dimensions, an additional difficulty emerges: ensuring sufficient regularity of the front. The first reason why such regularity is desirable is that the underlying fields are assumed to exhibit some regularity. The second reason concerns the orientation criteria that seek to exploit data in an area ahead of the crack. Indeed, these criteria require the areas associated with the different points of the front to be close enough to each other to have consistent associated directions. However, a zig-zag front could lead us to consider totally different areas for neighboring front points. It is then possible to resort to an additional step for smoothing the crack front, which corresponds to a correction of the initially predicted front. When only a few tangent vectors are used for the construction of the 3D surface, this smoothing is quite natural: the fewer these vectors are, the smoother the front will be, at the cost of a certain loss of information.

7.3.3.2. *Limiting perturbations*

Independently of the considered dimension, one of the main issues in choosing the insertion criterion is to limit the level and frequency of perturbations.

7.3.3.2.1. Origins

These perturbations are related to both the change of model and the change of discretization. Indeed, whether the discontinuity is introduced gradually using a cohesive zone model or more directly through an opening of the crack lips without any possible force transmission, the insertion of the discontinuity represents a change in the constitutive model at the crack location.

In addition, independently of the method chosen to represent discontinuity in the model, it will impose, compared to the continuous model, at least a change in discretization. This change causes a perturbation due to the errors associated with the transfer of the fields onto the new discretization (see section 7.4).

7.3.3.2.2. Effects and possible solutions

These perturbations cause difficulties when resuming computation after insertion. They result in jumps in the macroscopic force–displacement curve. The larger the inserted segment, the bigger the jumps (Broumand and Khoei 2013).

On the other hand, inserting a discontinuity element per element does not seem to be the ideal solution, as it would increase costs and errors associated with field transfer operations. A trade-off must therefore be found, namely to introduce increments that are short enough to avoid significant perturbations of the computation, but long enough to avoid too frequent insertions.

Let us consider the problem in two dimensions. If the considered increments are initially too short (e.g. of a only one element long), it is then possible to iteratively continue the search for increments from the new front, at the end of the same time step. This is known as the *exhaustion method*. This approach was in particular employed by Crété et al. (2014), Wolf et al. (2018) and Javanmardi and Maheri (2019). On the contrary, if the intended increments are already long enough, it may be preferable to proceed with the insertion of this increment and to consider inserting the following one once equilibrium has been restored. If necessary, it will be inserted during the next time step; and if the time steps are small enough, the delay may be negligible. The length of the increments can then be arbitrarily set by the user: this is the choice made by Feld-Payet (2010) and Feld-Payet et al. (2015). Another option, proposed by Seabra et al. (2013), is to consider a *variable length* increment (determined by the point exceeding the critical damage value in the determined crack propagation direction), but potentially inserted over several steps.

In three dimensions, the question of knowing how to transpose the notion of sufficient increment length, since crack increments are then surfaces and no longer segments arises.

7.3.4. *Outlook: strengthening the link with physics*

The evaluation of the insertion criterion is the step that has the strongest link with physics. However, proposals in the literature are often motivated by numerical reasons. This can be justified, on the one hand, by the fact that it is first necessary to have calculations that converge before being able to conduct a test-calculation comparison in a subsequent step. And, on the other hand, the situation can also be explained by the lack of *local* experimental data to identify the coefficients of the local approach model. At the moment, mainly *global* measurements, such as macroscopic force–displacement curves, are already compared.

Yet, some information is not accessible through global curves, such as the spatial distribution of strains. This type of information could be provided by analyses of digital image correlation results. This would be useful for insertion criteria based on a critical value of an equivalent strain, but it would also allow for better model identification. This type of analysis would in particular make it possible to study the relocalization of strains due to damage.

Moreover, global curves do not give access to the position of the crack front either. Again, to obtain this type of information, it would be interesting to resort to the analysis of numerical image correlation results. More specifically, methods used for finding crack orientation from numerical results could be employed for the analysis of experimental results. In this way, Feld-Payet et al. (2020) used the *Marching Ridges* technique to find potential crack paths from measured displacement fields. More spatial experimental data, but also temporal data, are thus needed: Crété et al. (2014) stress the need to have a sufficient amount of data over time to properly capture the all-too-often abrupt progression of the crack.

With the development of means of acquisition (not only in 2D, but also in 3D with tomography and laminography (Bouterf et al. 2020)), test image processing and means to compare experiments and calculation results (Garaud et al. 2017; Feld-Payet et al. 2022), we can hope to progress toward a validation of the insertion criteria and therefore of damage-to-crack transition strategies.

7.4. Resuming computations after inserting a discontinuity

7.4.1. *Issues*

From a numerical point of view, the main issue associated with the damage-to-crack transition is the introduction of a perturbation into the model.

This is a problem that has all the more impact as the way of representing the discontinuity implies a significant change in discretization. The impact is therefore very significant when choosing to represent the crack discretely, with X-FEM or mesh adaptation. In both cases, the change of discretization involves transferring data from the old to the new discretization. Indeed, while field transfer is often associated with the mesh adaptation technique, it is also necessary when using the X-FEM since the insertion of a discontinuity is accompanied by the adaptation of the integration scheme. These problems with transferring the fields stored at the integration points with X-FEM are mentioned in particular by Wells and Sluys (2002), Simone et al. (2003), Broumand and Khoei (2013), Seabra et al. (2013), Wang and Waisman (2016) and Javanmardi and Maheri (2019). It is therefore necessary to define a transfer operator for the data stored at the integration points, in the same way as for a strategy based on mesh adaptation.

It should be noted that the introduction of a perturbation highlights any instability in the problem. The first step to take is therefore to ensure that we start from a sound base. For example, in the presence of volumetric locking for an incompressible plastic flow, it is first necessary to make sure that we have provided a suitable solution (reduced/selective integration or mixed formulation as proposed by Feld-Payet et al. (2011)) before considering a change of discretization.

In contrast, let us point out that the presence of viscous effects in the model makes it possible to reduce to some extent the effect of perturbations.

7.4.2. *Field transfer*

This section is essentially taken from Chapter 2 in Feld-Payet (2010) for the operators dedicated to remeshing. It has been supplemented by operators that are more commonly used in association with the X-FEM technique. It is intended to provide an overview of the methods used for transfer, without however being exhaustive.

7.4.2.1. *Principle*

To resume computation after a change of discretization, it is necessary to transfer the values of the degrees of freedom (displacements, nonlocal variable, etc.), stored at the nodes, and the values of the internal variables (damage, cumulative plasticity, etc.), stored at the integration points. Nonetheless, the transfer of these variables remains difficult, because the transfer operator should, in theory, satisfy the following constraints (Perić et al. 1996):

– The relations between the transferred fields should always be respected: for example, the strain tensor should be equal to the symmetric part of the displacement gradient.

– The transferred fields should always satisfy the equilibrium (consistency of transferred stresses and external forces).

– The operator should minimize numerical diffusion: in particular, the damage band should not become wider.

– Boundary conditions should always be verified.

In practice, it is impossible to satisfy all of these constraints simultaneously. It is essentially the minimization of numerical diffusion that will guide the choice of a transfer operator (indeed, having fields respecting equilibrium and/or consistent with each other but that do not look anything like the starting fields does not seem reasonable). The problems of satisfying relations between the transferred variables will only be solved in a subsequent step, during a reequilibrium procedure.

7.4.2.2. *Nodal data transfer*

The transfer of nodal data is usually done by interpolating the values stored at the nodes of the old mesh $\mathcal{M}$. We simply have to identify to which element of the old mesh $\mathcal{M}$ each node of the new mesh $\mathcal{M}^*$ belongs and to use the shape functions associated with this element to obtain the value for the new node.

It should be noted, however, that Espinosa et al. (1998) proceeded in a slightly different way. Their transfer technique was based on the use of an auxiliary mesh with 3-node triangular elements connecting the nodes of the old mesh with six-node triangular elements. The shape functions associated with this auxiliary mesh were used to interpolate values at the nodes of the new mesh. The use of this type of method still remains anecdotal.

7.4.2.3. *Data transfer between integration points*

There are several possible techniques to transfer data between integration points.

7.4.2.3.1. Transfer via the nodes of the old and the new mesh

The procedure used by Lee and Bathe (1994), Perić et al. (1996) and Mediavilla (2005), for mesh adaptation, can be divided into three steps:

1) The values at the integration points are extrapolated at the nodes using the shape functions of the elements. The nodal values are then averaged in order to obtain a continuous field. Mediavilla acknowledges that this step is the source of significant diffusion and specifies that more elaborate techniques have been proposed to limit this phenomenon, such as resorting to an approximation based on least squares minimization (see the work of Hinton and Campbell (1974)) or using an iterative procedure following the work of Loubignac et al. (1977). A similar method inspired by the patch recovery procedure or SPR (*Superconvergent Patch Recovery*; see Zienkiewicz and Zhu 1992a, 1992b) has been proposed by Khoei and Gharehbaghi (2009) for 3D problems.

2) The nodal components are then transferred from the old nodes to the new ones by locating these nodes in the old mesh and using direct interpolation, as mentioned above.

3) The values at the new integration points are determined by simple interpolation in each element, using the shape functions of the new mesh.

This is the same technique that was used by Broumand and Khoei (2013) following the subdivision of the elements with X-FEM (the second step being then trivial since the nodes remain the same).

The problem with this method lies in the numerical diffusion that it creates: it is more significant than with a method using direct transfer from integration points to integration points (Perić et al. 1996).

7.4.2.3.2. Direct transfer from integration points to integration points

Constant values

A very simple transfer operator can be built by assigning constant values to the domains associated with the quadrature points of the old mesh. This construction is local. It is also possible to associate a new integration point with the value at the nearest integration point on the old mesh. This is, for example, the option chosen by Wang and Waisman (2016) (the closest integration point belonging to the same element). The authors then point out that the error decreases with the mesh size. Simone et al. (2003) have chosen to consider the maximum value in the element and, after transfer, to recalculate the increment that led to the critical damage. These authors all used this type of transfer operator after introducing a discontinuity with the X-FEM.

Direct interpolation

Ortiz and Quigley (1991) have used shape functions that are equal to one at the integration points. The method is based on interpolating the values when the new integration points are inside the group of the old integration points, and extrapolating the values when they are outside (while remaining in the old element). However, the field discontinuity along the boundaries between elements creates some problems, especially when extrapolating. Boroomand and Zienkiewicz (1999) have proposed to solve this problem by resorting to a particular type of remeshing, namely, the old element is divided into several elements so that the new integration points are always located within the group of old integration points. This transfer methodology is thus not suitable for any new arbitrary mesh.

The same method seems to be used by Seabra et al. (2011, 2013) with a bilinear extrapolation of the quantities from the old to the new integration points using shape functions associated with old integration points (following a change of model with the X-FEM technique). The authors describe this method as a *local smoothing technique* and attribute it to Hinton and Owen (1977).

A similar principle is used by Espinosa et al. (1998), but this time with an interpolation method only. In this chapter, the authors present a transfer operator that starts from a mesh with six-node triangular elements and full integration. The values at the new integration points are obtained using the shape functions of an auxiliary mesh. This mesh is composed of triangles with three nodes connecting the integration points. We then merely have to identify the auxiliary element to which the new integration point belongs, and then proceed by simple interpolation.

7.4.2.3.3. Transfer from integration points to integration points with smoothing

L^2 projection onto the nodes of the old mesh

The technique used by Camacho and Ortiz (1997) is intended to work similarly to the transfer from integration points to integration points. It is based on the work of Ortiz and Quigley (1991), but attempts to make the methodology more general by smoothing the data before transfer. This smoothing is obtained by projecting the values stored at three integration points to the nodes of the old mesh by means of an L^2 projection. The resulting smoothed field is then directly evaluated at the integration points of the new mesh.

However, Boroomand and Zienkiewicz (1999) pointed out that this method has one particular drawback. Since, for the L^2 projection, the integration is carried out over the whole domain, the information originating from a small area can spread over a relatively larger area.

Weighted average per element

Wells and Sluys (2002) proposed to estimate the internal variables at the new points thanks to a weighted average that takes into account the distance between the points. The authors highlight that the number of sampling points is limited because the transfer only takes place within an element (this technique is used after the introduction of a discontinuity by the X-FEM) and that the error decreases with the mesh size. This transfer method can be regarded as a technique for reconstructing the data by smoothing. We can note that with this technique, the new value can only be less than or equal to that of the sampling points.

Weighted average of least squares values per patch

Boroomand and Zienkiewicz (1999) proposed a transfer technique based on patch contributions, in order to minimize diffusion problems. The first step then consists in determining which element of the old mesh the new integration point belongs to. Each node of this element can be associated with a patch (defined by all of the elements sharing this node). On each of these patches, a polynomial approximation is then built from the values at the integration points thanks to a weighted least squares technique. The value sought for at the new integration point is obtained by calculating the weighted average of the smoothed values estimated at the nodes.

Projection after smoothing by least squares

Villon et al. (2002) proposed to transfer the fields at the integration points thanks to diffuse approximation techniques. These techniques enable to define a mechanical field with global continuity of any order that serves as an approximation for a given discrete mechanical field. This method is implemented in particular by Brancherie et al. (2006). It consists in locally decomposing, in the vicinity of the integration point of the new mesh, the scalar field to be transferred (or each scalar component of a tensor, for example) onto a polynomial basis centered at this point. The coefficients of the polynomial are obtained by a weighted least squares technique, from the values at the old integration points. Finally, the smoothed field is directly projected to the new integration point. This type of method is particularly efficient as long as the number of sampling points remains limited in order to avoid diffusion.

7.4.2.4. *Choice of transfer operator*

A comparative study by Patzák and Jirásek (2004) shows that the transfer of values stored at integration points using the nodes leads to significant numerical diffusion. This diffusion phenomenon must be avoided as it can cause the widening of the strain localization band. In contrast, direct transfer methods that rely on using the value of the nearest integration point or smoothing by least squares do not have this drawback. Direct value transfer from the nearest integration point represents a lower cost. However, least squares smoothing is more robust when the mesh size varies significantly during the remeshing process. It is also much simpler to implement and also less expensive than the transfer method proposed by Boroomand and Zienkiewicz (1999). By considering only a minimum of old integration points to build the smoothed function in the vicinity of the new integration point[14], it is possible to obtain better results, that is to say, with less diffusion limitation, than with smoothing by L^2 projection. Besides, this method is as easy to implement in 2D as in 3D.

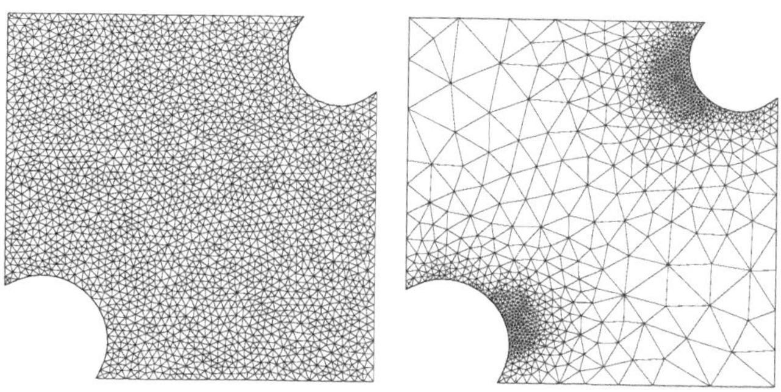

a) b)

Figure 7.16. *Initial mesh (a) and adapted mesh based on error estimation (b). The elements of the adapted mesh are concentrated in areas where nonlinear phenomena occur. The coarsest elements correspond to areas where behavior is linear*

14 Feld-Payet (2010) uses 13 sampling points in 2D for six-node triangles and 30 in 3D for tetrahedra with 10 nodes. The selection of these points takes into account the distance to the new integration point under consideration, but also the possible presence of an already open crack to only consider the points on the most relevant side of the crack.

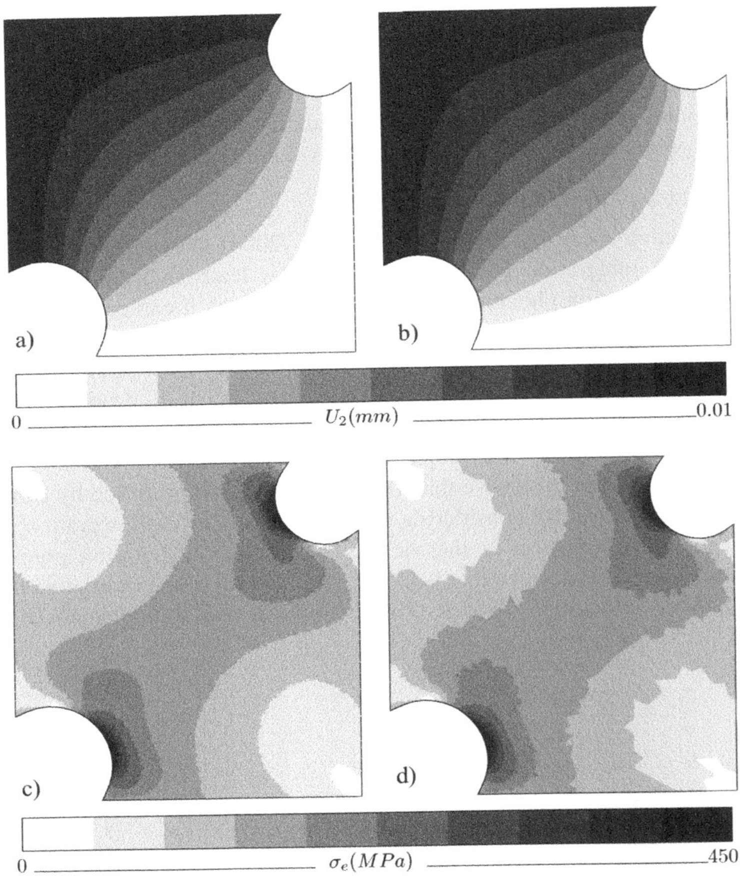

Figure 7.17. *Iso-values of the vertical displacement field U_2 before remeshing (a) and after transfer (b). Von Mises stress iso-values σ_e before remeshing (c) and after transfer (d)*

COMMENT ON FIGURE 7.17.– *The displacement distributions are almost identical between the two meshes. The von Mises stress distributions are very similar in areas where the stresses are most significant and globally consistent elsewhere. The transfer operator under consideration therefore gives qualitatively satisfactory results.*

An example of transfer between two meshes (shown in Figure 7.16) with direct interpolation for data stored at the nodes and direct projection after moving least squares smoothing for data stored at integration points is proposed in Figure 7.17. For the vertical displacement field, the conventional method gives visually identical distributions for both meshes. On the other hand, the distribution of the von Mises stress field shows that the considered transfer operator provides very good results in the area with the fine mesh size near the notches. The more significant the mesh size modification, the more affected the field transfer at integration points. However, this visual criterion can only be qualitative and not quantitative. Actually, the changes in the iso-values are not significant because the coarsening concerned undamaged areas, with low energy, and where the behavior of the material is still elastic (see Feld-Payet 2010; Feld-Payet et al. 2015 for information on the remeshing strategy based on an error estimation that led to an adapted mesh). This is why, an *improvement* in accuracy can be expected despite coarsening, in addition to a significant decrease in computational costs. The main concern should be to transfer the variables as well as possible in areas where the behavior is strongly nonlinear.

7.4.3. *Reequilibrium*

7.4.3.1. *Context*

Since the fields have been transferred in different ways for the variables stored at the nodes and at the integration points, regardless of the nonlinear relations between these variables, the fields on the new mesh are out of equilibrium. At this point, resuming the computation after insertion will be all the easier if the fields associated with the new mesh (after transfer) verify all of the equations of the problem. With this in mind, Brancherie et al. (2006) proposed transferring certain fields under constraint in order to better satisfy the static and kinematic admissibility conditions. Some authors, such as Camacho and Ortiz (1997), Mediavilla (2005) or Broumand and Khoei (2013), proposed to try to get a little closer to equilibrium by using certain local relations that link the variables together to transfer only a small number of variables and then calculating the others due to local relations. This type of transfer certainly facilitates the return to equilibrium. However, this requires a choice in the equations to be verified: it is therefore not easily generalizable in a finite element computation code. In addition, even with such precautions, transferred and calculated fields may still not exactly verify global equilibrium or the boundary conditions.

In all cases, there are various options to help resuming the computation after insertion of the discontinuity and data transfer. At the end of these operations, when calculation has converged, the fields on the new mesh are consistent and satisfy equilibrium. The analysis can therefore continue with the new discretization of the continuous problem with a crack.

7.4.3.2. *Time step division*

Boroomand and Zienkiewicz (1999) proposed to rely on the division of the following time step to facilitate the return to equilibrium. In other words, the applied forces are gradually imposed. This is a very simple technique, often easily accessible in finite element codes, and it can easily be combined with other measures.

7.4.3.3. *Elastic step*

In order for the fields on the new mesh to verify the discrete equilibrium equations, Mediavilla (2005) proposed to introduce an additional step to restore equilibrium for the boundary conditions of the current increment[15]. During these iterations, strains are assumed to be elastic in order to facilitate convergence of the solution. The arguments in favor of this choice are that the convergence rate of this step is quadratic and that it would be useless to resort to elastoplastic iterations when the material state is not supposed to change during this step.

7.4.3.4. *Compensation by vanishing residual forces*

Broumand and Khoei (2013) proposed to introduce a relaxation step during which vanishing residual forces associated with the insertion of a crack increment are applied in the opposite direction. This proposition was also implemented by Javanmardi and Maheri (2019). In both cases, field transfer was necessary due to the subdivision of rectangular elements with the X-FEM. It was specified that the total number of iterations depended on the size of the inserted crack increment and on the value chosen for maximal damage: the higher the damage value, the lower the residual forces.

Seabra et al. (2013) also chose to resort to an additional step. Before each enrichment of the element nodes, the displacements are stored. These displacements are then applied, as input to solve the global problem after nodal enrichment and transfer of the quantities stored at the integration points. Solving this problem makes it possible to calculate the associated forces at the enriched nodes. A decreasing percentage of these forces is then applied to the nodes until a given tolerance threshold is reached.

This type of technique can be interesting to facilitate return to equilibrium. Nevertheless, too few quantitative studies have been carried out on the subject until now to know its limitations.

7.4.3.5. *Choosing a reequilibrium method*

The choice of a reequilibrium method can be guided by the difficulties of implementation, the required level of knowledge of the problem equations and whether or not the method needs to be robust and applicable in various cases or adapted to a particular calculation. At the very least, it is necessary to allow the division of the time step for the following increment. This may be enough in the presence of viscosity where a distance from the loading surface can be tolerated. Further studies (preferably comparative) on these methods should enable better evaluating their benefits in terms of implementation costs and their limitations.

---

15 Mediavilla (2005) justified the introduction of this additional step by mentioning numerical stability issues. However, these may well be due to the choice of transfer operator that causes significant diffusion.

7.5. Conclusion

The field of damage-to-crack transition has been experiencing sustained growth since the early 2000s with many contributions attempting to answer these three questions:

1) Where should the discontinuity be placed? (a) Which orientation criterion should be chosen? (b) How can a continuous discrete crack representation be obtained?

2) When should the discontinuity be inserted? And just how far?

3) What measures needs to be taken to resume computation after insertion?

This chapter has proposed some guidelines based on each of these three questions to try to analyze the various contributions and highlight their benefits, similarities and differences. The choice of combination between the methods discussed here depends very much on the situation. It is important to know the tools available in the literature to make the most appropriate choice.

While many contributions have been made in the field of orientation criteria and *Crack Path Tracking*, there is now a need for more quantitative studies on the insertion criteria and the measures to be taken to enable computations to be resumed after inserting discontinuity.

Once the numerical challenges have been met, it is essential to validate the chosen methodology by comparing with experimental results. Only in this way can we hope to move towards a validation of the insertion criteria and therefore of the damage-to-crack transition strategies. This validation will lead to greater confidence in the use of local approaches for crack propagation and will enable the industry to consider more complex strategies than initiation criteria.

7.6. References

Annavarapu, C., Settgast, R.R., Vitali, E., Morris, J.P. (2016). A local crack-tracking strategy to model three-dimensional crack propagation with embedded methods. *Computer Methods in Applied Mechanics and Engineering*, 311, 815–837.

Areias, P. and Belytschko, T. (2005). Analysis of three-dimensional crack initiation and propagation using the extended finite element method. *International Journal for Numerical Methods in Engineering*, 63, 760–788.

Areias, P., Van Goethem, N., Pires, E. (2011). A damage model for ductile crack initiation and propagation. *Computational Mechanics*, 47(6), 641–656.

Askes, H. and Sluys, L.J. (2000). Remeshing strategies for adaptive ale analysis of strain localisation. *European Journal of Mechanics-A/Solids*, 19(3), 447–467.

Beese, S., Loehnert, S., Wriggers, P. (2018). 3D ductile crack propagation within a polycrystalline microstructure using XFEM. *Computational Mechanics*, 61(1), 71–88.

Belytschko, T. and Black, T. (1999). Elastic crack growth in finite elements with minimal remeshing. *International Journal for Numerical Methods in Engineering*, 45, 601–620.

Belytschko, T., Chen, H., Xu, J., Zi, G. (2003). Dynamic crack propagation based on loss of hyperbolicity and a new discontinuous enrichment. *International Journal for Numerical Methods in Engineering*, 58(12), 1873–1905.

Benvenuti, E. (2011). Mesh-size-objective XFEM for regularized continuous-discontinuous transition. *Finite Elements in Analysis and Design*, 47(12), 1326–1336.

Benvenuti, E. and Tralli, A. (2012). Simulation of finite-width process zone in concrete-like materials by means of a regularized extended finite element model. *Computational Mechanics*, 50(4), 479–497.

Benvenuti, E., Tralli, A., Ventura, G. (2008). A regularized XFEM model for the transition from continuous to discontinuous displacements. *International Journal for Numerical Methods in Engineering*, 74(6), 911–944.

Bernard, P., Moës, N., Chevaugeon, N. (2012). Damage growth modeling using the thick level set (TLS) approach: Efficient discretization for quasi-static loadings. *Computer Methods in Applied Mechanics and Engineering*, 233/236, 11–27.

Bobiński, J. and Tejchman, J. (2016a). Comparison of continuous and discontinuous constitutive models to simulate concrete behaviour under mixed-mode failure conditions. *International Journal for Numerical and Analytical Methods in Geomechanics*, 40(3), 406–435.

Bobiński, J. and Tejchman, J. (2016b). A coupled constitutive model for fracture in plain concrete based on continuum theory with non-local softening and extended finite element method. *Finite Elements in Analysis and Design*, 114, 1–21.

Boroomand, B. and Zienkiewicz, O. (1999). Recovery procedures in error estimation and adaptivity. Part II: Adaptivity in nonlinear problems of elasto-plasticity behaviour. *Computer Methods in Applied Mechanics and Engineering*, 176, 127–146.

Bottoni, M., Dufour, F., Giry, C. (2015). Topological search of the crack pattern from a continuum mechanical computation. *Engineering Structures*, 99, 346–359.

Bouchard, P., Bay, F., Chastel, Y., Tovena, I. (2000). Crack propagation modelling using an advanced remeshing technique. *Computer Methods in Applied Mechanics and Engineering*, 189(3), 723–742.

Boussaa, D. and Aravas, N. (2001). A numeric-symbolic approach to the problem of localization of plastic flow. *Computational Mechanics*, 27(4), 324–331.

Bouterf, A., Buljac, A., Hild, F., Jailin, C., Neggers, J., Roux, S. (2020). Digital volume correlation of laminographic and tomographic images: Results and challenges. In *Virtual Design and Validation*, Wriggers, P., Allix, O., Weißenfels, C. (eds). Springer, Cham.

Brancherie, D., Villon, P., Ibrahimbegović, A., Rassineux, A., Breitkopf, P. (2006). Transfert de champs par approximation diffuse avec conservation de l'énergie. *Revue européenne de mécanique numérique*, 15(1–3), 107–118.

Broumand, P. and Khoei, A. (2013). The extended finite element method for large deformation ductile fracture problems with a non-local damage-plasticity model. *Engineering Fracture Mechanics*, 112/113, 97–125.

Camacho, G. and Ortiz, M. (1997). Adaptive lagrangian modelling of ballistic penetration of metallic targets. *Computer Methods in Applied Mechanics and Engineering*, 142, 269–301.

Cao, T.-S. (2014). Numerical simulation of 3D ductile cracks formation using recent improved lode-dependent plasticity and damage models combined with remeshing. *International Journal of Solids and Structures*, 51(13), 2370–2381.

Cazes, F. and Moës, N. (2015). Comparison of a phase-field model and of a thick level set model for brittle and quasi-brittle fracture. *International Journal for Numerical Methods in Engineering*, 103(2), 114–143.

Cervera, M. and Chiumenti, M. (2006). Mesh objective tensile cracking via a local continuum damage model and a crack tracking technique. *Computer Methods in Applied Mechanics and Engineering*, 196(1), 304–320.

Cervera, M., Pelà, L., Clemente, R., Roca, P. (2010). A crack-tracking technique for localized damage in quasi-brittle materials. *Engineering Fracture Mechanics*, 77(13), 2431–2450.

Chen, B., Pinho, S., De Carvalho, N., Baiz, P., Tay, T. (2014). A floating node method for the modelling of discontinuities in composites. *Engineering Fracture Mechanics*, 127, 104–134.

Chiaruttini, V., Feyel, F., Chaboche, J. (2010). A robust meshing algorithm for complex 3D crack growth simulation. In *IV European Conference on Computational Mechanics*, 16–21 May. ECCM, Paris.

Comi, C., Mariani, S., Perego, U. (2007). An extended FE strategy for transition from continuum damage to mode I cohesive crack propagation. *Int. J. For Numerical Analytical Methods In Geomechanics*, 31, 213–238.

Crété, J., Longère, P., Cadou, J. (2014). Numerical modelling of crack propagation in ductile materials combining the GTN model and X-FEM. *Computer Methods in Applied Mechanics and Engineering*, 275, 204–233.

Cuvilliez, S. (2012). Passage d'un modèle d'endommagement continu régularisé à un modèle de fissuration cohésive dans le cadre de la rupture quasi-fragile. PhD Thesis, École Nationale Supérieure des Mines de Paris, Paris.

Cuvilliez, S., Feyel, F., Lorentz, E., Michel-Ponnelle, S. (2012). A finite element approach coupling a continuous gradient damage model and a cohesive zone model within the framework of quasi-brittle failure. *Computer Methods in Applied Mechanics and Engineering*, 237/240, 244–259.

Dias, I.F., Oliver, J., Lloberas-Valls, O. (2018). Strain-injection and crack-path field techniques for 3D crack-propagation modelling in quasi-brittle materials. *International Journal of Fracture*, 212(1), 67–87.

Duflot, M. (2007). A study of the representation of cracks with level sets. *International Journal for Numerical Methods in Engineering*, 70(11), 1261–1302.

Dufour, F., Legrain, G., Pijaudier-Cabot, G., Huerta, A. (2012). Estimation of crack opening from a two-dimensional continuum-based finite element computation. *International Journal for Numerical and Analytical Methods in Geomechanics*, 36(16), 1813–1830.

El Khaoulani, R.E. and Bouchard, P. (2012). An anisotropic mesh adaptation strategy for damage and failure in ductile materials. *Finite Elements in Analysis and Design*, 59, 1–10.

Elguedj, T., Gravouil, A., Combescure, A. (2007). A mixed augmented Lagrangian-extended finite element method for modelling elastic-plastic fatigue crack growth with unilateral contact. *International Journal for Numerical Methods in Engineering*, 71(13), 1569–1597.

Engelen, R., Geers, M., Baaijens, F. (2003). Nonlocal implicit gradient-enhanced elasto-plasticity for the modelling of softening behaviour. *International Journal of Plasticity*, 19, 403–433.

Espinosa, H., Zavattieri, P., Emore, G. (1998). Adaptive FEM computation of geometric and material nonlinearities with application to brittle fracture. *Mechanics of Materials*, 29, 275–305.

Essongue-Boussougou, S. (2017). Augmented finite element method for quasi-brittle fracture: Application to woven ceramic matrix composites. PhD Thesis, Université de Bordeaux, Bordeaux.

Feist, C. and Hofstetter, G. (2006). An embedded strong discontinuity model for cracking of plain concrete. *Computer Methods in Applied Mechanics and Engineering*, 195(52), 7115–7138.

Feld-Payet, S. (2010). Amorçage et propagation de fissures dans les milieux ductiles non locaux. PhD Thesis, École Nationale Supérieure des Mines de Paris, Paris [Online]. Available at: http://pastel.archives-ouvertes.fr/pastel-00583011.

Feld-Payet, S., Besson, J., Feyel, F. (2011). Finite element analysis of damage in ductile structures using a non local model combined with a three-field formulation. *International Journal of Damage Mechanics*, 20, 655–680.

Feld-Payet, S., Chiaruttini, V., Besson, J., Feyel, F. (2015). A new marching ridges algorithm for crack path tracking in regularized media. *International Journal of Solids and Structures*, 71, 57–69.

Feld-Payet, S., Le Besnerais, G., Bonnand, V., Pacou, D., Thiercelin, L. (2020). Crack path tracking from full field measurements: A novel empirical methodology. *Strain*, 56(2), e12333.

Feld-Payet, S., Fourest, T., Bouda, P., Garaud, J.-D., Bettonte, F. (2022). Escale : une plateforme ouverte pour le dialogue essai-modèle-calcul. In *CSMA 2022 – 15e colloque national en calcul des structures*, 16–20 May, Giens [Online]. Available at: https://hal.science/hal-03687482/.

Fries, T.-P. and Baydoun, M. (2012). Crack propagation with the extended finite element method and a hybrid explicit-implicit crack description. *International Journal for Numerical Methods in Engineering*, 89(12), 1527–1558.

Furst, J.D. and Pizer, S.M. (1998). Marching optimal-parameter ridges: An algorithm to extract shape loci in 3D images. In *International Conference on Medical Image Computing and Computer-Assisted Intervention – MICCAI'98*, 11–13 October, Cambridge, MA. Springer, Berlin, Heidelberg.

Gao, X., Faleskog, J., Shih, C., Dodds, R. (1998). Ductile tearing in part-through cracks: Experiments and cell-model predictions. *Engineering Fracture Mechanics*, 59(6), 761–777.

Garaud, J.-D., Sylvia, F-P., Francesco, B., Tireira, A., Le Sant, Y., Le Besnerais, G., Belon, S. (2017). Escale : plateforme logicielle pour le dialogue essai-calcul. In *13e colloque national en calcul des structures*, May. Université Paris-Saclay, Giens [Online]. Available at: https://hal.science/hal-01908509/.

Garzon, J., O'Hara, P., Duarte, C.A., Buttlar, W.G. (2014). Improvements of explicit crack surface representation and update within the generalized finite element method with application to three-dimensional crack coalescence. *International Journal for Numerical Methods in Engineering*, 97(4), 231–273.

Geelen, R.J., Liu, Y., Dolbow, J.E., Rodríguez-Ferran, A. (2018). An optimization-based phase-field method for continuous-discontinuous crack propagation. *International Journal for Numerical Methods in Engineering*, 116(1), 1–20.

Geers, M., de Borst, R., Brekelmans, W., Peerlings, R. (1998). Strain-based transient-gradient damage model for failure analyses. *Computer Methods in Applied Mechanics and Engineering*, 160(1), 133–153.

Gupta, P. and Duarte, C.A. (2014). Simulation of non-planar three-dimensional hydraulic fracture propagation. *International Journal for Numerical and Analytical Methods in Geomechanics*, 38(13), 1397–1430.

Hadamard, J. (1902). Sur les problèmes aux dérivées partielles et leur signification physique. *Princeton University Bulletin*, XIII(4), 49–52.

Hambli, R. (2001). Finite element simulation of fine blanking processes using a pressure-dependent damage model. *Journal of Materials Processing Technology*, 116(2), 252–264.

Hansbo, A. and Hansbo, P. (2004). A finite element method for the simulation of strong and weak discontinuities in solid mechanics. *Computer Methods in Applied Mechanics and Engineering*, 193(33/35), 3523–3540.

Hill, R. (1958). A general theory of uniqueness and stability in elastic-plastic solids. *Journal of the Mechanics and Physics of Solids*, 6(3), 236–249.

Hinton, E. and Campbell, J. (1974). Local and global smoothing of discontinuous finite element functions using a least squares method. *Int. J. Num. Meth. Eng.*, 8, 461–480.

Hinton, E. and Owen, D.P. (1977). *Finite Element Programming*. Academic Press, London.

Huespe, A.E., Needleman, A., Oliver, J., Sánchez, P.J. (2009). A finite thickness band method for ductile fracture analysis. *International Journal of Plasticity*, 25(12), 2349–2365.

Huespe, A.E., Needleman, A., Oliver, J., Sanchez, P.J. (2012). A finite strain, finite band method for modeling ductile fracture. *International Journal of Plasticity*, 28(1), 53–69.

Jäger, P., Steinmann, P., Kuhl, E. (2008a). Modeling three-dimensional crack propagation – A comparison of crack path tracking strategies. *International Journal for Numerical Methods in Engineering*, 76(9), 1328–1352.

Jäger, P., Steinmann, P., Kuhl, E. (2008b). On local tracking algorithms for the simulation of three-dimensional discontinuities. *Computational Mechanics*, 42(3), 395–406.

Jäger, P., Steinmann, P., Kuhl, E. (2009). Towards the treatment of boundary conditions for global crack path tracking in three-dimensional brittle fracture. *Computational Mechanics*, 45(1), 91–107.

Javani, H.R. (2011). A computational damage approach towards three-dimensional ductile fracture. PhD Thesis, Technische Universiteit Eindhoven, Eindhoven.

Javani, H.R., Peerlings, R.H.J., Geers, M.G.D. (2016). Three-dimensional finite element modeling of ductile crack initiation and propagation. *Advanced Modeling in Simulation in Engineering Sciences*, 3, 1–25.

Javanmardi, M. and Maheri, M.R. (2019). Extended finite element method and anisotropic damage plasticity for modelling crack propagation in concrete. *Finite Elements in Analysis and Design*, 165, 1–20.

Jirásek, M. (2000). Comparative study on finite elements with embedded discontinuities. *Computer Methods in Applied Mechanics and Engineering*, 188(1/3), 307–330.

Jirásek, M. (2004). Non-local damage mechanics with application to concrete. *Revue française de génie civil*, 8(5/6), 683–707.

Jirásek, M. and Zimmermann, T. (2001). Embedded crack model. Part II: Combination with smeared cracks. *International Journal for Numerical Methods in Engineering*, 50, 1291–1305.

Khoei, A. and Gharehbaghi, S. (2009). Three-dimensional data transfer operators in large plasticity deformations using modified-SPR technique. *Applied Mathematical Modelling*, 33, 3269–3285.

Labergère, C., Rassineux, A., Saanouni, K. (2014). Numerical simulation of continuous damage and fracture in metal-forming processes with 2D mesh adaptive methodology. *Finite Elements in Analysis and Design*, 82, 46–61.

Lee, N. and Bathe, K. (1994). Error indicators and adaptive remeshing in large deformation finite element analysis. *Finite Elements in Analysis and Design*, 16, 99–139.

Lemaitre, J. and Chaboche, J.-L. (2004). *Mécanique des matériaux solides,* 2nd edition. Dunod, Paris.

Ling, D., Yang, Q., Cox, B. (2009). An augmented finite element method for modeling arbitrary discontinuities in composite materials. *International Journal of Fracture*, 156(1), 53–73.

Lloberas-Valls, O., Huespe, A., Oliver, J., Dias, I. (2016). Strain injection techniques in dynamic fracture modeling. *Computer Methods in Applied Mechanics and Engineering*, 308, 499–534.

Loubignac, G., Cantin, G., Touzot, G. (1977). Continuous stress fields in finite element analysis. *AIAA Journal*, 15, 1645–1647.

Ma, Z., Yang, Q., Su, X. (2019). A conforming augmented finite element method for modeling arbitrary cracking in solids. *Journal of Applied Mechanics*, 86(7), 071002.

Mandel, J. (1966). Conditions de stabilité et postulat de drucker. In *Rhéologie et mécanique des sols*, Kravtchenko, J. and Sirieys, P.M. (eds). Springer, Berlin, Heidelberg.

Mediavilla, J. (2005). Continuous and discontinuous modelling of ductile fracture. PhD Thesis, Technische Universiteit Eindhoven, Eindhoven.

Mediavilla, J., Peerlings, R.H.J., Geers, M.G.D. (2006). Discrete crack modelling of ductile fracture driven by non-local softening plasticity. *International Journal for Numerical Methods in Engineering*, 66(4), 661–688.

Melenk, J.M. and Babuška, I. (1996). The partition of unity finite element method: Basic theory and applications. Research Report, ETH Zürich, Zürich.

Moës, N., Dolbow, J., Belytschko, T. (1999). A finite element method for crack growth without remeshing. *International Journal for Numerical Methods in Engineering*, 46(1), 131–150.

Moës, N., Stolz, C., Bernard, P.-E., Chevaugeon, N. (2011). A level set based model for damage growth: The thick level set approach. *International Journal for Numerical Methods in Engineering*, 86(3), 358–380.

Oliver, J., Huespe, A., Samaniego, E., Chaves, E. (2002). On strategies for tracking strong discontinuities in computational failure mechanics. In *Fifth World Congress on Computational Mechanics*.

Oliver, J., Huespe, A., Samaniego, E., Chaves, E. (2004). Continuum approach to the numerical simulation of material failure in concrete. *International Journal for Numerical and Analytical Methods in Geomechanics*, 28, 609–632.

Oliver, J., Huespe, A.E., Sánchez, P.J. (2006). A comparative study on finite elements for capturing strong discontinuities: E-FEM vs X-FEM. *Computer Methods in Applied Mechanics and Engineering*, 195(37/40), 4732–4752.

Oliver, J., Dias, I., Huespe, A. (2014). Crack-path field and strain-injection techniques in computational modeling of propagating material failure. *Computer Methods in Applied Mechanics and Engineering*, 274, 289–348.

Ortiz, M. and Quigley, J. (1991). Adaptive mesh refinement in strain localization problems. *Computer Methods in Applied Mechanics and Engineering*, 90, 781–804.

Patzák, B. and Jirásek, M. (2003). Process zone resolution by extended finite elements. *Engineering Fracture Mechanics*, 70(7), 957–977.

Patzák, B. and Jirásek, M. (2004). Adaptive resolution of localized damage in quasi-brittle materials. *Journal of Engineering Mechanics*, 130, 720–732.

Peerlings, R., de Borst, R., Brekelmans, W., de Vree, J. (1996). Gradient-enhanced damage for quasi-brittle materials. *International Journal for Numerical Methods in Engineering*, 39, 3391–3403.

Peerlings, R., Javani, H., Mediavilla, J., Geers, M. (2008). Modelling of damage initiation and propagation in metal forming. *International Journal of Material Forming*, 1, 1123–1126.

Pereira, J.P., Duarte, C.A., Guoy, D., Jiao, X. (2009). hp-generalized fem and crack surface representation for non-planar 3-D cracks. *International Journal for Numerical Methods in Engineering*, 77(5), 601–633.

Perić, D., Hochard, C., Dutko, M., Owen, D. (1996). Transfer operators for evolving meshes in small strain elasto-plasticity. *Computer Methods in Applied Mechanics and Engineering*, 137, 331–344.

Pourmodheji, R. and Mashayekhi, M. (2012). Improvement of the extended finite element method for ductile crack growth. *Materials Science and Engineering: A*, 551, 255–271.

Rabczuk, T. and Belytschko, T. (2007). A three-dimensional large deformation meshfree method for arbitrary evolving cracks. *Computer Methods in Applied Mechanics and Engineering*, 196(29), 2777–2799.

Rabczuk, T., Zi, G., Bordas, S., Nguyen-Xuan, H. (2010). A simple and robust three-dimensional cracking-particle method without enrichment. *Computer Methods in Applied Mechanics and Engineering*, 199(37–40), 2437–2455.

Riccardi, F., Kishta, E., Richard, B. (2017). A step-by-step global crack-tracking approach in E-FEM simulations of quasi-brittle materials. *Engineering Fracture Mechanics*, 170, 44–58.

Rodríguez-Ferran, A. and Huerta, A. (2000). Error estimation and adaptivity for nonlocal damage models. *International Journal of Solids and Structures*, 37(48/50), 7501–7528.

Roth, S.-N., Léger, P., Soulaimani, A. (2015). A combined XFEM-damage mechanics approach for concrete crack propagation. *Computer Methods in Applied Mechanics and Engineering*, 283, 923–955.

Saanouni, K. (2008). On the numerical prediction of the ductile fracture in metal forming. *Engineering Fracture Mechanics*, 75(11), 3545–3559.

Saloustros, S., Cervera, M., Pelà, L. (2018a). Challenges, tools and applications of tracking algorithms in the numerical modelling of cracks in concrete and masonry structures. *Archives of Computational Methods in Engineering*, 26(4), 961–1005.

Saloustros, S., Cervera, M., Pelà, L. (2018b). Tracking multi-directional intersecting cracks in numerical modelling of masonry shear walls under cyclic loading. *Meccanica*, 53(7), 1757–1776.

Seabra, M.R.R., Cesar de Sa, J.M.A., Andrade, F., Pires, F. (2011). Continuous-discontinuous formulation for ductile fracture. *International Journal of Material Forming*, 4, 271–281.

Seabra, M.R.R., Šuštarič, P., Cesar de Sa, J.M.A., Rodič, T. (2013). Damage driven crack initiation and propagation in ductile metals using XFEM. *Computational Mechanics*, 52(1), 161–179.

Simatos, A. (2010). Méthode XFEM pour la modélisation de grandes propagations de fissure en déchirure ductile : transition d'un milieu continu vers une fissure via un modèle de zone cohésive pour le modèle de Rousselier. PhD Thesis, Institut National des Sciences Appliquées de Lyon, Lyon.

Simone, A., Wells, G., Sluys, L. (2003). From continuous to discontinuous failure in a gradient-enhanced continuum damage model. *Computer Methods in Applied Mechanics and Engineering*, 192(41–42), 4581–4607.

Song, J.-H., Areias, P.M., Belytschko, T. (2006). A method for dynamic crack and shear band propagation with phantom nodes. *International Journal for Numerical Methods in Engineering*, 67(6), 868–893.

Song, J.-H., Wang, H., Belytschko, T. (2008). A comparative study on finite element methods for dynamic fracture. *Computational Mechanics*, 42(2), 239–250.

Stolarska, M., Chopp, D., Moës, N., Belytschko, T. (2001). Modelling crack growth by level sets in the extended finite element method. *International Journal for Numerical Methods in Engineering*, 51, 943–960.

Strouboulis, T., Babuška, I., Copps, K. (2000). The design and analysis of the generalized finite element method. *Computer Methods in Applied Mechanics and Engineering*, 181(1/3), 43–69.

Tamayo-Mas, E. and Rodríguez-Ferran, A. (2015). A medial-axis-based model for propagating cracks in a regularised bulk. *International Journal for Numerical Methods in Engineering*, 101(7), 489–520.

Tamayo-Mas, E., Feliu-Fabà, J., Casado-Antolin, M., Rodríguez-Ferran, A. (2019). A continuous-discontinuous model for crack branching. *International Journal for Numerical Methods in Engineering*, 120(1), 86–104.

Teng, X. (2008). Numerical prediction of slant fracture with continuum damage mechanics. *Engineering Fracture Mechanics*, 75(8), 2020–2041.

Thomas, T.Y. (1957). Extended compatibility conditions for the study of surfaces of discontinuity in continuum mechanics. *Journal of Mathematics and Mechanics*, 6(3), 311–322.

Tvergaard, V. (1982). Influence of void nucleation on ductile shear fracture at a free surface. *Journal of the Mechanics and Physics of Solids*, 30(6), 399–425.

Tvergaard, V. and Needleman, A. (2006). Three dimensional microstructural effects on plane strain ductile crack growth. *International Journal of Solids and Structures*, 43(20), 6165–6179.

Villon, P., Borouchaki, H., Saanouni, K. (2002). Transfert de champs plastiquement admissibles. *C.R. Mécanique*, 330, 313–318.

Wang, Y. and Waisman, H. (2016). From diffuse damage to sharp cohesive cracks: A coupled XFEM framework for failure analysis of quasi-brittle materials. *Computer Methods in Applied Mechanics and Engineering*, 299, 57–89.

Wells, G. and Sluys, L. (2002). A new method for modelling cohesive cracks using finite elements. *International Journal for Numerical Method in Engineering*, 50, 2667–2682.

Wolf, J., Longère, P., Cadou, J.M., Crété, J.P. (2018). Numerical modeling of strain localization in engineering ductile materials combining cohesive models and X-FEM. *International Journal of Mechanics and Materials in Design*, 14(2), 177–193.

Wu, L., Becker, G., Noels, L. (2014). Elastic damage to crack transition in a coupled non-local implicit discontinuous galerkin/extrinsic cohesive law framework. *Computer Methods in Applied Mechanics and Engineering*, 279, 379–409.

Xue, L. and Wierzbicki, T. (2009). Numerical simulation of fracture mode transition in ductile plates. *International Journal of Solids and Structures*, 46(6), 1423–1435.

Yang, F., Rassineux, A., Labergere, C., Saanouni, K. (2018). A 3D h-adaptive local remeshing technique for simulating the initiation and propagation of cracks in ductile materials. *Computer Methods in Applied Mechanics and Engineering*, 330, 102–122.

Yun, K., Kim, T.-J., Jang, P.S., Wang, Z., Ronald, S. (2019a). An improved crack tracking algorithm with self-correction ability of the crack path and its application in a continuum damage model. *International Journal for Numerical Methods in Engineering*, 117(2), 249–269.

Yun, K., Wang, Z., Chang, M., Liu, J., Kim, T.-J., Son, N., Ji, K., Ronald, S. (2019b). A computational methodology for simulating quasi-brittle fracture problems. *Computers & Structures*, 215, 65–79.

Zienkiewicz, O.C. and Zhu, J.Z. (1992a). The superconvergent patch recovery and a posteriori error estimates. Part 1: The recovery technique. *International Journal for Numerical Methods in Engineering*, 33(7), 1331–1364.

Zienkiewicz, O.C. and Zhu, J.Z. (1992b). The superconvergent patch recovery and a posteriori error estimates. Part 2: Error estimates and adaptivity. *International Journal for Numerical Methods in Engineering*, 33(7), 1365–1382.

List of Authors

Pierre ALART
LMGC
Université de Montpellier
France

Jacques BESSON
CNRS
Mines Paris PSL
France

Franz CHOULY
Université Bourgogne-Franche-Comté
Dijon
France
and
University of Chile
Santiago
Chile

Sylvia FELD-PAYET
DMAS-ONERA
Université Paris-Saclay
Châtillon
France

Patrick HILD
CNRS
Université de Toulouse
France

Frédéric LEBON
LMA
Aix-Marseille Université
France

Éric LORENTZ
EDF Lab Paris-Saclay
EDF R&D/ERMES
Palaiseau
France

Nicolas MOËS
École centrale de Nantes
France

Yves RENARD
Institut Camille Jordan
Université de Lyon
and
CNRS
INSA Lyon
France

Vladislav A. YASTREBOV
CNRS
Mines Paris PSL
France

Index

A, B, C

D, E

F

G, H, I

L

M, N

P, R

S

T, U, W

Printed and bound by CPI Group (UK) Ltd, Croydon, CR0 4YY

13/09/2026

14960996-0008